The Complete Guide to MicroStation 3D

David Wilkinson

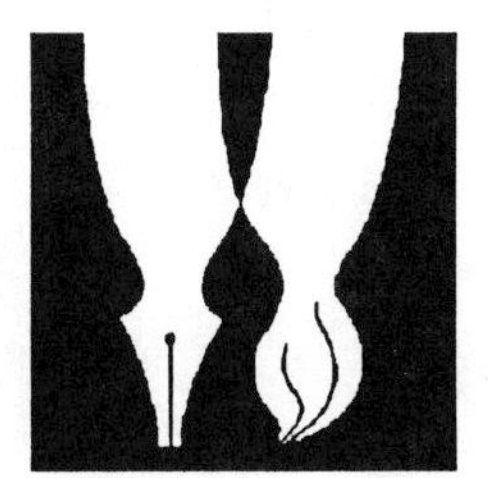

Cover photography by Duke Dinh of Black Art Photography.

First Printing February 1991

2nd Floor, 94 Flinders Street,
Melbourne, Victoria 3000.
Australia.
Fax: (03) 818 3704. International 61 3 818 3704.

ISBN 0-646-01678-4
Published in Australia. Printed in the United States of America.

To Lena

Acknowledgements

In April 1990, Mach Dinh-Vu first suggested that I should write this book and my initial reaction was 'No, I just don't have the time'. In a matter of days, however, I resigned from my job and was looking forward to devoting my time to 'the book'.

This was possible only because my wife, Lena, gave me her full support and allowed me to become a full-time 'dependant' for however long the project took to complete.

I have had encouragement and support from many other people, also. In particular, the CAD department at Minenco Pty. Ltd., my former work place. David Hill, Alan Ashton and Carole Lowe have helped with proof-reading at various stages. Alan also has come to my aid with items of hardware and software when needed. I am grateful to them all for their time and effort.

Many thanks as well to Bentley Systems Inc. who, through Mach, kept me up to date with Beta releases of version 4. These were invaluable, as was the hardware lock they supplied.

About the Author

David Wilkinson is a former civil designer/draftsman in Melbourne, Australia. As a civil designer he commenced working with VAX based IGDS in 1985, after many years 'on the board'. His over-riding interest with CAD from the first was with the 3D aspects. He was involved in a number of projects using CAD, both as an electronic drawing aid and as a design tool. Among these was the 3D modelling of a section of an underground mine. In 1987 he joined the CAD support department of a consulting engineering firm in Melbourne. Among his tasks, in this position, were trouble-shooting, user support/training and development of 3D applications. He held this position till May 1990, when he resigned to become a freelance consultant, and to write this book. He has been involved with MicroStation on PC's from the first release in Australia.

Introduction

Before the arrival of 3D CAD most designers/draftsmen converted their ideas into two-dimensional drawings of plans, elevations etc. They carried out design checks using a combination of drawings and calculations. Part of the time consuming checking process was ensuring that the drawings were correct.

With three-dimensional CAD, especially MicroStation, we can create a full scale three-dimensional computer model of our designs. Many of the design checks, such as clearance corridors and the horizontal and vertical layout schemes can be carried out using the model. This removes one possible cause of mistakes, that of the translation of the design into two-dimensional drawings, by the draftsman.

Displayed on the screen is a 'graphical' representation of our calculations. CAD allows us to see the results of our calculations as we perform them. The computer processes the data and shows us the results as a three-dimensional image. Errors are more easily picked up when seen in a model, rather than a maze of calculations on a piece of paper. From the 3D model we can produce 2D drawings of plans, elevations, isometrics and the like. The computer, through the MicroStation software, does the translations into 2D drawings for us. MicroStation allows us to view the model from any direction and produce a line drawing at any scale, or a more life-like rendered image if required. With experience and practice, most designs can be produced as full size, three-dimensional, computer models.

As with most things associated with CAD, both 2D and 3D, there is usually more than one way to get the required result. This book is essentially for MicroStation operators, experienced in 2D, who want to advance into 3D CAD. The book starts with the important ideas, or concepts, that need to be understood. It takes the operator from the basics of creating a simple model through to the more complex problems that arise in 3D modelling. Some exercises may seem simple but they help to introduce ideas that, if fully understood, simplify the mechanics of creating models.

Which Version of MicroStation ?

This book has been written for the two current versions of MicroStation - version 4 and version 3.3. There are significant differences, in some areas, between these two versions. This has necessitated separate discussions on several topics in the book.

Version 3.3 will run on earlier model PC's and is commonly referred to as the 286 PC version.

Version 4 for PC requires a minimum configuration of a 386 machine. Version 4 also will look and function similarly on other platforms. That is, the graphical user interface (GUI) has been standardised on the PC, Macintosh and Intergraph Unix systems.

No matter which platform you are using, the procedures are similar, when working in a MicroStation design file.

How to use this book

For the first-time user of MicroStation 3D, this book should be read from front to back. The introduction of concepts and the various tools and key-ins is progressive. Successive chapters build on knowledge gained from the previous chapters and topics. The exercises should be completed, no matter how simple they may appear to be.

Other users, who have had some previous experience with MicroStation, still should read through the early chapters. Many 3D CAD operators can construct simple models but are unable to solve more complex issues. This often is due to a lack of understanding of basic concepts. In particular, this applies to views and view rotations.

Special emphasis should be placed on the chapters discussing Display Depth, Active Depth and View Manipulations. These are the basic tools for working in a three-dimensional file.

Basic Requirements

Readers should be competent users of MicroStation 2D. They should know the meanings of basic terms such as TENTATIVE POINT, DATA POINT, RESET. Also, they should be familiar with the normal 2D methods and tools for creating the various elements.

A knowledge of editing text files with a text editor, outside MicroStation, is required for some of the later topics. However, this is not essential to the learning of how to create 3D models in MicroStation.

Terminology and Formats Used

Terminology used in this book is consistent with that used in the documentation of the new version 4 of MicroStation. We use 'tools' to do things rather than 'commands'. A tool is both an icon in a palette and the operation represented by the icon. When a tool is chosen in a palette, its name appears in the command window command field. This name is similar to the command description in earlier versions of MicroStation.

Where key-ins are used, the full key-in is shown. Normally, key-ins can be reduced to a minimum of three characters for each word involved. For example 'PLACE LINE' could be reduced to 'PLA LIN'.

Symbols, in the left margin, are used to indicate various points:

A floppy disk symbol is used to signify that there is a file supplied on the accompanying disk. Location (directory) on the disk, and name of the file is given directly below the symbol.

Points of special significance have this symbol in the left margin.

This symbol makes the beginning of an Exercise, or a separate section of an exercise. Where exercises are provided, they are in a format similar to the following:

Make the Active Level 5 and FIT each view

- Select the Place Line tool.
- Place a line from XY=0,0,0 to XY=5,020,10

Version 4

Where the exercise or section of the exercise refers specifically to either version 3.3 or version 4, this is noted below the symbol.

Accompanying Disk

There is an accompanying disk that can be purchased with this book. It contains copies of design files and models that are discussed throughout the book. As well, it contains other files that are mentioned in the text.

In the root directory of this disk is an ASCII file named INFO.TXT which contains a description of the files that are contained on the disk.

Table of Contents

2 : Working in 3D

3 : Project & Surface of Revolution

4 : Cells & Reference Files

5 : Placing & Manipulating Elements

6 : Creating 3D Models

7 : Advanced Techniques

8 : B-spline Surfaces version 4

9 : Introduction to Rendering

10 : Rendering version 3.3

11 : Rendering version 4

12 : Drawing Production

13 : Tips & Tricks

14 : Practice Examples

Index

1 : The 3D Environment

In this chapter you will be introduced to the MicroStation 3D environment and shown, step by step, how it operates. Placing elements in a file is easy, whether in 2D or 3D. Understanding what the screen is displaying is simple in 2D. We are familiar with drawings, and 2D CAD is like an electronic drawing. With 3D, the display on the screen may not be as easy to understand at first. Once you have been working in 3D for a time, it will become as familiar as the more mundane 2D 'drawings'. We have to forget the old methods of projecting three-dimensional ideas and designs onto a flat drawing. The computer can do that for us later, when we want a drawing from our model. We will be working in a three-dimensional world, creating our models as they really are.

When we design anything, from a mechanical part to a freeway, we have design rules and restrictions that we work with. With experience these rules and restrictions become second nature to us and we accept them without question. With CAD we also have rules and restrictions. With experience and practice, these also become second nature to us.

Some of the following exercises may seem very simple, but they are used to demonstrate the way we move in the 3D world of MicroStation. As was mentioned in the Introduction there are quite significant differences between version 3.3 and version 4 of MicroStation, but the basics for creating models are unchanged. It is extremely important to learn the basics of the 3D environment. Many 3D operators are confused by the concepts of VIEWS, VIEW ROTATIONS and DISPLAY DEPTH. Once these, and other basic ideas are understood, the rest is straight forward. Working in 3D is *not* difficult, particularly with MicroStation.

Ideas or concepts that need to be understood are:-

- **THE DESIGN CUBE**
- **WIREFRAME DISPLAY**
- **VIEWS**
- **BASIC ELEMENT PLACEMENT**
- **ACTIVE DEPTH**
- **DISPLAY DEPTH**

Procedures that need to be mastered are:-

- **PRECISION INPUTS (especially DL and DX)**
- **VIEW ROTATION**
- **ELEMENT MANIPULATION**
- **FENCE MANIPULATION**
- **PROJECTED SURFACES**
- **SURFACE OF REVOLUTION**

The order in which the various topics are listed and/or appear in the following chapters does not indicate greater or lesser importance. They are introduced as they are needed in the tutorials.

File Creation

There are differences between 3D and 2D design files which require different seed files to be used for each. Apart from this, there is no other difference in the way we create a 3D design file. The same applies for cell libraries.

3D Design Files

With PC's using version 3.3 and the MCE menu system, selecting 'Create 3D Design File' in the utilities menu produces a list of seed files from which to choose. A similar list is displayed with version 4 when the 'SEED' button is selected in the 'New' dialog box chosen in the 'File' menu.

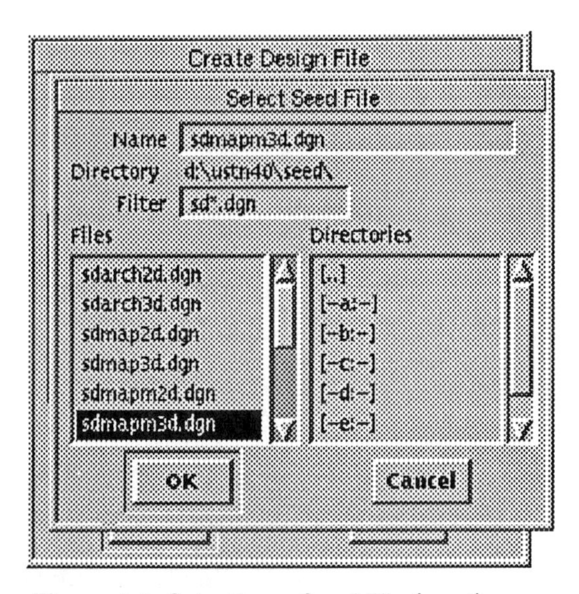

Figure 1.1 Selecting a Seed File (ver.4)

When we create a file, the selected seed file is copied to be the new file that we nominate. We don't have to use menu systems to create new design files. We can get the same result by using the respective operating systems' copy command and copy the seed file to be our new file. The seed files are located in the MicroStation DATA directory with version 3.3 and in the SEED directory with version 4.

With MicroStation (prior to version 4) on the Intergraph Unix workstations, there isn't a comparable menu system. We have to know the commands to create a new file, after typing MCE. To create a 3D file we specify the file 'dimension' to be 3D instead of 2D. The one seed file is used, by default, for all 3D files.

3D Cell Libraries

As with design files we create a 3D cell library by copying a seed file to be the new file. The seed file in this case is a file named SEED3D.CEL. It is located in the MicroStation DATA directory with version 3.3 and in the SEED directory with version 4. With version 3.3 on the PC we select 'Create 3D Cell Library' in the Utilities menu of the MCE menu. With version 4 we select the 'SEED' button in the 'Create Cell Library' dialog box. On the Intergraph Unix workstations the file dimension '3C' specifies that the new file is to be a 3D cell library.

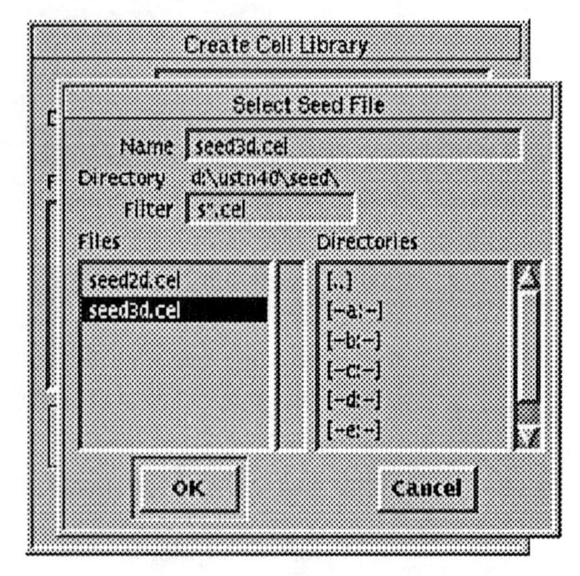

Figure 1.2 Selecting Seed Cell Library (ver 4)

Preparation for 3D Exercises

In preparation, create a 3D design file and a 3D cell library for the exercises. It is a good idea to give your design files a name that is indicative of what they contain. For instance, you may call your training files TRAIN3D.DGN and TRAIN3D.CEL.

3D Tools

Here we will look at where, on the menus, to find the new 3D tools that we will be using. While many of these tools will not mean anything to you now, we will be discussing all of them as we advance through the exercises. The following figures show their locations on the various standard menus. Use these diagrams for quick reference, till you become familiar with them. Maybe you can photo-copy the diagrams relative to your system. The sidebar menu referred to is the standard USTN.SBM, which can be used on the PC's and the Intergraph Unix workstations. EMENU is the version 3.3 digitizer menu which is illustrated. Following this is a description of locations of the equivalent tools for version 4 menus including settings boxes, dialog boxes and tool palettes.

Figure 1.3 shows the location of the 3D tools on the sidebar menu. Figure 1.4 shows the location of the same tools on the paper menu. You will notice that some appear on the sidebar menu only. Where this occurs, you will be given an alternate key-in to use as well. Many of the 3D tools are 'buried' three and four sub-menus deep on the sidebar menu. In chapter 14 (Tips and Tricks) are listings of simple '3D' sidebar menus, for versions 3.3 and 4, that you may like to use. These menus contain tools that we commonly use with 3D work, but are 'buried' in the standard sidebar menus, dialog boxes etc. They are simple, single level menus which can be used in conjunction with the standard menus.

While using more than one sidebar menu with version 3.3 takes extra screen space, the convenience of having commonly used tools readily available more than compensates for this. Version 4 gives us other options. We can 'tear off' parts of menus for later use, which is fine for creating customized menus 'on the fly'. These, as well as any sidebar menus, can be 'stacked' on top of each other, in one area of the screen, leaving the rest available for graphics. With a dual screen configuration, this problem of screen space is greatly reduced.

Digitizer, or tablet menus, don't obscure the screen at all, and tools are accessed with one press of a cursor button. They are generally much quicker to use, particularly when going from one tool to another (e.g., from place line to setting key-point snap, to chaining elements to form a shape).

We can use a combination of menus also, tablet style and on screen. Whichever style/s of menu you are most comfortable with is the best for you. You may find also, that the alternative key-in is sometimes the most efficient way to select a tool.

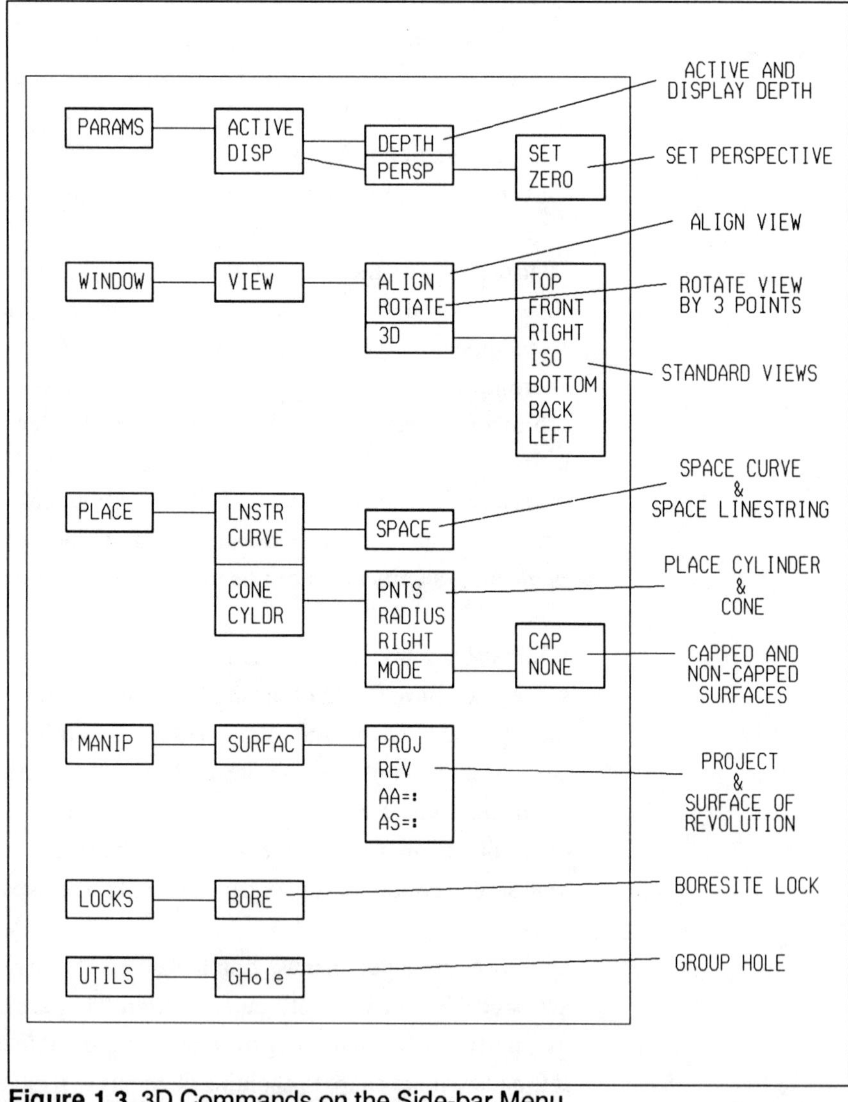

Figure 1.3 3D Commands on the Side-bar Menu

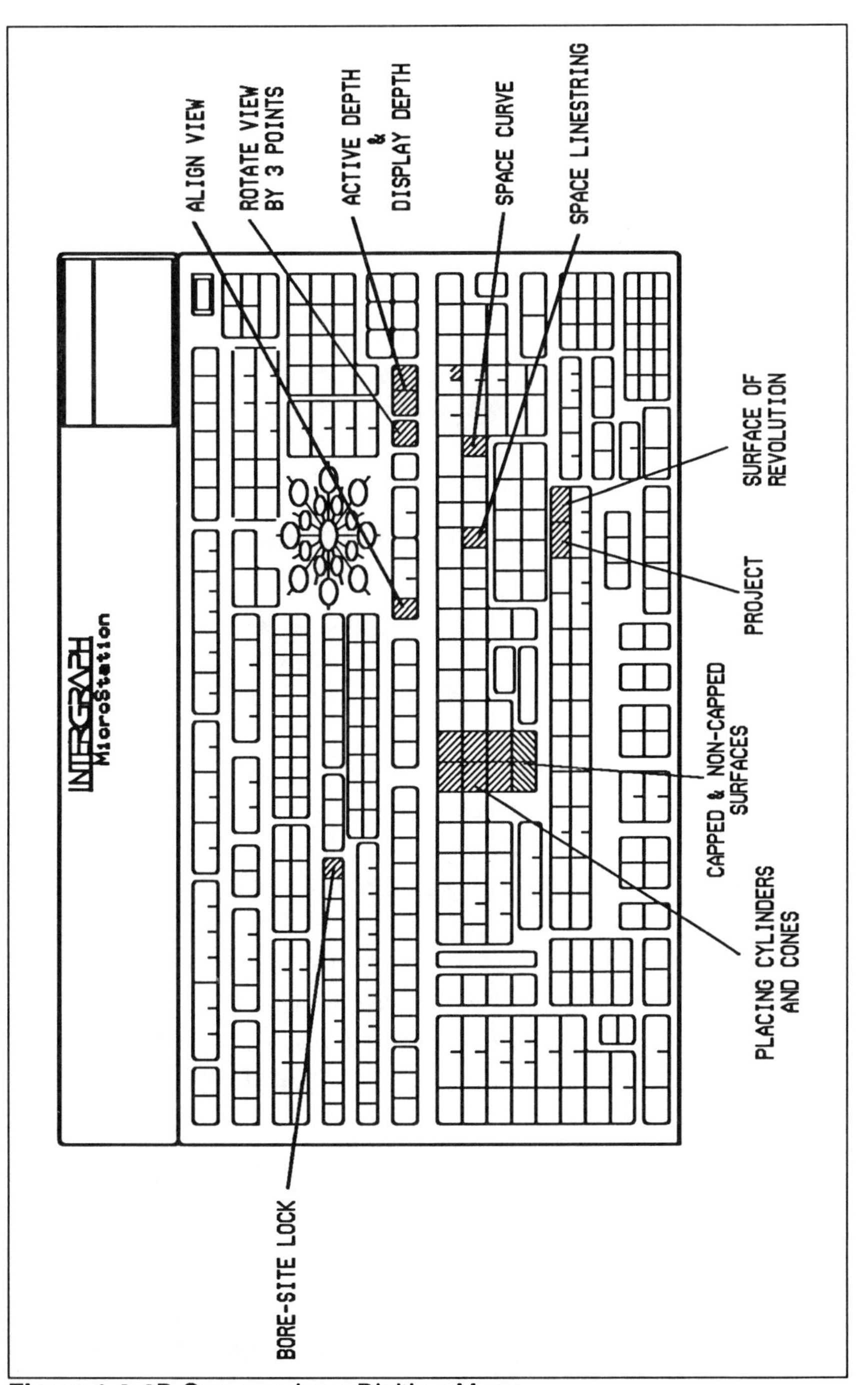

Figure 1.4 3D Commands on Digitizer Menu

Version 4 Alternative Menus

With version 4 we can use its standard pull-down menus, settings boxes, tool palettes and dialog boxes. They can be used exclusively, or along with one or more of the other menus. The various pull-down menus, palettes etc. are accessed from the command window menu bar.

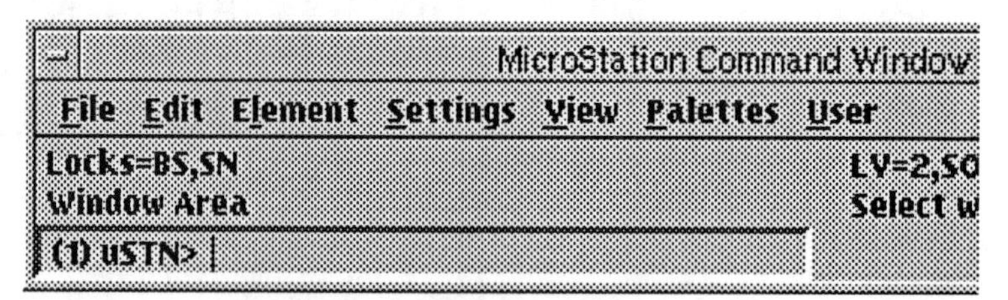

Figure 1.5 The Command Window Menu Bar

You will be shown where to locate all the relevant tools as you advance through the exercises. Here, we will briefly discuss the location of the version 4 equivalents to those mentioned on the previous pages.

Selecting '3D' in the 'Palettes' pull-down menu gives us the main 3D palette with 16 icon selections. They cover:

- Placing SPACE LINESTRINGS and CURVES.
- Placing CYLINDERS and CONES.
- PROJECTING surfaces.
- SURFACE of REVOLUTION.
- Setting and displaying DISPLAY DEPTH
- Setting and displaying ACTIVE DEPTH.
- Placing SPHERES and SLABS (new with version 4).

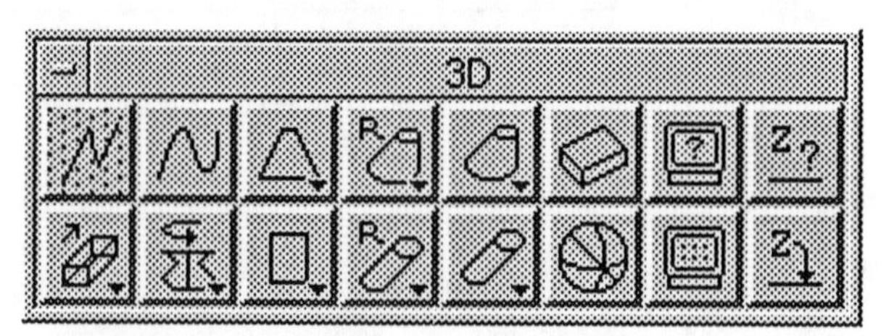

Figure 1.6 The 3D Tool Palette

Version 4 refers to capped or non-capped surfaces as SOLIDS or SURFACES. The option to select either is presented in a pop-down menu, each time we select particular 3D palette tools. For example, when we select tools to project surfaces or place elements such as cylinders and cones, we are given the opportunity to choose SURFACE or SOLID.

Standard views can be selected in the 'View Rotation' settings box. This is selected in the 'View' menu. This same settings box is used to 'dynamically' rotate views, if required.

GROUP HOLE is located in the 'Chain' sub-palette selected in the 'Main' palette or the 'Main' sub-menu.

BORESITE LOCK is set in the 'LOCKS' settings box. This is chosen in the 'Settings' pull-down menu.

Version 3.3 uses the SET PERSPECTIVE key-in to present views in a realistic fashion with perspective. With version 4 we use the SET CAMERA key-in which allows us to define a camera position, target point and type of lens being used. All the camera setup tools are found in the 'View' pull-down menu of the command window bar menu.

Digitizing/Tablet Menu version 4

In keeping with the format of icons used with the on-screen menus, the digitizing menu for version 4 also uses icons. These icons are identical to the screen menu icons.

The Design Cube

With 2D MicroStation we carry out our work on a large plane (like a flat sheet of paper). The plane is defined by X and Y axes that each have a dimension of 2^{32} (approx. 4.29 billion) Units Of Resolution (UOR's). We can 'draw' within the limits of the design plane, like we can on a sheet of paper.

3D design files have a third dimension, as is reflected in their coordinate readout. The third dimension is depth. A 3D design file is a large cube. This design cube is a finite volume of space we work in. It is like a large box in which we can create three-dimensional models of our designs. We can move the box around and view it, and its contents, from any direction. We can manipulate the contents inside the box. We can move inside the box also, and view the components from close quarters when needed.

X, Y and Z axes define the design cube and each, like 2D design files, has a dimension of 2^{32} UOR's. The origin (0,0,0) is at the very center of the design cube. The allocation of UOR's to the design units determines the 'physical' size of the design cube. The more UOR's that we assign to a physical dimension (e.g., for greater accuracy), the smaller in physical dimensions the design cube becomes.

For the exercises we will accept the default values of the seed file. We will be working in generic 'units' instead of feet or meters or any other system of measurement.

In terms of 'real-world' coordinates, the X, Y, and Z axes may be related to length, width and height when working with objects, such as mechanical parts. For site-works the X, Y, and Z axes correspond to Easting, Northing, and Elevation respectively.

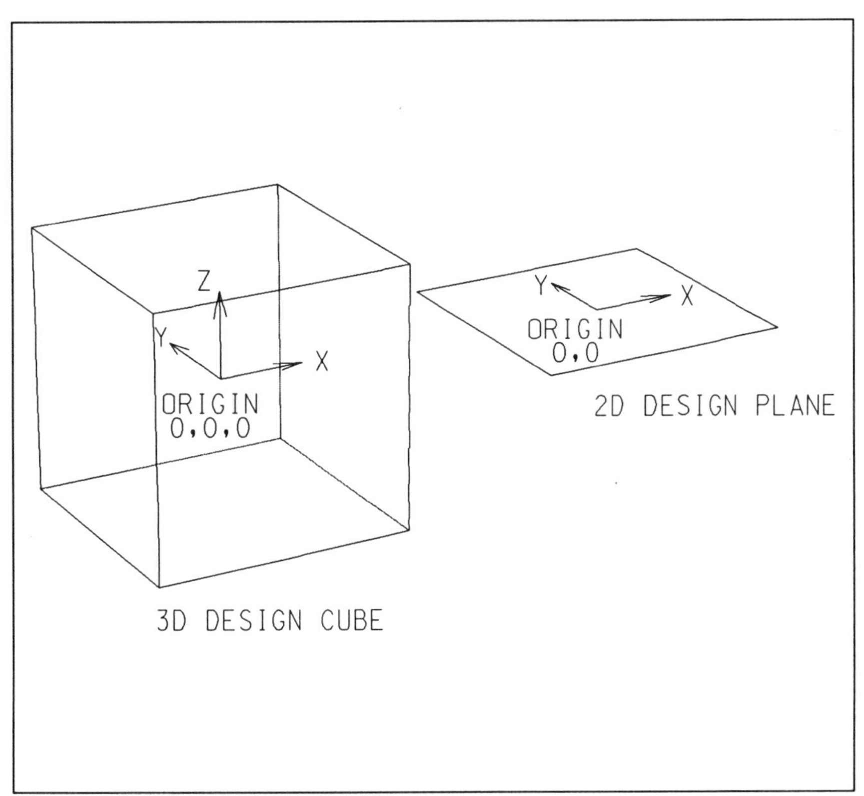

Figure 1.7 3D and 2D Design Files

To remember the orientation of the axes, do the following:

- Hold your right hand with the thumb, fore-finger and middle finger pointing at right angles to each other as in figure 1.8.
- Consider the thumb to be pointing in the positive direction of the X axis.
- The fore-finger points in the positive direction of the Y axis and the middle finger in the positive direction of the Z axis.

No matter how you move or rotate your hand the axes stay perpendicular to one another. Similarly, when we rotate our design cube, its axes stay in their relative orientations to each other. This concept will become more important when we discuss views, and view manipulations.

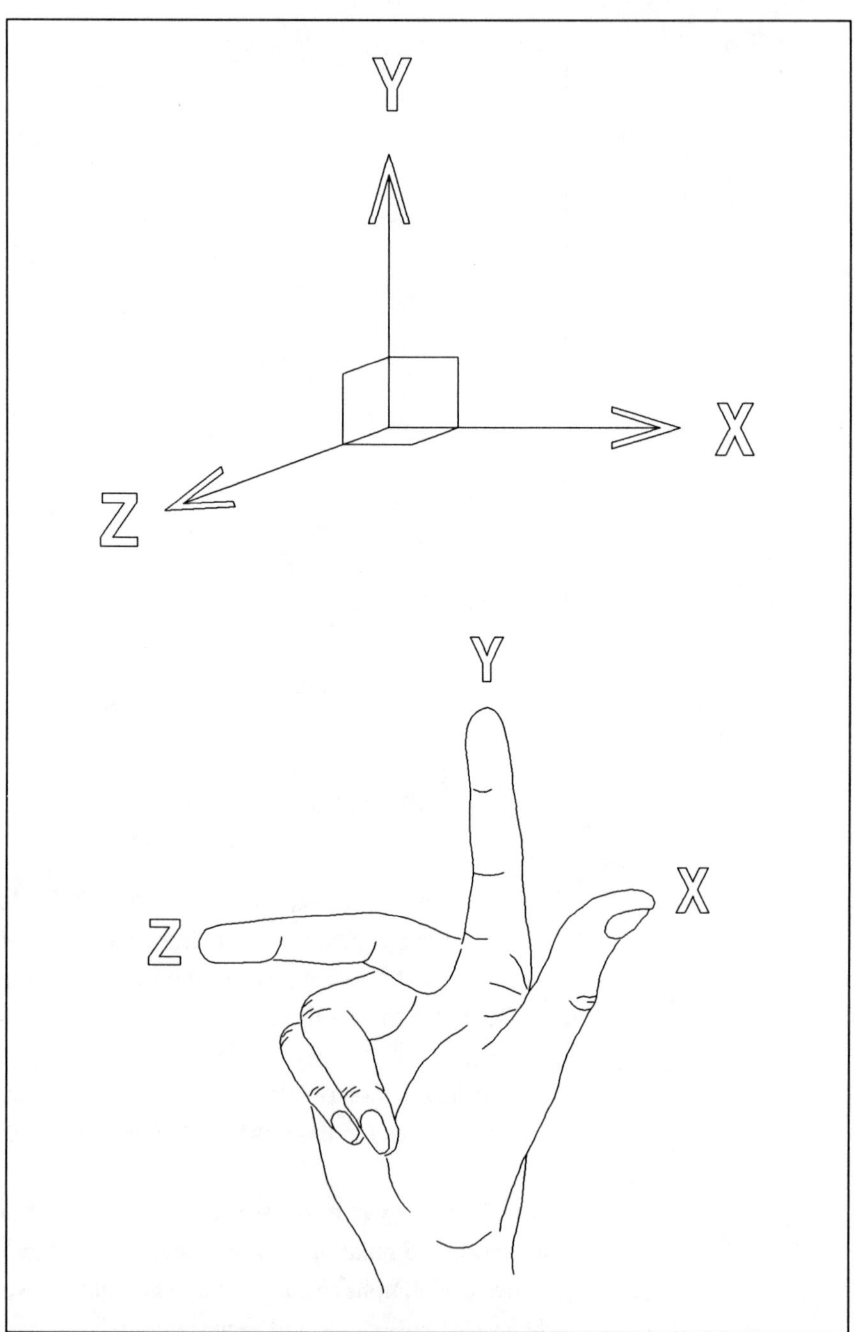

Figure 1.8 Orientation of Design File Axes

Wireframe Display

As we construct a model in the design cube, it displays by default in 'wireframe' form. That is, we can see through the model and only the outlines or boundaries of surfaces display. To get a more realistic picture, we can use the hidden line removal, rendering, and edges utilities. These are discussed in later chapters. First, we have to be capable of creating a model.

With version 4 of MicroStation, we may optionally set a view or views to display constantly with hidden lines removed or as rendered view/s. As we add elements to the design any views set to be rendered or have hidden lines removed update as well. Having a view or views set to other than wireframe mode slows the system because of the many calculations required. We will discuss this in the chapter on rendering. For our normal design work, we use the standard wireframe mode.

Figure 1.9 shows a simple model in both wireframe mode and after performing hidden line removal.

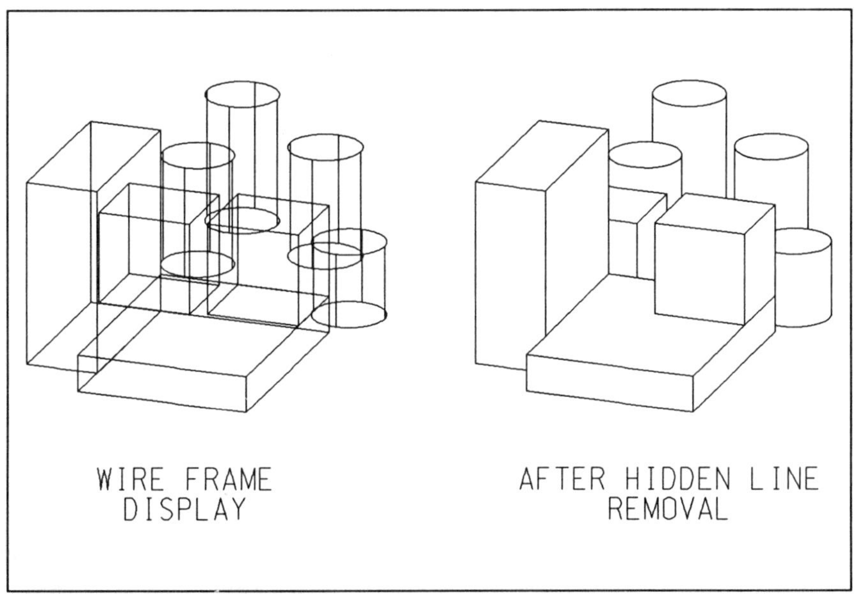

Figure 1.9 Screen Displays

Views

To work successfully in 3D we must be able to look at, or view, our design from any direction. This is possible with a physical model. While a physical model is usually a scaled version of the real thing, with CAD we create the model at full size. To change the direction, from which we are looking at our model, we rotate the DESIGN CUBE to the orientation we require. We do not rotate the model about its co-ordinates, we rotate the 'world' of the model, co-ordinate system and all. The world is the DESIGN CUBE. Later, we will see how to rotate the design cube about any of the axes. Now we will look at the standard viewing parameters available to us.

MicroStation has 7 standard views. Six of these views - TOP, BOTTOM, FRONT, BACK, RIGHT and LEFT - are perpendicular to planes of the Design Cube (i.e., orthogonal). Additional to these, there is a seventh standard view that we call ISO. The ISO view sets our eye point to a location south-west (front-left) and about 35 degrees above horizontal. It does not line up with any of the design cube axes or planes.

Figure 1.10 shows a wireframe model of a cube with the viewing locations of the standard views and how they appear on the screen. (The cube is similar to the model that is present in the standard 3D seed files as delivered.)

In this example, all views are looking at the same cube from different directions.

A parallel in real life could be a T.V. producer at a telecast of a football game. He, or she, would normally have several monitors simultaneously showing scenes from camera locations around the ground. From these monitors, scenes are selected to transmit. All scenes are of the same football game, but from different viewing locations. Similarly, we can view our model from up to eight locations at once using views in the quadrants. With version 3.3 we need a dual screen configuration to display eight views simultaneously. Version 4 allows us to have all eight views on the one screen, if we wish. We select the view that is easiest to work in at any time. We can use the other views to assist us when necessary. We can zoom in or out of any view when required.

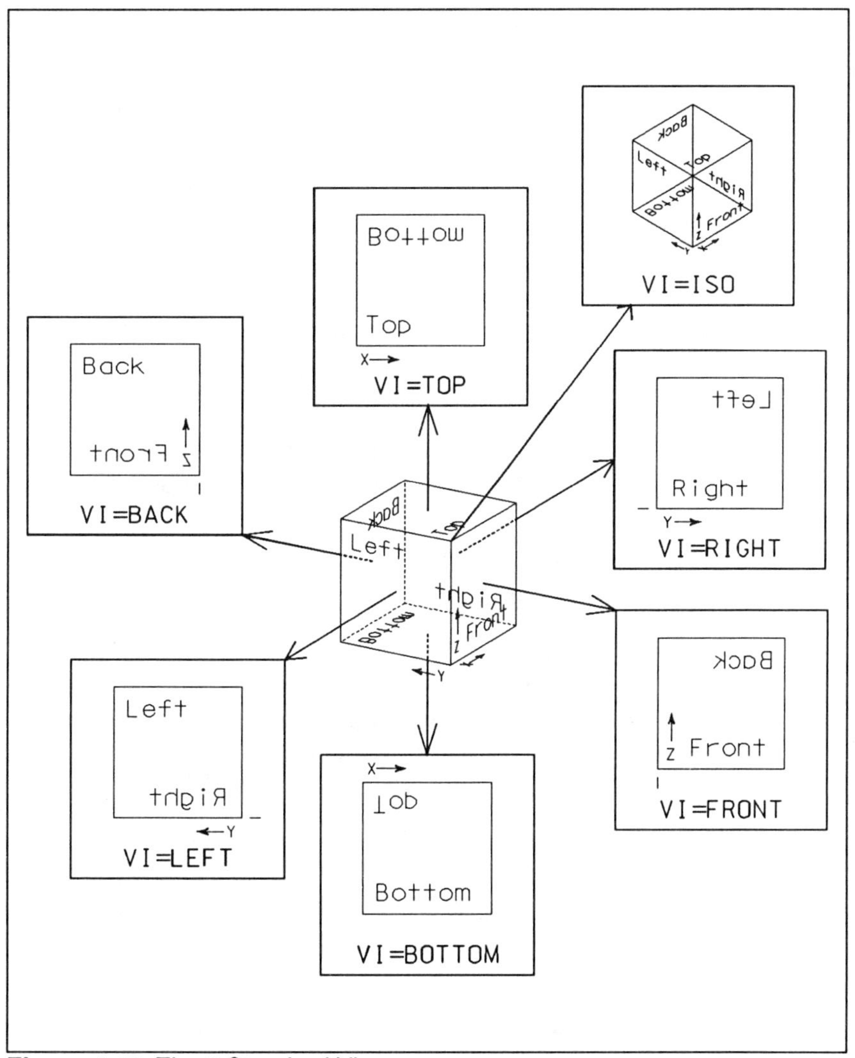

Figure 1.10 The 7 Standard Views

Names of the views refer to the position from which we are looking at the design cube.

For example (refer to figure 1.10), when we look at our screen:
In a **TOP** view

- We are looking down on the design cube from the top.
- The XY plane is parallel to our screen (like a 2D file).
- X is positive from left to right (horizontally).
- Y is positive from bottom to top (vertically).
- Z is positive toward the viewer and perpendicular to the screen.

In a **FRONT** view

- We are looking into the design cube from the front.
- The XZ plane is parallel to our screen.
- X is positive from left to right (horizontally).
- Z is positive from bottom to top (vertically).
- Y is positive away from the viewer and perpendicular to the screen.

In a **RIGHT** view

- We are looking into the design cube from the right.
- The YZ plane is parallel to our screen.
- Y is positive from left to right (horizontally).
- Z is positive from bottom to top (vertically).
- X is positive toward the viewer and perpendicular to the screen.

To recall the standard views, the key-in is the same as that for saved views, namely 'VI=viewname'. Keying in 'VI=TOP' followed by a data point in the required view or quadrant of the screen would set that view so that we see the DESIGN CUBE from the TOP.

While the key-in is the same, there is an important difference between STANDARD VIEWS and SAVED VIEWS.

A SAVED VIEW remembers levels that were on and off, also the location of the eye-point and how far zoomed in or out we were. A STANDARD VIEW only relocates the eye-point to be in the various positions, relative to the design cube, as described previously. Our eye-point remains the same distance from the model.

STANDARD VIEWS cannot be deleted. If we accidentally create a SAVED VIEW with the same name as a STANDARD VIEW, then the SAVED VIEW takes priority. We disable the particular STANDARD VIEW. To regain the STANDARD VIEW simply delete the SAVED VIEW (using DV=viewname).

'Hands-on' is the best way to learn so enter your 3D training file. Probably you will see the cube mentioned earlier. To commence with a clean file, delete all elements and then compress the file. Regularly compressing your files is a good habit to adopt. Keeping them to a minimum size, reduces search times etc.

In the following exercises, reference will be made to one screen only. Those with dual screen configurations may use either or both, but the exercises will be directed to using a single screen with four views or quadrants. Version 4 allows users to vary the size of their views. Initially, for these lessons, make the four views equal sizes.

Set up your screen with 4 quadrants/views such that views 1-4 are ISO, TOP, FRONT and RIGHT respectively:

- Enter 'VI=ISO' then data point in top right view.
- Enter 'VI=TOP' then data point in top left view.
- Enter 'VI=FRONT' then data point in lower left view.
- Enter 'VI=RIGHT' then data point in lower right view.

Note!

With version 4, we are told the orientation of the view in the window title bar. That is, if the window title bars are enabled, and the window is displaying a standard view. Note also that, if we 'Tile' the screen, the position of views 1 and 2 reverses. View 1 becomes the top left window, and view 2 the top right window. There is no indication of the view orientation with version 3.3

Further, in preparation for the exercises:

- Make the active level 1 and the active weight 5.
- Place an active point at 0,0,0 as follows:

Select *Place Active Point* tool or key in PLACE POINT.

- Key in XY=0,0,0 for the location of the point.
- FIT each view.

The third figure of the 'XY=' key-in is for the Z value. In 2D we only had X and Y coordinates. 3D has a third coordinate value to consider, that of the Z axis. As with 2D, if a value is 0, it may be omitted. For example, to copy an element 20 units in the Z direction, the key-in 'DL=,,20' is the same as 'DL=0,0,20'.

Basic Placement of Elements

Before going any further we will briefly look at two very important concepts - DISPLAY DEPTH and ACTIVE DEPTH. A detailed discussion of both can be found, commencing page 1-26.

Each view in a 2D file displays a selected area of the overall design plane. We can change the amount displayed with ZOOM and WINDOW AREA. In a 3D file, each view displays a selected volume of the overall design cube. The volume is bounded by the WINDOW AREA of the view and the DISPLAY DEPTH of the view. The hatched area, in figure 1.11, indicates the WINDOW AREA of the view. As you can see, the 3D design file also has a depth factor. For example, the design file might be 20,000 units deep, but the view may only display the first 100 units of that, or any other slice. It is possible to be looking at a TOP view, with all the levels on, for instance, and not be able to see our model. Even when we zoom out to the maximum window, the model does not appear. The reason would be that the DISPLAY DEPTH is set such that the model is not contained within it. A FIT will change the display parameters, both window area and depth of display, to include all the elements.

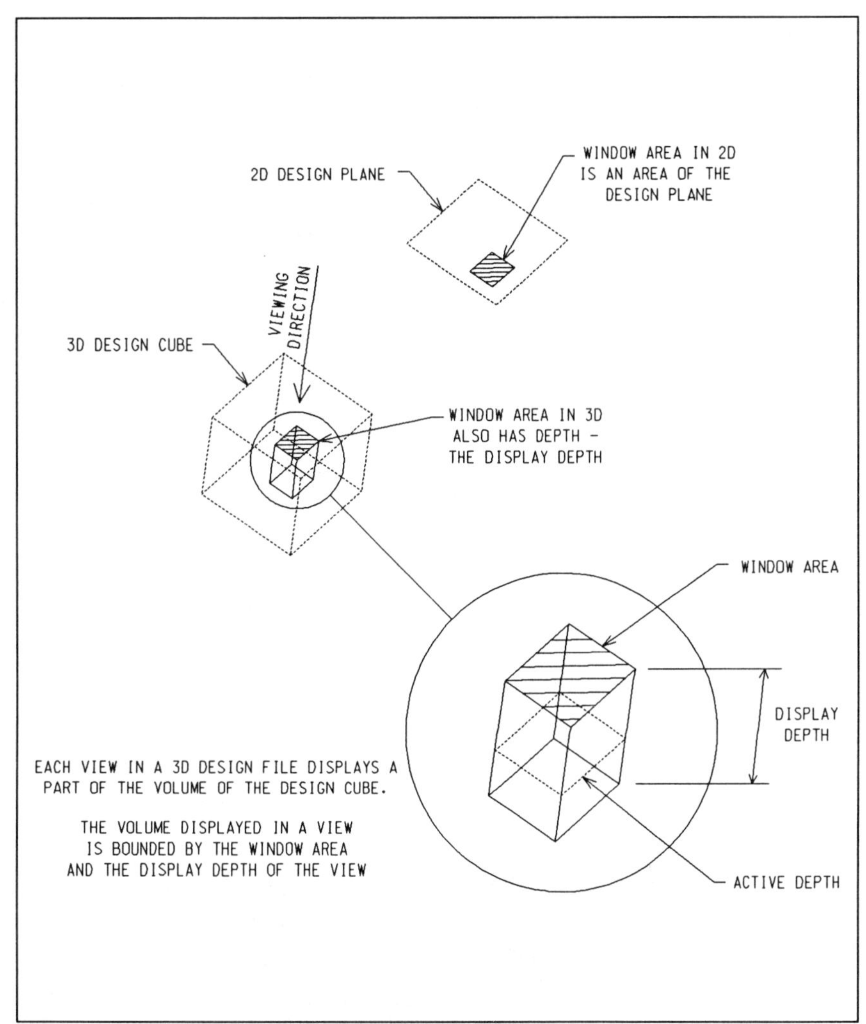

Figure 1.11 Display Depth and Active Depth

Any elements we place in a view, with data points, will fall somewhere within the depth parameters of the view (i.e., the DISPLAY DEPTH). Where the elements fall, is on a plane known as the ACTIVE DEPTH. The plane of the ACTIVE DEPTH is always parallel to the view being used.

Note!

ACTIVE DEPTH is always within the limits of the DISPLAY DEPTH.

Placing elements in a 2D file is like drawing on a sheet of paper. The elements all appear on the same plane or sheet of paper. In 3D we place elements in space. We can place them horizontally (like floors), vertically (like walls), or at any other angle or direction we choose. The problem is how to tell the system (i.e., MicroStation), the exact orientation in which we want the element placed. This is where the views again come into significance. We could key in values (i.e., X,Y, and Z co-ordinates) for each point of every element. That would solve the problem but would be extremely laborious. The answer is that MicroStation takes as a default, a plane parallel to the view we are using.

When we place elements such as BLOCKS, CIRCLES WITH RADIUS, TEXT etc. they are placed parallel to the plane of the view we are using. If you find it easier to remember, this is parallel to the screen in the view we are using. To prove this, do the following:

First, make the active color yellow, the active weight 0, the text size 50 (TX=50) and the text justification bottom left. Now:

- Select the *Place Text* tool from the menu. Make sure that you don't select VIEW INDEPENDENT text by mistake.
- Place the word 'TOP' in the TOP view (view 2 - top left quadrant) using a tentative and accept to the active point in the view. FIT each view to show the text.
- With version 3.3, you may need to ZOOM OUT in the FRONT and RIGHT views to see the text, if it gets lost in the borders. When we use FIT with a single element the element appears at the bottom of the view for a horizontal line and along the left edge for a vertical line. With version 4, when there is a single line element, a FIT puts the element in the center of the view.

Notice that the word 'TOP' appears normally in the TOP view (where we placed it). In the FRONT (quadrant 3) and RIGHT (quadrant 4) views it appears as three dashes and one dash respectively (refer figure 1.12). Because we placed the word 'TOP' in the TOP view, it is in the XY plane of the design cube. The XY plane in a TOP view is parallel to the screen.

In the FRONT view we are looking along the XY plane, in the Y direction. We see the three letters, edge-on, as three dashes. Think of a sheet of transparent paper, with the word TOP printed on it. If we look at this sheet of paper, along its plane from the bottom edge, we would see the word as three dashes.

In the RIGHT view we are again looking along the XY plane, but this time in the X direction. The three letters here, are one behind the other and appear to us as a single dash.

In the ISO view (top right quadrant) you will see that the text appears skewed but legible. Remember that the ISO view does not align with any of the design cube's planes.

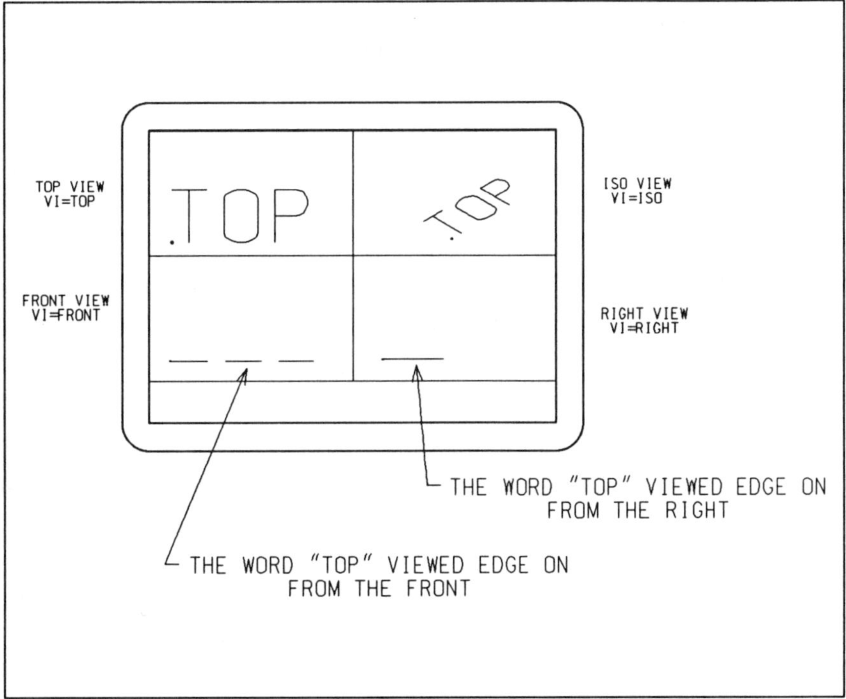

Figure 1.12 Viewing the Text Element

To continue:

° Change the active color to red and place the word 'FRONT' in the FRONT view (view 3 - bottom left quadrant).
° Use a tentative and accept on the active point in that view. If it is obscured by the word TOP, tentative to the left end of the 3 dashes.
° FIT each view.

Again, look at the other views. Notice that this time the word 'FRONT' appears as five dashes in the TOP view and a single vertical dash in the RIGHT view. We placed it in a FRONT view therefore it is vertical.

To finish the exercise:

° Change the active color to blue and place the word 'RIGHT' in the RIGHT view (view 4 - bottom right quadrant), at the active point.
° FIT each view.

Your screen should resemble figure 1.13. All four views are displaying the same three words but from different directions. Tentative to the dashes in any view. By checking in the other views you will see the relevant element highlight in each. Because text is two-dimensional, it appears as dashes when viewed edge-on. We have the four quadrants of our screen set up with different views, or viewing directions. We can simultaneously see the three words from different directions. Words that aren't seen clearly from one direction, are clearly visible when viewed from another location. In the ISO view all three words are both visible, and legible.

Be sure that you understand what we have just discussed. It is important to understand what is being displayed on your screen, and why.

Leave the text as a reminder for which view is which, in the following exercises. This applies mainly to version 3.3 users. With version 4, we can see which of the standard views a particular window is, by checking the window title bar.

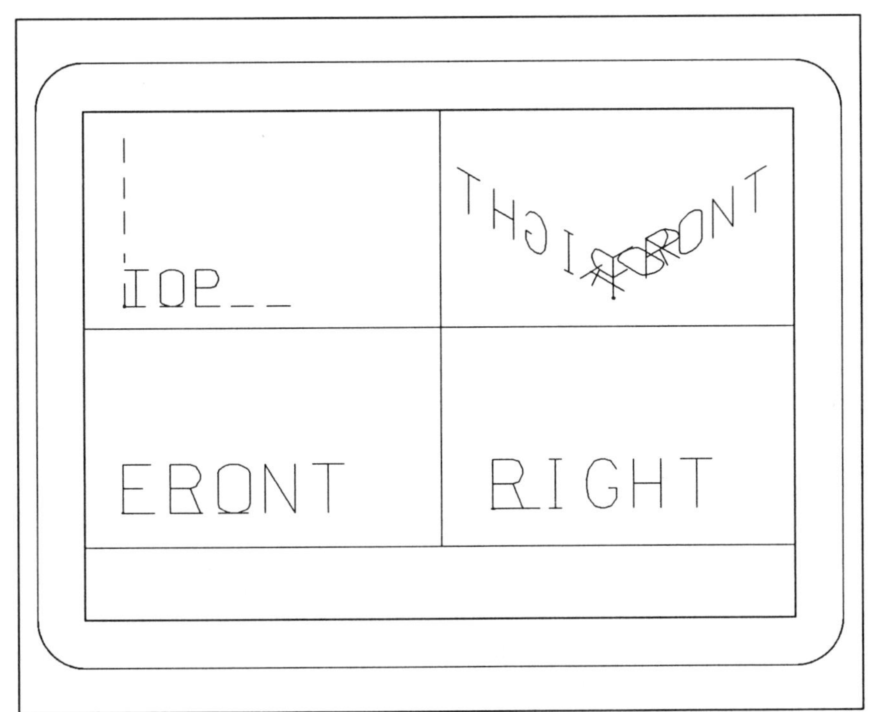

Figure 1.13 Placing Text in 3D

Before moving on, we will use a technique to help us in the exercises:

- Place another active point at XY=1000,1000,1000.
- Fit each view.
- File Design

Our reason for placing this second active point is to ensure that each view displays the same part of the design file. We know that a view in a 3D file is looking at part of the overall volume of our design file. The second active point is, in effect, a diagonally opposite point of an imaginary cube within the design cube (figure 1.14). This means that each view, when FITted, is looking at the same volume within the design cube. Our initial exercises will be carried out within this smaller 'cube'. Therefore, we will not have to FIT or ZOOM OUT to see elements in each view, after placing them.

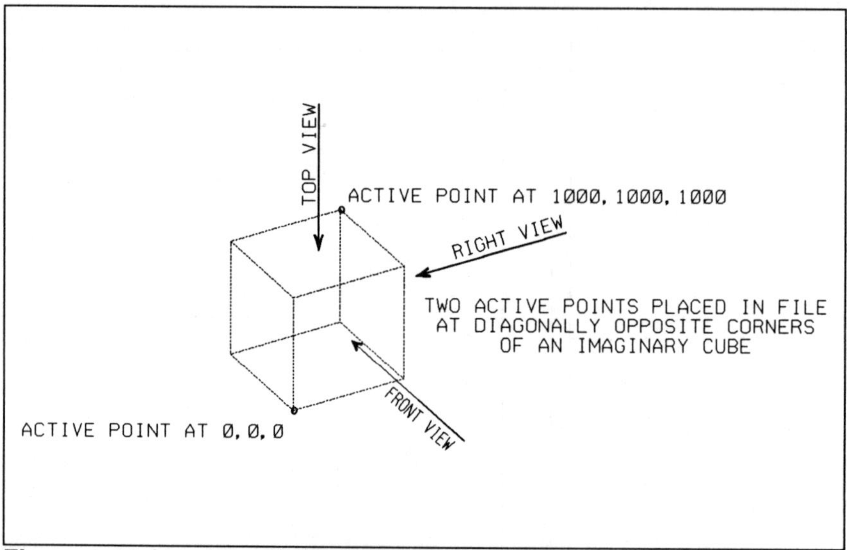

Figure 1.14 Imaginary Cube Between Two Points

Your screen should resemble one of those illustrated in figure 1.15. With the version 4 screen, note that the viewing orientation of each window is displayed in the window title bar (e.g., View 3 - Front).

At this point, COMPRESS your file. This will allow you to use the UNDO function to clear away unwanted elements, without losing the text or active points.

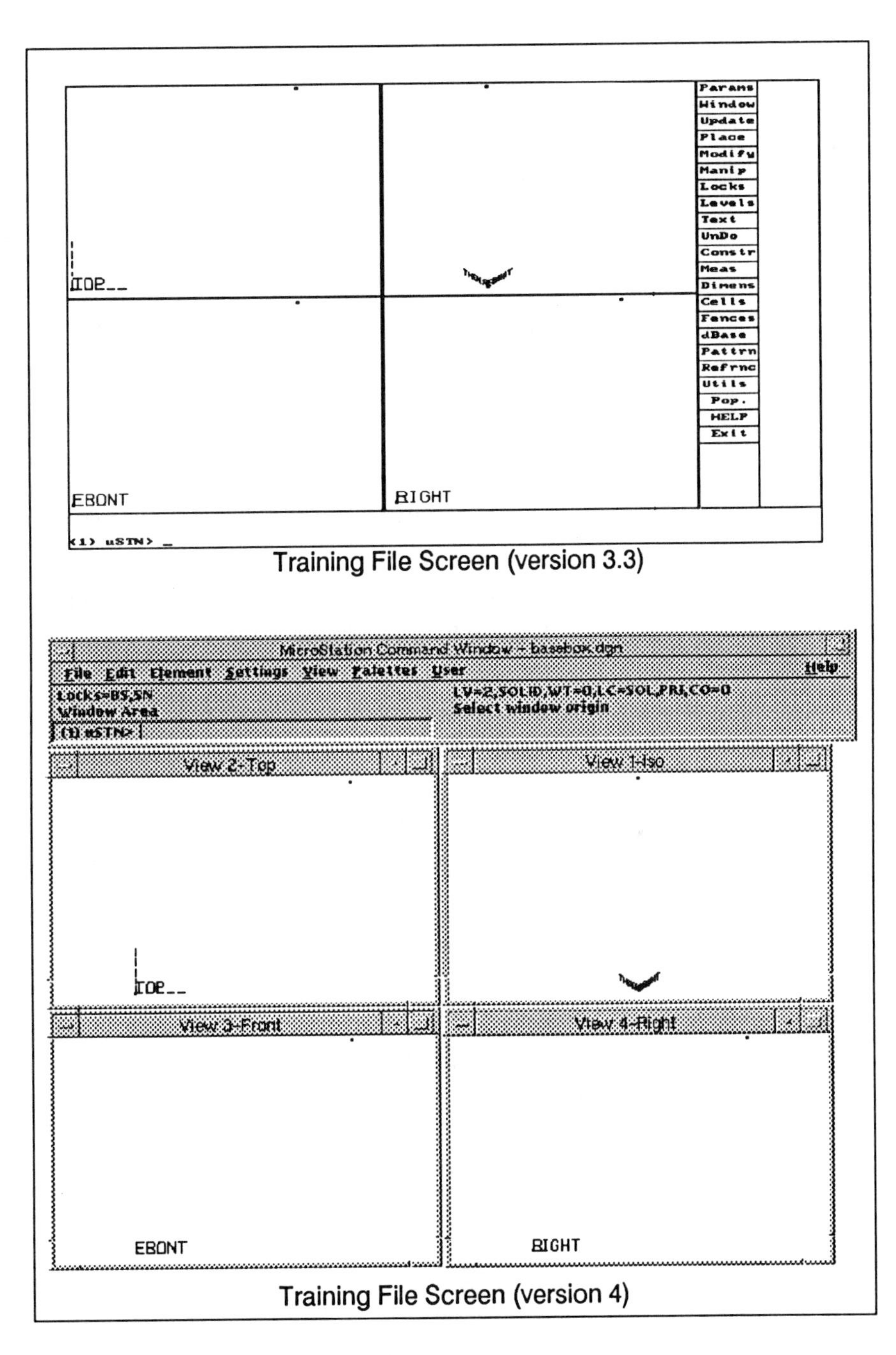

Training File Screen (version 3.3)

Training File Screen (version 4)

Figure 1.15 Training File Ready to Commence

We will now look at placing other elements in 3D

Do the following:

- Make the active level 2 and try placing an element, such as a block, in the various views. Do this with DATA POINTS only. Don't tentative to other elements.

Notice that as you place elements in a particular view, they appear as 'lines' in the other orthogonal views. Like a sheet of paper, a block, when viewed edge-on, appears as a line. The same applies for circles, shapes, polygons etc. They are flat, or two-dimensional, surfaces. They are parallel to the screen, in the view that we used to place them.

- Delete, or UNDO the elements just placed.

To further investigate how elements are placed in a 3D file:

- Place a block or shape in the TOP view with DATA POINTS only (don't tentative to any other elements).
- Change the active color each time and place other shapes, in the TOP view.

Check in the FRONT and RIGHT views. Notice that all the shapes are in the same plane. In these views they line up exactly and would appear to be collinear if they overlapped (figure 1.16).

Repeat the process in the RIGHT and/or FRONT views and notice how the elements that you place are on a particular plane in each view. This applies when we place elements with data points, without a tentative to an existing element. This plane, in each view, we call the ACTIVE DEPTH.

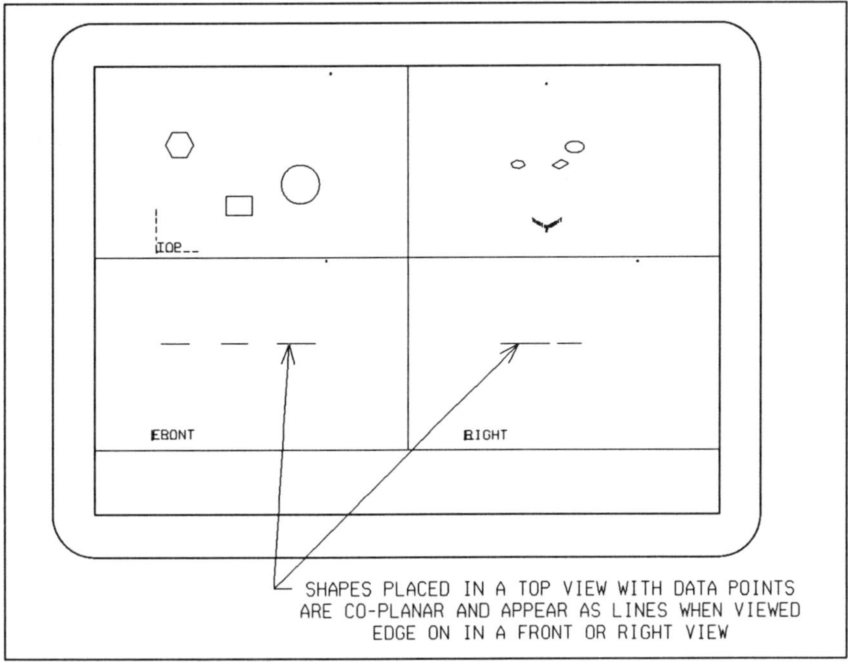

Figure 1.16 Placing Shapes with Data Points

Active Depth

ACTIVE DEPTH is a plane, parallel to the screen in each view, where elements will be placed by default.

Elements are placed at the ACTIVE DEPTH if you don't tentative to an existing element, or use a precision input. Snapping to an existing element over-rides the ACTIVE DEPTH. Because the ACTIVE DEPTH plane is variable for each view, we call it VIEW DEPENDENT. The ACTIVE DEPTH plane depends on which view we use.

ACTIVE DEPTH for the TOP and BOTTOM views is in the XY plane of the design cube with a depth value along the Z axis.

For the FRONT and BACK views it is in the XZ plane of the design cube with a depth value along the Y axis.

For the RIGHT and LEFT views it is in the YZ plane of the design cube with a depth value along the X axis.

The active depth plane is always parallel to the plane of the view, which is, in turn, parallel to the screen. Consider that your screen has X, Y, and Z axes. The origin of the screen's axes is the same as that of the design cube.

- The X axis is horizontal and positive from left to right.
- The Y axis is vertical and positive from bottom to top.
- The Z axis is perpendicular to the screen and positive toward the viewer.

These axes never change. We rotate the design cube when we choose the various views (TOP, FRONT, RIGHT etc.), but the axis system of the screen stays the same.

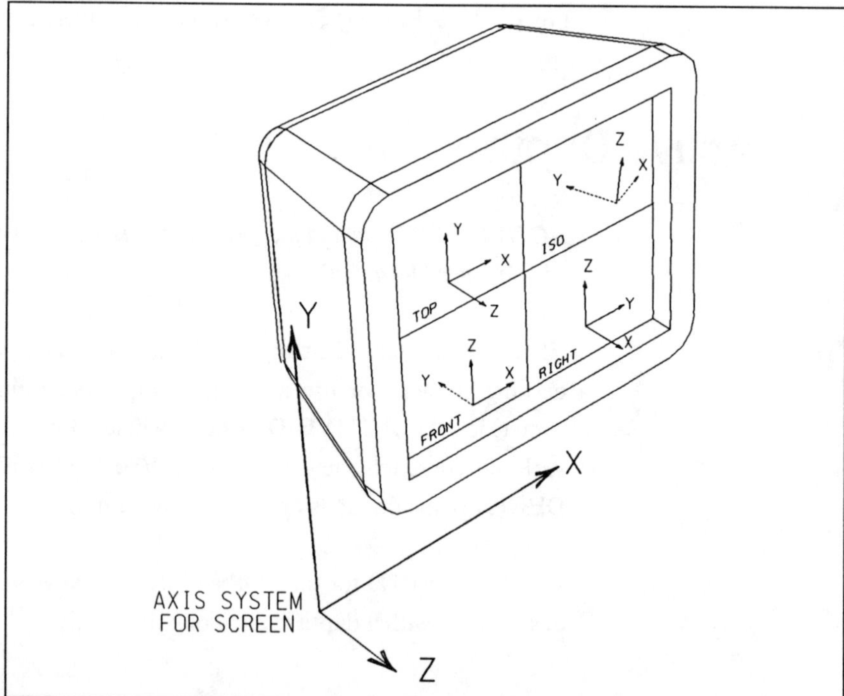

Figure 1.17 The Screen or View Axis System

ACTIVE DEPTH, or ACTIVE Z VALUE, is always in the XY plane of the screen with a depth value along the Z axis of the screen. That is, it is a plane parallel to our screen, no matter which view we are working with.

Figure 1.17 shows how the axes system of the design cube is located, relative to the screen's axes, for the views as you should have them set. Only in a TOP view are the two axes systems completely aligned.

ACTIVE DEPTH may be specified either with a key-in, or graphically with data points.

Setting Active Depth by key-in

We can specify our own value for the ACTIVE DEPTH with a key-in. The key-in is 'AZ='. Think of it as '*A*ctive *Z* value ='. Remember that it is the Z value relative to the coordinates or axis system of the screen. It always refers to a plane parallel to your screen. As an exercise:

Delete the elements on level 2. FIT the TOP view.

- Key in AZ=500
- Select the TOP view when prompted.
- Place a red block in the TOP view with DATA POINTS.

Now key in AZ=200

- Select the TOP view when prompted.
- Place a green circle in the TOP view, again with DATA POINTS.

For the third one, key in AZ=800

- Again select the TOP view when prompted.
- Place a blue hexagon in the TOP view, also with DATA POINTS only.

Your screen now should look similar to figure 1.18

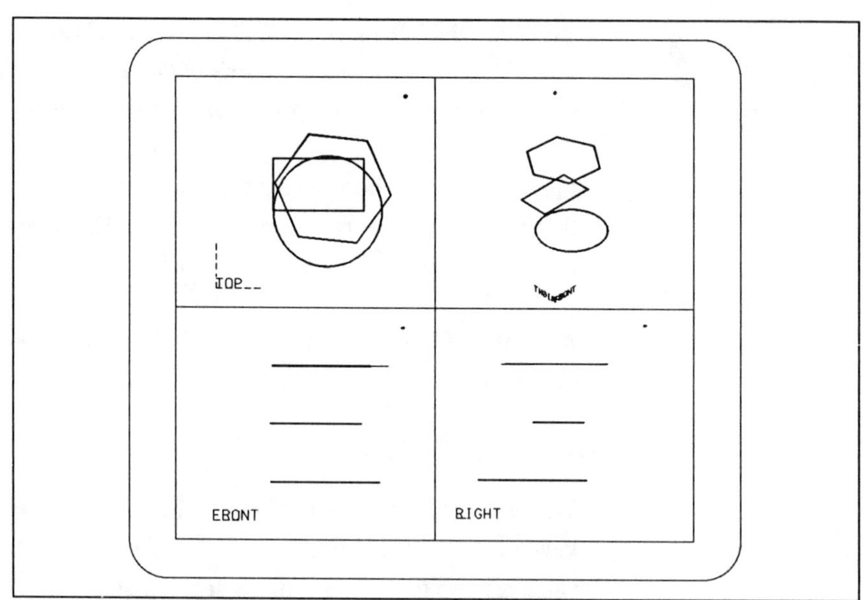

Figure 1.18 Placing Elements at an Active Depth

In the FRONT and RIGHT views we see that the 3 elements are separated vertically though we placed them in the same view (TOP). Changing the ACTIVE DEPTH in the view prior to placing each element caused them to be placed on the different planes as specified. To check, tentative to each element. Check the Z value in each case (i.e., the third figure of the co-ordinate readout). We placed the elements in a TOP view, so the Z value for the screen is the same as the Z value for the design cube.

It is not obvious, looking at the TOP view, that the elements are separated vertically. The FRONT and RIGHT views, though, show the separation clearly. Using the various views helps us to visualize our design more clearly. In particular, this helps when working in wireframe mode.

Setting Active Depth graphically

Once we have commenced a design, there may be an existing element already at the required ACTIVE DEPTH for a particular view. To place another element at the same depth we could tentative to the element to find the coordinates for it. We would then have to determine which axis is aligned with the screen Z axis in order to key in the correct value with the 'AZ=' key-in. This would work but it is cumbersome. MicroStation has another tool which we can use to achieve the same result. We can set the ACTIVE DEPTH graphically.

For this exercise we will place another green circle, in the TOP view, at the same ACTIVE DEPTH as the existing green circle.

While the key-in is the same for both versions 3.3 and 4, there are differences in the operation of the tool. We will deal with each version separately.

Version 3.3

Version 3.3

Make the active color green. Now:

- Select SET ACTIVE DEPTH from the menu or use the alternative key-in 'DEPTH ACTIVE'.
- When prompted to 'ENTER ACTIVE DEPTH POINT', tentative to and accept the green circle. This may be done in *any* view (including the view you are working in). Other views should be checked to see that the correct element highlights if there is any doubt.
- Place a data point in the TOP view when prompted to 'SELECT VIEW'.

We have now set the ACTIVE DEPTH in the TOP view to the same value as the existing green circle.

- Place a circle in the TOP view with data points. Check in the other views to ensure that it is in the same plane as the first circle.

Version 4

We can select the *Set Active Depth* tool from one of the standard menus, or use the key-in 'DEPTH ACTIVE'. As well, we can select the tool in the '3D' palette, which we choose in the 'Palettes' pull-down menu.

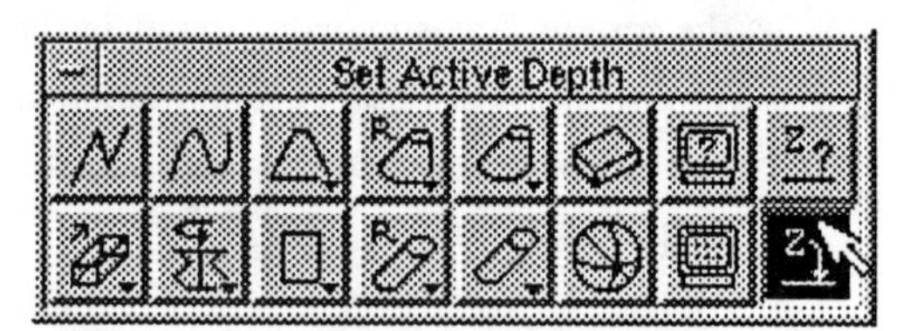

Figure 1.19 Set Active Depth Tool

Version 4

To demonstrate how this works, do the following:

- FIT each view and then WINDOW AREA into a smaller section in the central area of the TOP view.
- In the 3D palette, select the *Set Active Depth* tool, (lower left hand corner - figure 1.19).
- When prompted for the view, select the TOP view.

You will notice dashed lines, in all views, indicating the viewing parameters of the selected TOP view. Both the display volume of the view and the active depth plane are dynamically displayed, with different style dashed lines, as shown in figure 1.20.

We can see in each view, just which part of our design cube is being displayed in the selected view. The ISO view, in figure 1.20, shows the whole display volume for the selected TOP view. If your display has parts of these lines missing in any of the views, don't be concerned. It simply means that the display depth of the other view/s is not big enough to contain the whole display volume of the selected view. This does not affect the operation of the tool.

We are now prompted to 'SET THE ACTIVE DEPTH POINT':

- Move your cursor to a view other than the selected view.

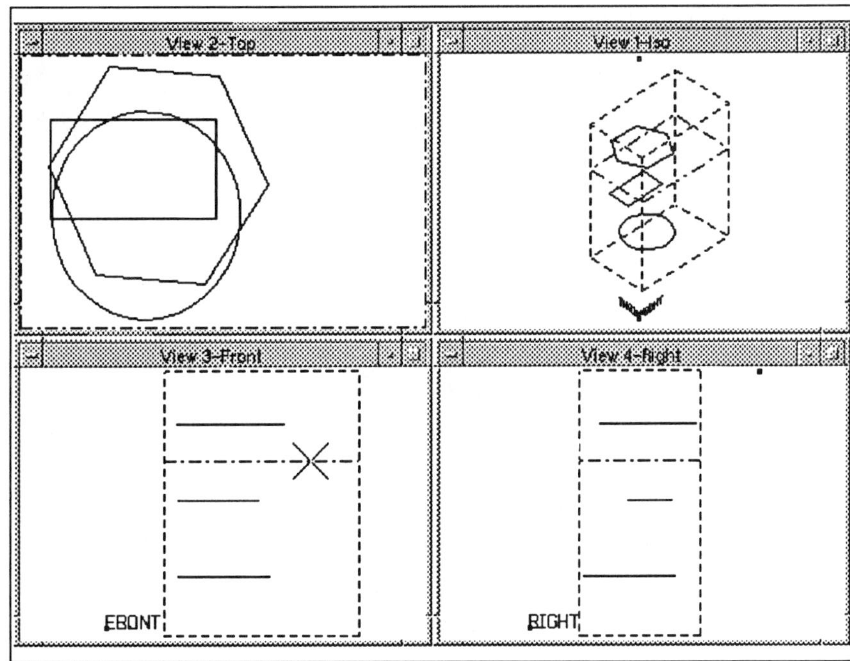

Figure 1.20 Setting Active Depth with Version 4

Notice that as you move the cursor in the other views, the dashed line representing the ACTIVE DEPTH plane moves also. We are being shown dynamically, just where the active depth plane will be located if we place a data point. Unlike the version 3.3 method, *we must use a view other than the selected view* to set ACTIVE DEPTH.

- Tentative to and accept the green circle in one of the other views. This will set the ACTIVE DEPTH in the selected view (TOP) to the same value as the existing green circle.
- Place a circle in the TOP view with data points. Check in the other views to ensure that it is in the same plane as the first circle.

Both Versions

By using previously placed elements to set ACTIVE DEPTH, we avoid the chance of making errors with keying in values. This assumes that the existing element is in its correct location.

ACTIVE DEPTH also may be specified for the other views. It will *always* be a Z value as far as the screen co-ordinates are concerned. Its value, related to the design cube, will vary in each view.

> In a RIGHT view the ACTIVE DEPTH will be an X design cube value. The X axis of the design cube, in a RIGHT view, aligns with the Z axis of the screen.
>
> In the FRONT view the ACTIVE DEPTH will be a Y design cube value, The Y axis of the design cube, in a FRONT view, aligns with the Z axis of the screen.

Note:

> In the FRONT view a positive Z value for the screen is equivalent to a negative Y value for the design cube. Y, for the design cube, is positive into the screen. Z, for the screen, is negative into the screen. Refer again to figure 1.17 if this is unclear.

With an ISO view the Z value for the screen does not align with any plane of the design cube.

To check the ACTIVE DEPTH in any view we type 'AZ=$' followed by a data point in the required view. The system typically responds with:

> 'View 1: Active Depth= 475.306'

Another method is to tentative to a vacant area of the view and check the relevant value of the co-ordinate readout. For example, ACTIVE DEPTH will be the Y value in a FRONT view, and the X value in a RIGHT view as explained above. A tentative point will default to the active depth plane if it finds no elements.

Using the *Show Active Depth* tool, is another option for version 4 users. Its icon is located directly above that for setting active depth. We are prompted to select a view, which gives a response is similar to that from the key-in 'AZ=$'.

ACTIVE DEPTH goes hand in hand with another depth parameter, DISPLAY DEPTH.

Display Depth

WINDOW AREA allows us to select an area of our file to work in. DISPLAY DEPTH allows us to select a 'slice' of our file parallel to the screen or view. Each view has a window area and a DISPLAY DEPTH that define a volume of the design cube.

In each view we see a selected volume out of the total DESIGN CUBE volume. Figure 1.21 depicts a Design Cube with example viewing parameters for TOP, FRONT and ISO views. Each view has a Window Area (shown hatched), and a Display Depth. For the illustration, each view is looking at a different part of the Design Cube.

ACTIVE DEPTH is a default plane, in each view, that we work on. DISPLAY DEPTH is a depth, in each view, that we see. It is a display parameter and is defined by FRONT and BACK clipping planes in each view. The FRONT clipping plane is the one nearest the viewer and the BACK clipping plane is the plane furthest from us in each view.

Note:

ACTIVE DEPTH cannot be set outside the DISPLAY DEPTH limits. If we attempt to set the ACTIVE DEPTH to a value that is outside these limits the system responds with 'ACTIVE Z SET TO DISPLAY DEPTH'. When this occurs, MicroStation instead sets the ACTIVE DEPTH to the nearest DISPLAY DEPTH value.

Take for example, a view with a DISPLAY DEPTH of -100 to +100. If we attempted to make the ACTIVE DEPTH 175 in that view, then the system would set it to 100 and display the error message. If we keyed in 'AZ=-175' and selected this same view, then ACTIVE DEPTH would be set to -100 and the error message displayed. In each case the ACTIVE DEPTH is set to the nearest DISPLAY DEPTH value.

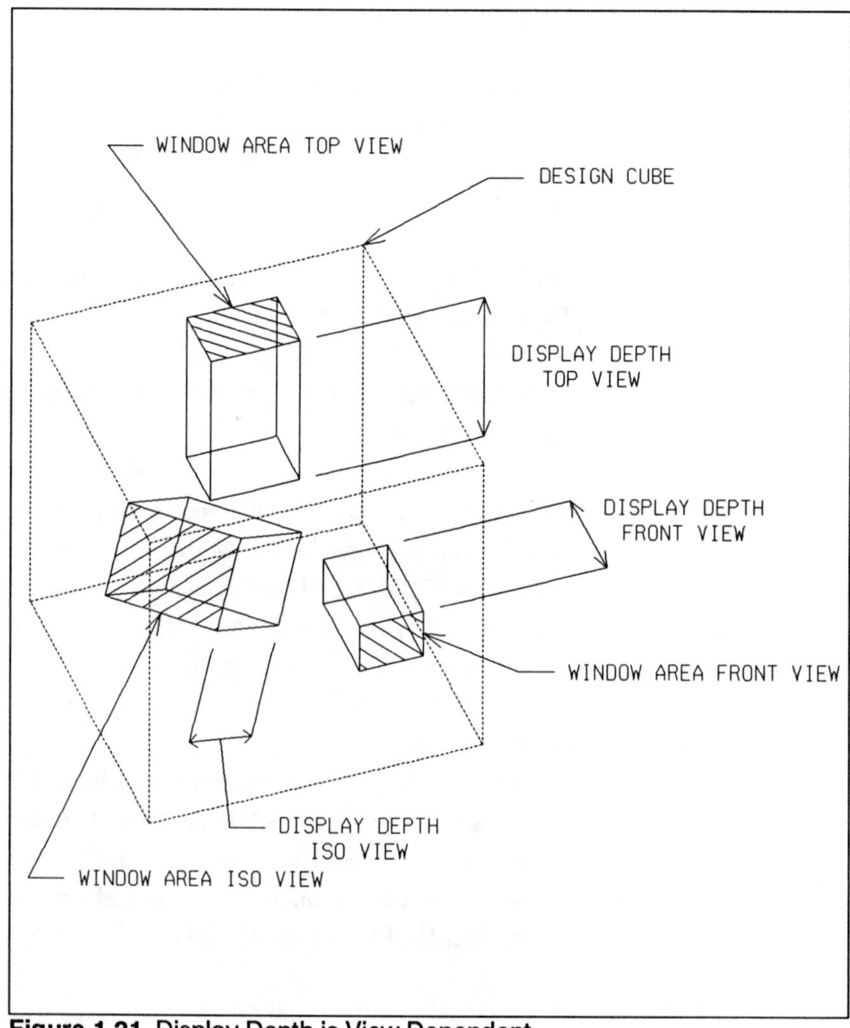

Figure 1.21 Display Depth is View Dependent

DISPLAY DEPTH, like ACTIVE DEPTH, is view dependent and may be specified with a key-in, or graphically.

Setting Display Depth by key-in

As an illustration, we will work in the TOP view. From the previous exercises, there should be a block, two circles and a hexagon displayed in the view. By checking other views we can see that they are separated in the Z axis of the design cube. In a TOP view the Z axis of the design cube aligns with the Z axis of the view or screen.

We will set the DISPLAY DEPTH so that the two circles only, display. These circles were placed at the design cube Z value of 200, while the other shapes were placed at Z values of 500 and 800.

We will set the DISPLAY DEPTH to be from 190 to 210. This will include the elements we require at depth 200, and exclude the other elements in front of, or behind our DISPLAY DEPTH.

The key-in we use for selecting DISPLAY DEPTH is 'DP = Zvalue1,Zvalue2' where the two Z values are relative to the view or screen Z axis.

It can be remembered as '*D*is*P*lay Depth = '

To display the two circles only:

° Key-in DP = 190,210 and select the TOP view when prompted.

The TOP view now only displays those elements that lie between 'view' Z values of 190 and 210 (refer to figure 1.22). Here the 'view' Z values are the same as the design cube Z values because we are using a TOP view.

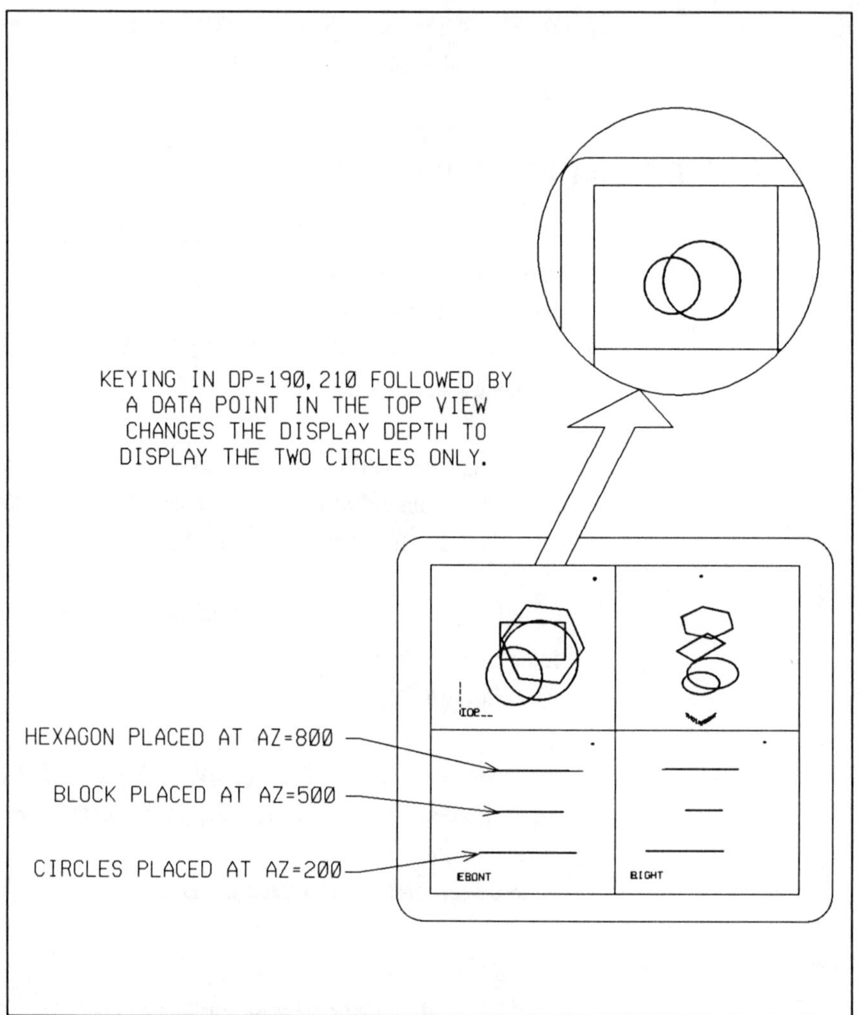

Figure 1.22 Setting Display Depth by Key-in

Now change the DISPLAY DEPTH again:

° Key-in DP = 490,510 and select the TOP view when prompted.

You should only see the element placed at an active depth of 500 in the TOP view. That is, the element you placed after keying in AZ = 500 - the red block. By using DISPLAY DEPTH we can selectively work on parts of the design without having other components behind and/or in front, confusing the issue.

Changing Display Depth Relative

We have another associated key-in that we can use to vary the display depth in a view. This key-in changes the minimum and maximum display depth settings relative to the current values.

Syntax for the key-in is 'DD = delta Z min, delta Z max'. Here, the two values - 'delta Z min' and 'delta Z max' - are the required changes to the minimum and maximum display depth parameters.

For example, your TOP view should have a DISPLAY DEPTH of 490 to 510. We could change this to be 790 to 810 using the 'DD =' key-in, as follows.

Set the DISPLAY DEPTH in the TOP view to be 490 to 510, if it is not already set. (Key in DP = 490,510 followed by data point in TOP view).

- Key in DD = 300,300.
- Select the TOP view when prompted.

Your TOP view should be displaying the element that was placed after keying in AZ = 800 - the blue hexagon. We have changed the DISPLAY DEPTH to be 790 to 810. That is, it has changed by plus 300 in both the minimum and maximum values.

The first value always alters the minimum display depth parameter, and the second value alters the maximum parameter. Omitting one of the values is equivalent to entering a 0. For example, if we wanted to increase the DISPLAY DEPTH of the TOP view to include the elements at a Z value of 200, we could key in 'DD = -600'. This would set the minimum DISPLAY DEPTH value to be 190. That is, 600 less than at present (790-600).

Using this key-in, we can shift or alter the size of our viewing 'slice' without knowing the existing Z values.

Setting Display Depth Graphically

As with ACTIVE DEPTH, it is possible to set DISPLAY DEPTH graphically. This is the way we would usually do it because it is quicker, and does not require us to key in values. We normally use a view that is perpendicular to the view in which we want the DISPLAY DEPTH set. This makes it easy to exclude elements that we don't want to appear in our working view.

We will set the DISPLAY DEPTH in the TOP view, which is along the Z axis of the design cube. The Z axis of the design cube appears as the *screen's* vertical (Y) axis in the FRONT and RIGHT views. We can use either of these views to set the front and back clipping planes for the DISPLAY DEPTH in the TOP view.

Again with setting ACTIVE DEPTH there is a variation in the procedure between version 3.3 and version 4. As before, we will deal with each version separately.

Version 3.3

Version 3.3

- FIT the TOP view so that all elements display.
- Select DISPLAY DEPTH from the menu, or use the alternative key-in 'DEPTH DISPLAY'.
- When prompted for the FIRST and then the SECOND DISPLAY DEPTH POINTS place data points just either side of the blue 'line' (figure 1.23). Do this in either the FRONT or RIGHT views (so that you include no other elements between the points). The blue line in the FRONT and RIGHT views, is an edge-on view of the hexagon.
- Place a data point in the TOP view when prompted to SELECT VIEW FOR DISPLAY DEPTH.

When we placed data points for the first and second display depth points, we were defining the front and back clipping planes. The third data point instructed the system in which view to apply these limits. Only the blue hexagon should be displayed in the TOP view as figure 1.23 shows.

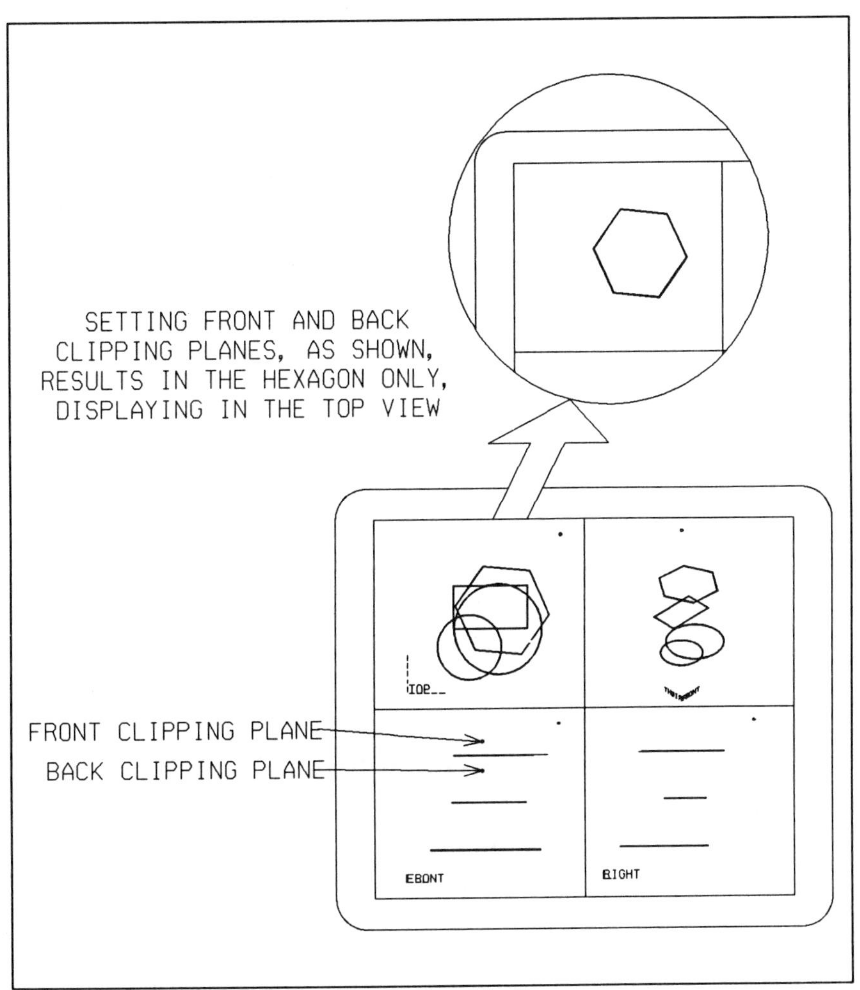

Figure 1.23 Setting Display Depth Graphically

Version 4

We can select the *Set Display Depth* tool from one of the standard menus, or use the key-in 'DEPTH DISPLAY'. As well, we can select it in the '3D' palette where it is located adjacent to the *Set Active Depth* tool (refer figure 1.24).

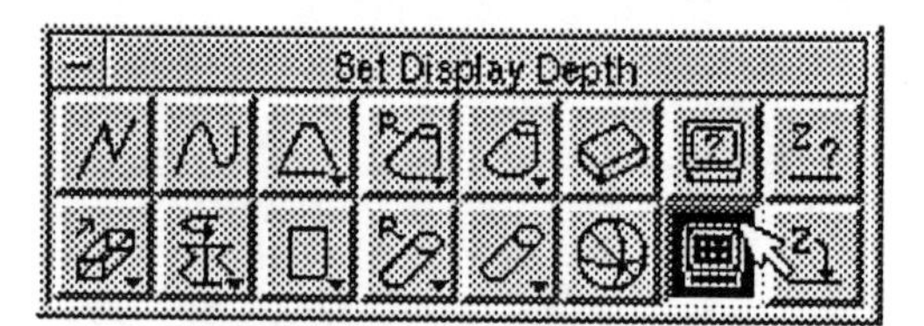

Figure 1.24 Set Display Depth Tool

Version 4

We will use a similar method to that adopted when we set the active depth.

- FIT each view and then WINDOW AREA into a smaller section in the central area of the TOP view.
- Select *Set Display Depth* tool from one of the menus, or key in 'DEPTH DISPLAY'.
- When prompted for the view, select the TOP view. As with the *Set Active Depth* tool, you will notice dashed lines appear indicating the viewing parameters and active depth plane of the selected TOP view.
- We are prompted to select the front clipping plane first.
- Move the cursor to a view, other than the TOP view. As you move the cursor notice that it moves the front clipping plane (figure 1.25). Place a data point just above the blue line.
- Now, you will see that the cursor moves the back clipping plane. Place a point just below the blue line. This causes the DISPLAY DEPTH in the selected view to change to the new parameters.

Note:

We used WINDOW AREA, in the TOP view to make its display volume less than that of the other views. This was so that the dynamic display graphics would appear in the other views. The tool will operate whether or not these graphics can display. The graphics are an aid to us. They are not necessary to the operation of the tool.

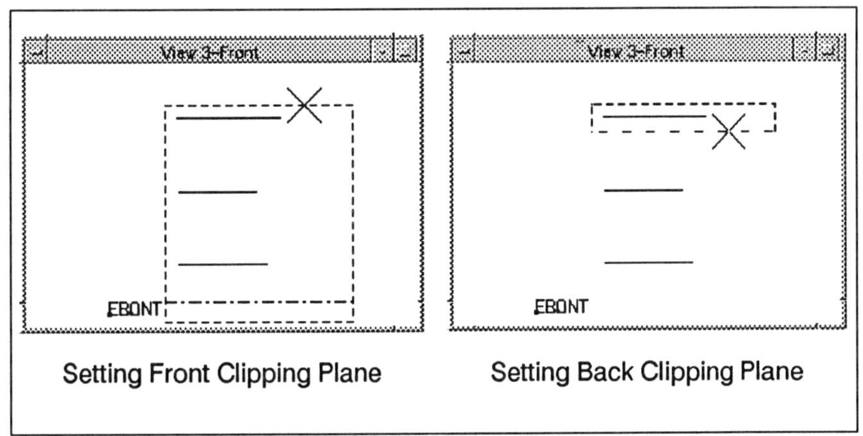

Figure 1.25 Setting Display Depth with version 4

Both Versions

Try the same procedure again. This time place the data points, so that the top and middle elements only, display. Refer to figure 1.26.

FITting a view resets the display parameters for the selected view. In 2D this gave the minimum window area required to display all the elements on the currently enabled levels. With 3D files there is also DISPLAY DEPTH. When we FIT a view, we get the minimum volume required to display all elements on the currently enabled levels. That is, the minimum window area and the minimum display depth that includes all the elements. Initially, we placed two active points in our file at opposite corners of an imaginary cube. This ensured that a FIT would display at least the same volume in each view.

If we want to check the DISPLAY DEPTH in any view we can do this by entering 'DP=$'. We then select a view when prompted and the system typically responds with:

'View 1: Display Depth= 1005.354,256.705'

With version 4 we can also select the *Show Display Depth* tool in the '3D' palette where it is located directly above the SET DISPLAY DEPTH icon.

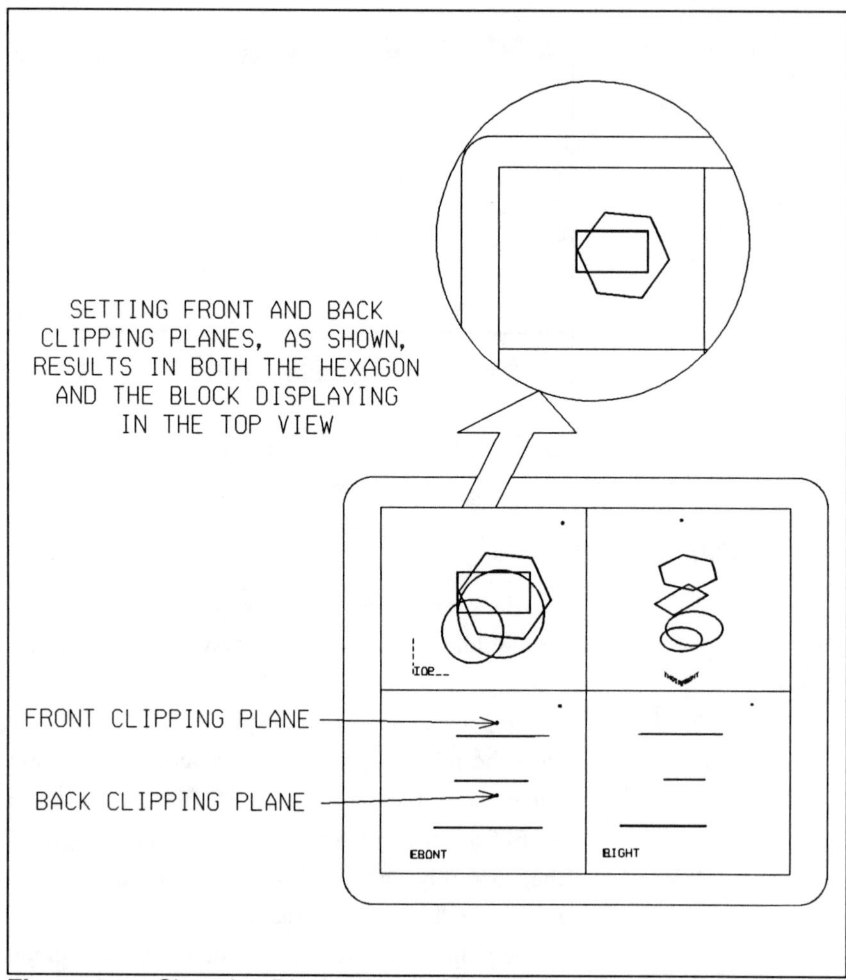

Figure 1.26 Changing Display Depth Graphically

So far we have worked mainly in the TOP view when discussing ACTIVE DEPTH and DISPLAY DEPTH. They may be used in any view, and as you work through the exercises, this will become more obvious.

Bore Site Lock

In 3D we have a new lock - BORESITE LOCK. When boresite lock is OFF, data points will find only those elements at, or very near, the ACTIVE DEPTH of a view. Elements still will be located with a tentative point, however. Tentative points over-ride BORE SITE LOCK. When BORE SITE LOCK is ON we can manipulate elements, using DATA POINTS only, without changing their ACTIVE DEPTH. To see this in operation do the following:

- Set the active depth in the TOP view to 650 (enter AZ=650 followed by a data point in the TOP view).
- Set bore site lock on. Use the locks menu, the settings box (version 4) or the key-in 'LOCK BORESITE ON'.
- Select the *Move Element* tool and move one of the circles in the TOP view, using a data point only to identify it.

When it highlights, notice in the other views that it remains at its current view Z value. It may be moved in the TOP view plane but it stays at its current Z value in that view (figure 1.27). So, with bore site lock on, we can move elements in a view without changing their depth value for the view.

We will now see what happens when we tentative to an element first.

- Select the *Move Element* tool
- Tentative to the circle and accept it.

Notice in the other views that, when accepted, it now 'jumps' to the current ACTIVE DEPTH. If you accept with a data point, not only does the circle move in the plane of the TOP view, it moves also to the current active depth of the view you are using (figure 1.28). Any other elements to which you tentative, before accepting, also will move to the current active depth of the view.

So, we can see that a tentative point over-rides the bore site lock. With a tentative point any element can be identified, whether it is near the active depth or not, and whether or not bore site lock is active.

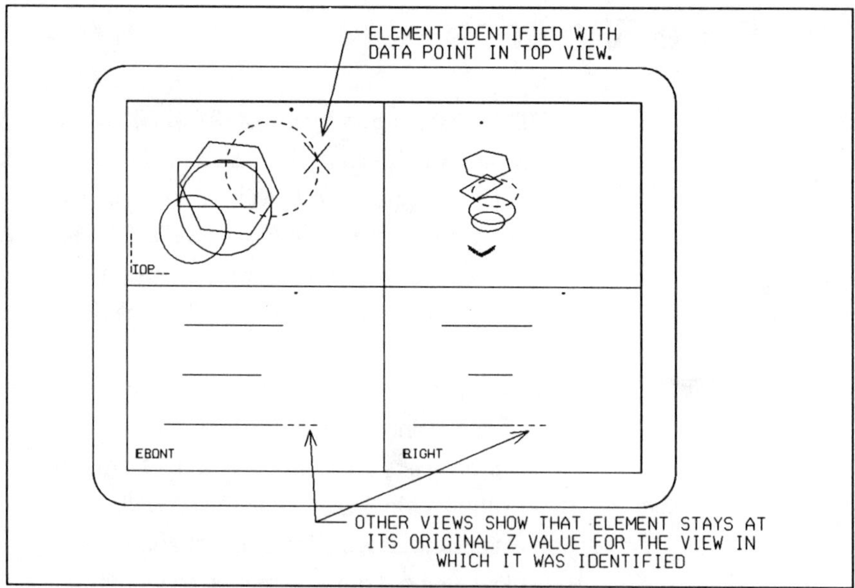

Figure 1.27 Using Data Points with Bore Site Lock

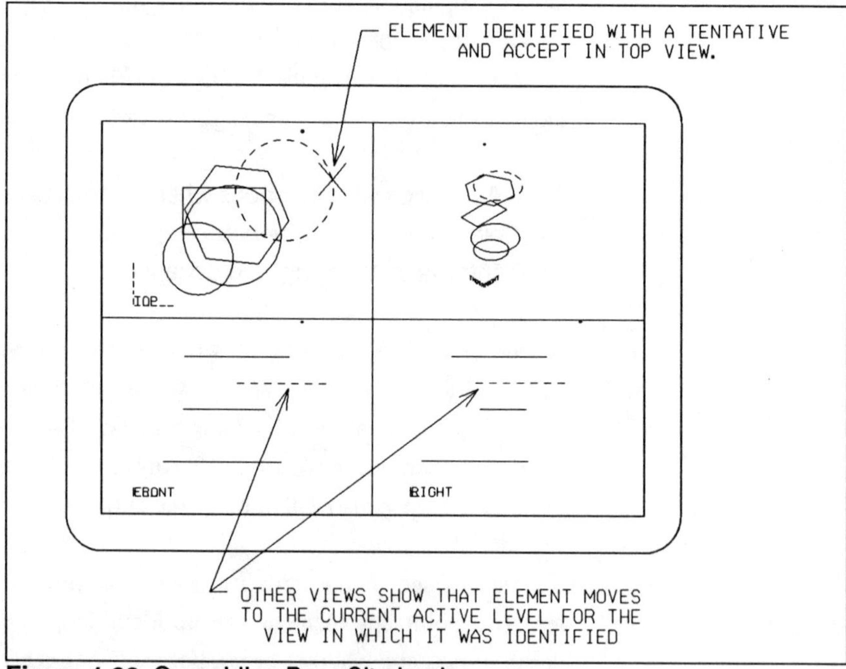

Figure 1.28 Over-riding Bore Site Lock

Normally, to check which locks are active, we can key in LOCK. This results in a list, in the message area of our screen, indicating the currently active locks. Boresite lock is not one of the locks that displays in earlier versions of MicroStation. For these versions, to check whether or not BORESITE LOCK is active, we can key-in 'LOCK BORESITE'. The system responds with either 'Boresite - ON' or 'Boresite - OFF'.

Version 4 does show the status of boresite lock when we key in LOCK. The abbreviation 'BS', in the list, indicates that BORESITE LOCK is active. We can also check in the settings boxes for LOCKS, either Full or Toggles, as shown in figure 1.29.

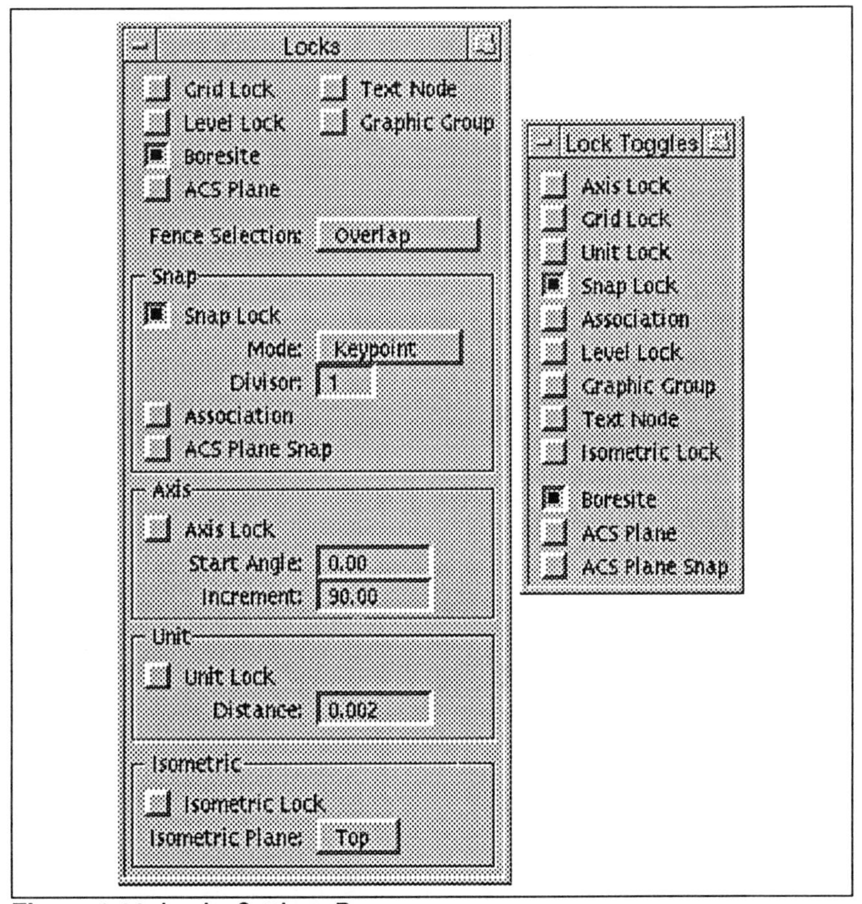

Figure 1.29 Locks Settings Boxes

What We Have Covered So Far:

We have now looked at the 3D world that we work in with MicroStation. In the next chapter, we will be building on the basics learned in this chapter. Before continuing, be sure that you are clear on what we have discussed so far. The following is a brief summary of the points we have covered:

- With 3D we work in a volume of space we call the design cube.
- We can look at our model from various directions, using the standard views - TOP, BOTTOM, FRONT, BACK, RIGHT, LEFT and ISO.
- Each view has a DISPLAY DEPTH and an ACTIVE DEPTH.
- Elements are placed, by default, at the ACTIVE DEPTH of the view being used.
- We can specify the DISPLAY DEPTH and ACTIVE DEPTH parameters for each view. This can be done with key-ins, or graphically.
- With BORESITE LOCK we can manipulate elements without them moving to the ACTIVE DEPTH of the view.

2 : Working in 3D

In the previous exercises we have been placing elements with data points only. Elements placed in this fashion fall on the active depth plane of the view we are using. In the 'real world' we want to specify exact locations for elements. For site-works, this may be as coordinate values. For structures and mechanical plant or components it might be locations relative to existing parts. From your 2D work you would know that we specify points with precision inputs.

Precision Inputs

Precision inputs that are most commonly used are 'XY=', 'DL=', 'DX=' and 'DI='. All of these key-ins work similarly to the way they do in 2D files.

The syntax for each is as follows:

XY = X value, Y value, Z value - Design Cube Axes.
DL = X value, Y value, Z value - Design Cube Axes.
DX = X value, Y value, Z value - View or Screen Axes.
DI = Distance value, Degrees from Horizontal - View or Screen Axes.

As you can see, the 'XY=' and 'DL=' key-ins relate to the design cube. The view being used at the time has no effect on them. The others, 'DX=' and 'DI=', relate to the view or screen axes. They are both VIEW DEPENDENT. They depend on the orientation of the view for their direction.

In particular, it is very important to understand the difference between DX and DL.

DX always refers to the axes of the view (i.e., the screen).
DL always refers to the axes of the design cube or model, no matter which view you are using. A way to remember this is that DL refers to the axes of the mo*D*e*L*.

Generally, DX key-ins are easier to use because they remain constant relative to the screen. As you look at your screen, no matter which view you are using, the axes for the DX key-ins are as follows:

X axis is always horizontal and positive from left to right.
Y axis is always vertical and positive from bottom to top.
Z axis is always perpendicular to the screen and positive toward the viewer.

Once the plane that we want to work in is parallel to the screen or view, we can use DX key-ins. Alternatively, the DL key-in allows us to work in any view and use the co-ordinate system of the model. Only when using a TOP view are the axes of the design cube and the view or screen aligned. Therefore, key-ins for DX and DL are identical only in a TOP view.

Two important points to remember when working in 3D are:

- Many basic elements we construct in 3D are placed, by default, parallel to the screen in the particular view. For example, a block created in a TOP view is placed horizontally - in a FRONT or RIGHT view, vertically.

- With key-in precision inputs, MicroStation assumes that the view we want to use is the one that we last worked in. That is, the view in which the last tentative or data point was placed. We will call this the 'current' view. A simple way to make a view current is to place a tentative point and then reset. UPDATE or FITting the view are two other methods. A tentative to an existing element in a view automatically makes that view current.

\DGN\TRAIN3D.DGN

In the following exercise we will use both DX and DL key-ins. We will construct a simple box that is 500 units long, 600 units wide and 400 units high (i.e., X=500, Y=600, Z=400). Its origin is at XY=250,300,200.

Start with the base:

- Make the TOP view 'current', active level 2 and weight 0.
- Select the *Place Block* tool
- Key in XY=250,300,200 for the first point.
- Key-in DL=500,600 (or DX=500,600) for the second point. Because we are using a TOP view DL and DX are identical.

It is not necessary to specify a value for Z, because we are placing the block in a horizontal plane and the Z offset is zero.

Now construct the front face:

- Make the FRONT view 'current' and active level 3.
- Select the *Place Block* tool.
- Key in XY=250,300,200 for the first point.
- For the second point we may use either DX or DL, but they differ.

  DL refers to the axes of the model. The front face is 500 units long in the X direction and 400 units high in the Z direction. Therefore, the key-in is DL=500,0,400 or DL=500,,400 (omitting the zero).

  DX refers to the axes of the view or screen. The front face, here, is 500 units long horizontally (X) and 400 units high vertically (Y) relative to the screen. Thus the key-in is DX=500,400.

Next we construct one of the sides:

- Make the RIGHT view 'current'.
- Select the *Place Block* tool.
- Key in XY=250,300,200 for the first point.
- For the second point the key-in is either of the following:

  DL=0,600,400 or DL=,600,400 (600 units in the Y direction, 400 units in the Z direction of the model).

  DX=600,400 (600 units horizontally (X) and 400 units vertically (Y) relative to the screen).

Your screen should now resemble figure 2.1

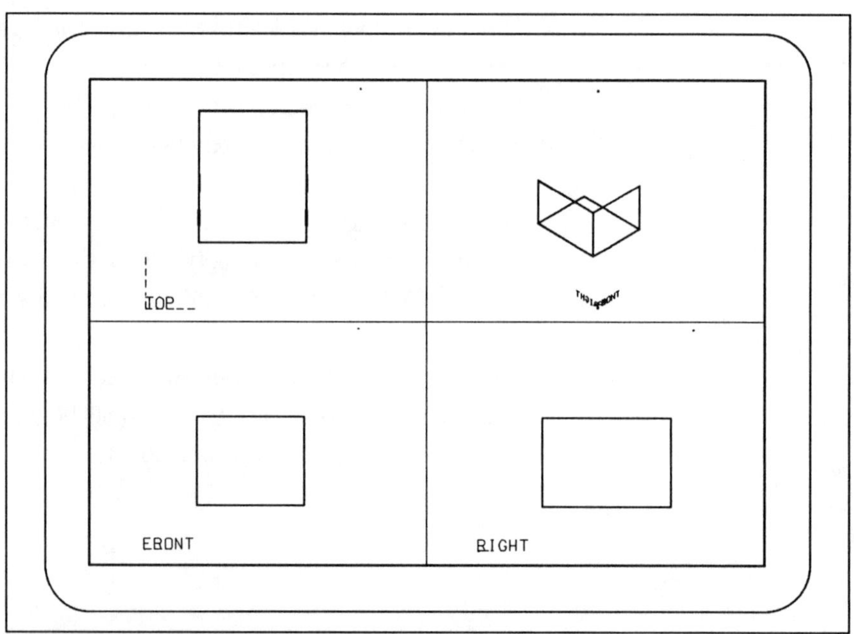

Figure 2.1 Commencing the Box

Looking at the ISO view it shows that we have constructed the BASE, FRONT and LEFT faces of the box.

For each element we used XY=250,300,200 as the first point. The orientation of the faces is different because we used different views for placing each one. We made the view with the required orientation 'current' before using precision inputs to create the blocks or faces.

We will copy the base 400 units in the model's Z direction now, to form the top of the box:

° Make the active level 2 and turn off level 3, so that only the base displays in all views.

Now, select the *Copy Element* tool:

° Tentative to and accept the base in the TOP view. (This automatically makes the TOP view 'current').

° To specify the distance, key in DX=,,400. This will copy the element 400 units in the Z direction of the view (i.e., 400 units toward you).

° Check in FRONT and RIGHT views, that the copy is directly above the base.

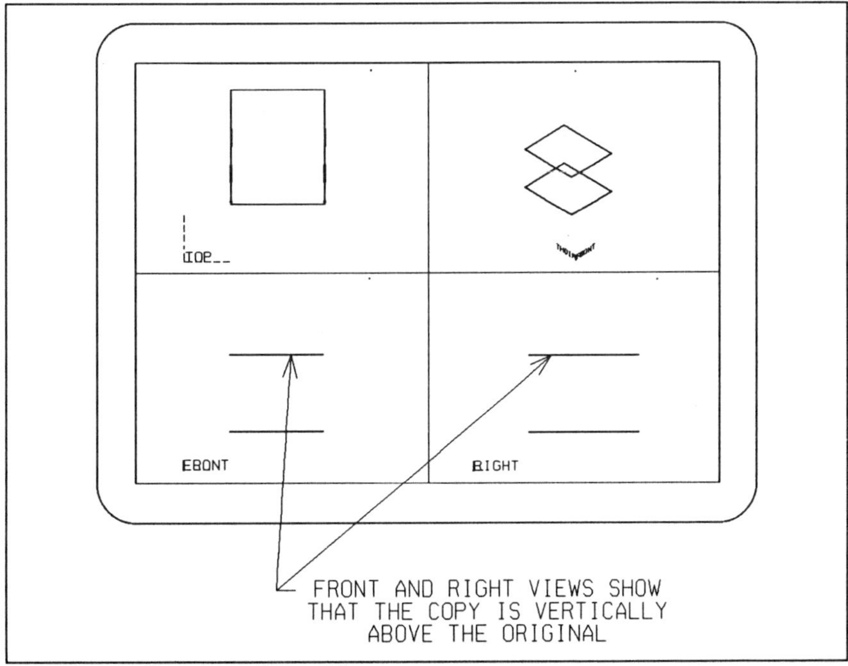

Figure 2.2 Checking Other Views

DELETE or UNDO the copy and this time copy the base using either the FRONT or RIGHT view.

Select the *Copy Element* tool:

- ° Identify the element in either the FRONT or RIGHT view.
- ° Key in DX=,400 to specify the distance. This will copy the element 400 units in the Y direction of the view. In both the FRONT and RIGHT views, the Y direction is vertical from bottom to top of the view or screen.

Again, delete or undo the copy. Now we will perform the same operation using the ISO view. Because the ISO view does not align with any of the axes of the model, the DX key-in is not appropriate. The DL key-in, however, may be used in *any* view. It relates to the axes of the design cube (or model).

Select the *Copy Element* tool:

- ° Identify the element in the ISO view.
- ° Key in DL=,,400 to specify the distance. This will copy the element 400 units in the Z direction of the model.

Looking at the other views will verify that the copy is directly above the original.

- ° Make the active level 3 and check that the top face is in the correct location, sitting on top of the front and side faces.
- ° Now, turn level 2 off in all views and update the screen. Only the front and side faces of the box should be visible .

To create the other two faces we could calculate the co-ordinates and place the elements with 'XY=' and 'DX=' or 'DL=' key-ins. We also could construct them by copying the existing faces using a 'DX=' or 'DL=' key-in as we did for the TOP face. Both these methods involve key-ins and/or calculations.

Another form of precision input, is the TENTATIVE POINT, or snapping to existing elements in the file.

Select the *Copy Element* tool:

° Using the ISO view, tentative to and accept the front face of the model at the bottom corner, where it meets the side face.

° Copy it to the other end of the side face. Refer to figure 2.3.

Now copy the existing side face to the other side.

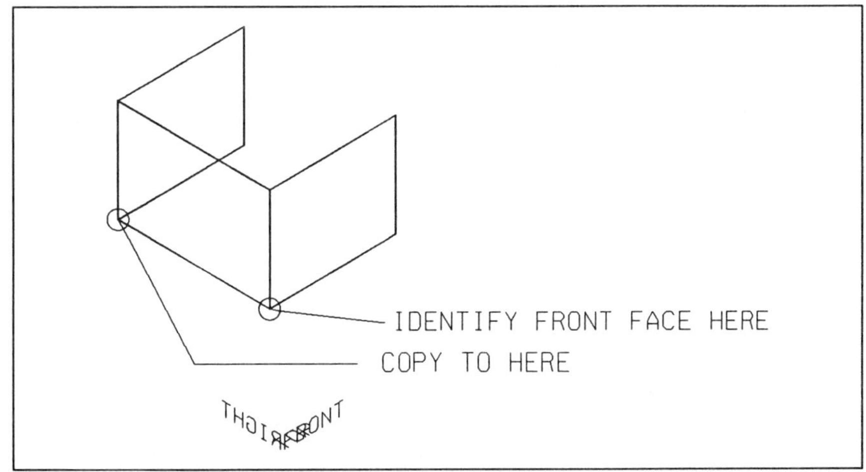

Figure 2.3 Copying using Existing Elements

Note!

When we copy or move elements in 3D files they retain their same orientation relative to the design cube. When we copied the front and side faces, the copies remained parallel to the original faces. This is true no matter what view or views we use. Elements may be identified in one view and another view used for the actual manipulation. Once we place elements, they retain their orientation, relative to the design cube, unless we rotate them.

This allows us to use the ISO view for such operations as copying and moving. We are no longer restricted to working in a two-dimensional fashion with plans, elevation and sections. We are creating a three-dimensional model in the three-dimensional world of the design cube. Normally, the model is easier to visualize in the ISO view. We can see elements that would be hidden, or confused with other elements, in the orthogonal views.

We will now do an exercise to consolidate the new concepts and procedures you have learnt. Our task is to place text describing the location of each face of the box. That is, TOP, BOTTOM, FRONT, BACK, RIGHT and LEFT. The text is to be placed in about the middle of each face (using data points).

In preparation:

° Set the text size to 50 and the Active Level to 63 for the text.

Each face of the box has a corresponding standard view. We can use these to get the correct orientation for placing the text. Think of how you would complete the task if you had the box in front of you. Probably, you would rotate each face in turn toward you to write on it. Setting view orientation is equivalent to this. You would then place your pen on the face and write the word. In the design file, we set the ACTIVE DEPTH and PLACE TEXT.

Some hints to get you started:

- Use quadrant 1 (presently the ISO view) as your working view, for those views not already displayed (i.e., LEFT, BACK and BOTTOM).
- Set the ACTIVE DEPTH to the face of the box that is to have the text placed on it, before placing the text.
- Use the TOP, FRONT, or RIGHT views to select graphically the ACTIVE DEPTH point for the view that you are working in.

For example, to place the word BACK on the back face of the box we could do the following:

° Key in 'VI=BACK' followed by a data point in quadrant 1 (then FIT if necessary).

° Set the ACTIVE DEPTH to the back face in the BACK view. Use the RIGHT or TOP views to select the ACTIVE DEPTH point. In the TOP view, the back face is the top edge. In the RIGHT view the back face is the right edge. Refer to figure 2.4.

° Place the word 'BACK' in about the middle of the back face, with a data point, using the back view.

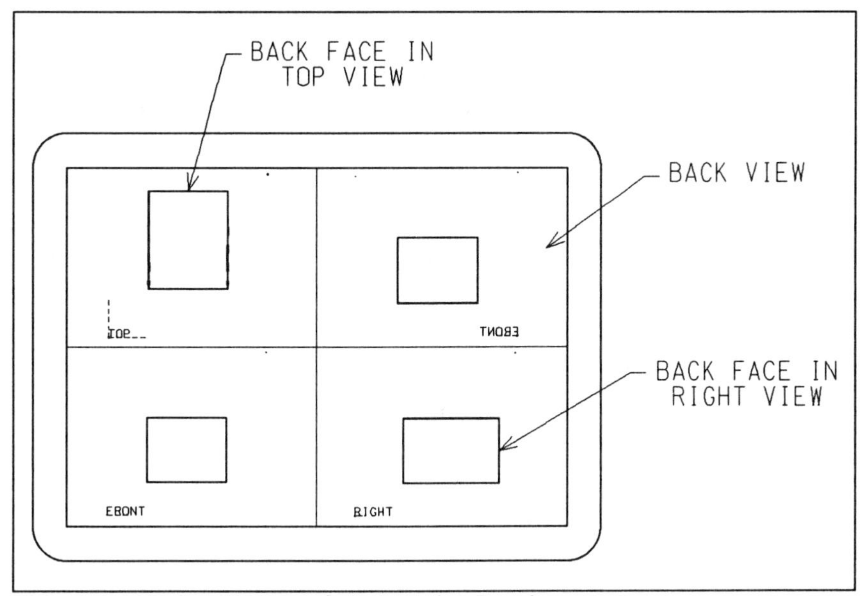

Figure 2.4 Using Views to Select Active Depth

Systematically follow the same procedure for each of the other faces. When you have finished, your model, in an ISO view, should look similar to that in figure 2.5.

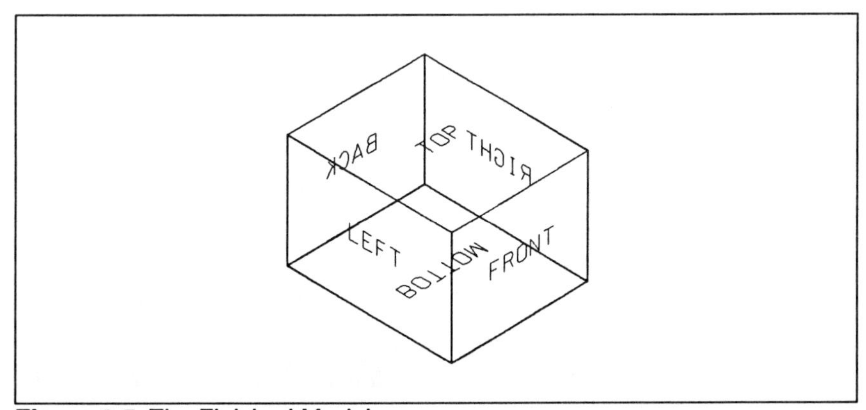

Figure 2.5 The Finished Model

Before continuing, we will look again at the effects of changing the DISPLAY DEPTH in a view. We will set the DISPLAY DEPTH in the FRONT view so that we can see the words TOP and BOTTOM.

Window area in closer to the model in the FRONT view, then:

- Select the *Set Display Depth* tool from the menu. Alternatively, key in 'DEPTH DISPLAY'.

Version 3.3

- Use the TOP view to select the front and back limits for the DISPLAY DEPTH (refer figure 2.6).
- Select the FRONT view when prompted.

Version 4

- Select the FRONT view when prompted.
- Use the TOP view to select the front clipping plane, first, then the back clipping plane (refer figure 2.6).

As figure 2.6 shows, the FRONT view now displays the words TOP and BOTTOM as three and six horizontal dashes respectively. In the FRONT view, we are looking at the words edge on. As well, parts of the words LEFT and RIGHT appear as single vertical dashes. Notice also, the dots around the text. Setting the DISPLAY DEPTH has effectively taken a slice through the model. This slice, defined by front and back clipping planes, is parallel to the screen. The dots are the lines of the TOP, BOTTOM, RIGHT and LEFT faces, viewed edge on, along the line. Tentative to a face in another view and the corresponding 'dots', in the FRONT view, will highlight also. DISPLAY DEPTH only allows us to see the elements, or parts of elements, contained within its front and back limits. Parts of elements that protrude through the DISPLAY DEPTH limits do not display on the screen.

In the lower right corner of figure 2.6 we are looking at an illustration of the FRONT view but from an ISO view point. This is to show more clearly, just what is being displayed in the FRONT view, with the restricted display depth. Elements shown dotted do not display in the FRONT view because they are outside the range of the set display depth. They are, however, still present in the design file and would be seen in other views having different display parameters.

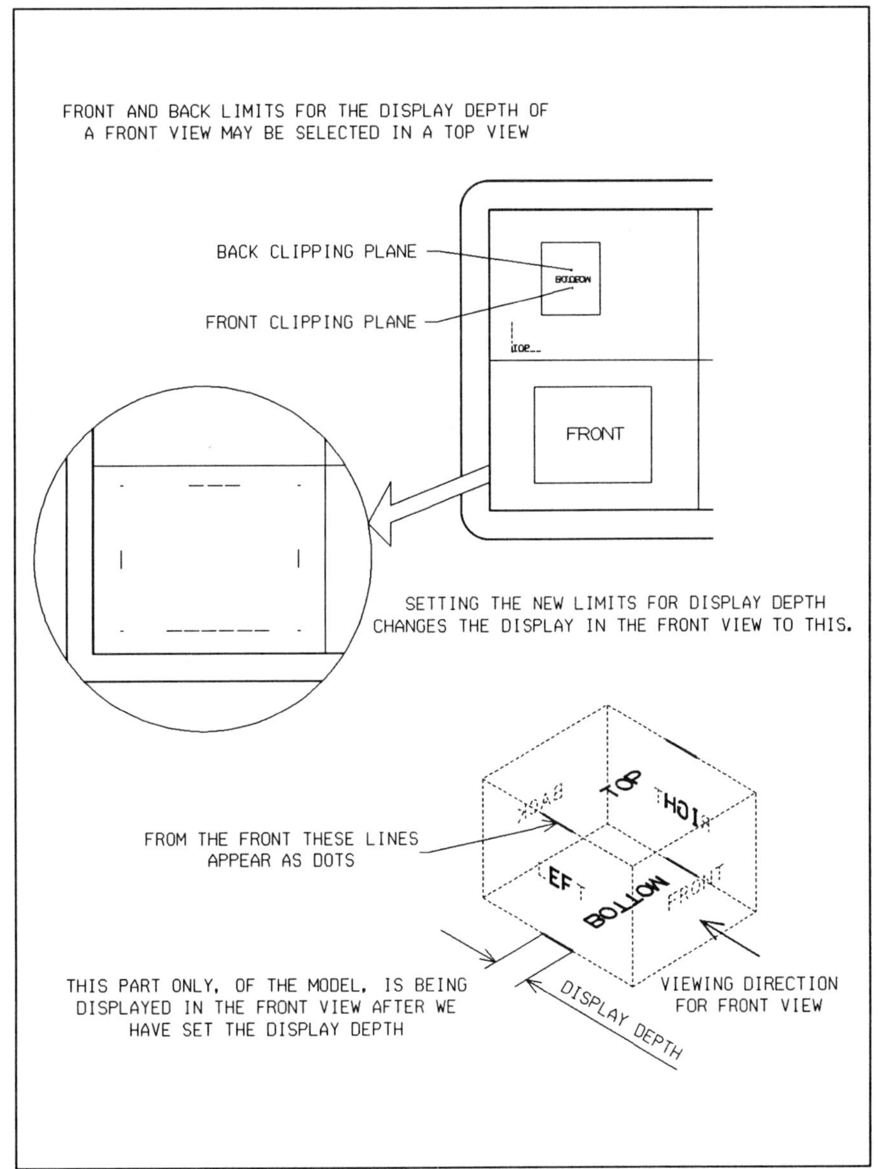

Figure 2.6 Setting Display Depth

With version 4 we are shown graphically, in the other views, what the new display parameters will be. Figure 2.7 shows the display in the four views as the DISPLAY DEPTH is set for the FRONT view. We can see in each of the other views that we are setting the DISPLAY DEPTH, for the FRONT view, to be a narrow slice.

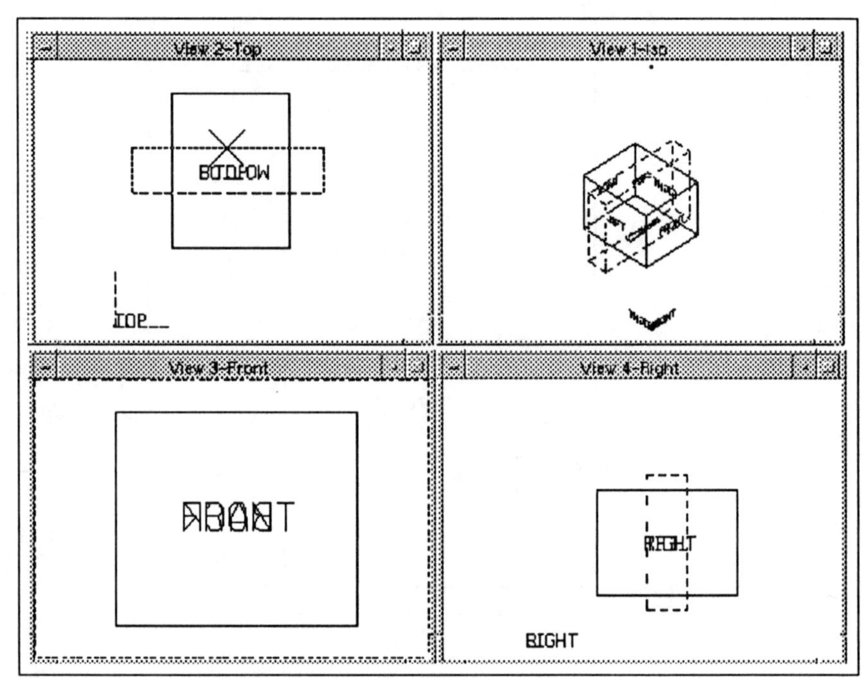

Figure 2.7 Setting Display Depth with version 4

We have seen that we can view our model from various directions - TOP, BOTTOM, FRONT, BACK, RIGHT, LEFT and ISO. These are the directions or views that are standard with MicroStation. By using these views we can place elements with different orientations in the three-dimensional space of the design cube. This is fine for regular shapes where everything is at right angles or perpendicular to one another. Quite often we have models that include parts that cannot be created in one of the standard views. We need to be able to cater for these situations also.

We will be looking at how to manipulate our views to other than the standard orientations next. Before going on to this section, however, it is a good idea to practice what has been learnt so far. Several examples of simple models, along with solutions, can be found in chapter 14 (Practice Examples).

View Rotation

We have looked at the 7 standard views that are available. Also, we may rotate views about the X, Y, or Z axis of the view or screen. This allows us to look at our model from any direction. Sometimes we may only need to rotate the view a few degrees to place an element or make something more clear.

We can compare rotating views as we work in 3D, to a technician sitting down and working on the inner components of a T.V. set. When he wants to see the connections on the back of the tube, for instance, he turns the set so that the back of the tube is facing him. If he wants to see something that is obscured by another component, he may tilt the set or turn it slightly. In this example, the outer casing of the T.V. can be related to our design cube, while the components inside are equivalent to our model. When the outer case (the design cube) is rotated, the components (our model) are rotated as well. The components stay in the same position, relative to the outer casing. Likewise, our model remains in the same position relative to the design cube coordinates.

With our computer model, while we can't physically touch it, we can manipulate it. We have key-ins that we can use to rotate the design cube about the X, Y, or Z axis of our view or screen.

With all versions of MicroStation, the following rule applies:

Note!

ALL VIEW ROTATIONS ARE RELATIVE TO THE AXES OF THE SCREEN, NOT THE MODEL OR DESIGN CUBE.

Version 4 has two types of view rotation RELATIVE and ABSOLUTE, while version 3.3 has RELATIVE rotation only.

Rotate View Relative (version 3.3 and version 4)

With RELATIVE view rotations:

ROTATIONS ARE RELATIVE TO THE PRESENT VIEW ORIENTATION.

From the starting point of a TOP view, for example, the other views can be duplicated with rotations about the axes of the view or screen. We know that in a TOP view, the axes system of our design cube aligns with the axes system of our screen (X horizontal, Y vertical and Z perpendicular to the screen).

- A FRONT view is the same as a TOP view, with a RELATIVE ROTATION of 90 degrees backward about the horizontal (X) axis of the screen

- A RIGHT view is the same as a FRONT view, with a RELATIVE ROTATION of 90 degrees clockwise about the vertical (Y) axis of the screen.

In each of the above we would rotate the view from its present orientation to the new orientation. Imagine that you are holding the model box in front of you, looking at the TOP face.

Rotating it backward about a horizontal axis would bring the FRONT face into view. Rotating it from this orientation, by 90 degrees clockwise about a vertical axis, would expose the RIGHT face (figure 2.8).

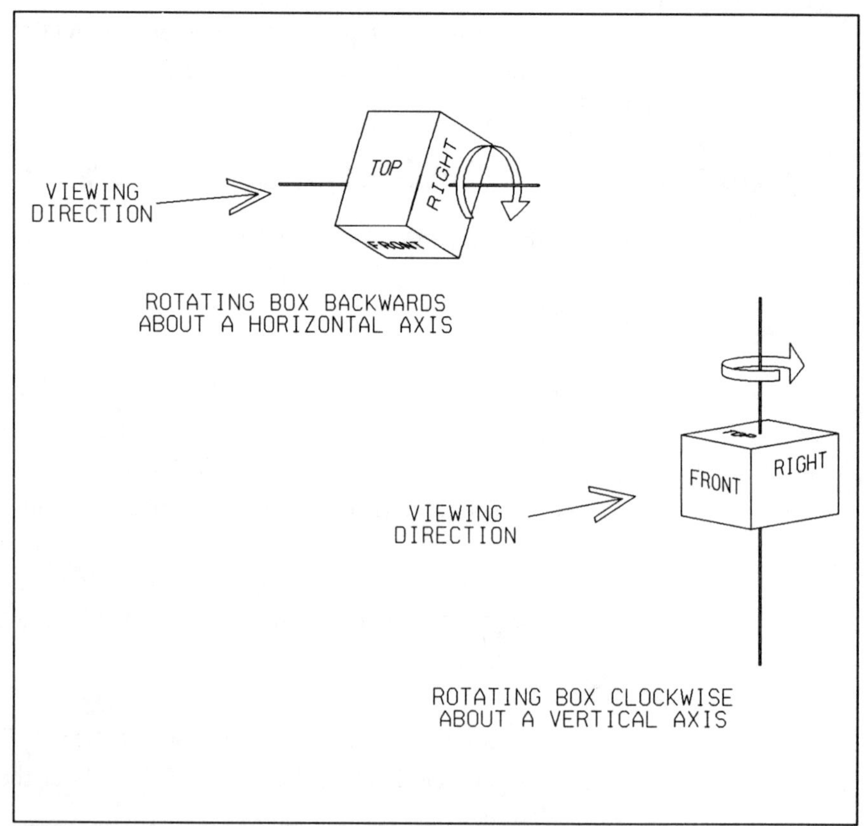

Figure 2.8 Rotating a Physical Model

We can rotate our design cube and model (i.e., our views) in a similar fashion using the ROTATE VIEW RELATIVE key-in (RV=). There is a protocol for determining which direction is positive for view rotation. Figure 2.9 shows the positive directions for rotating about the 3 axes.

Figure 2.9 also shows a simple way to remember the positive direction for rotating views. Hold your right hand with the outstretched thumb pointing in the positive direction of the required axis. The curled fingers point in the positive direction for the view rotation angle.

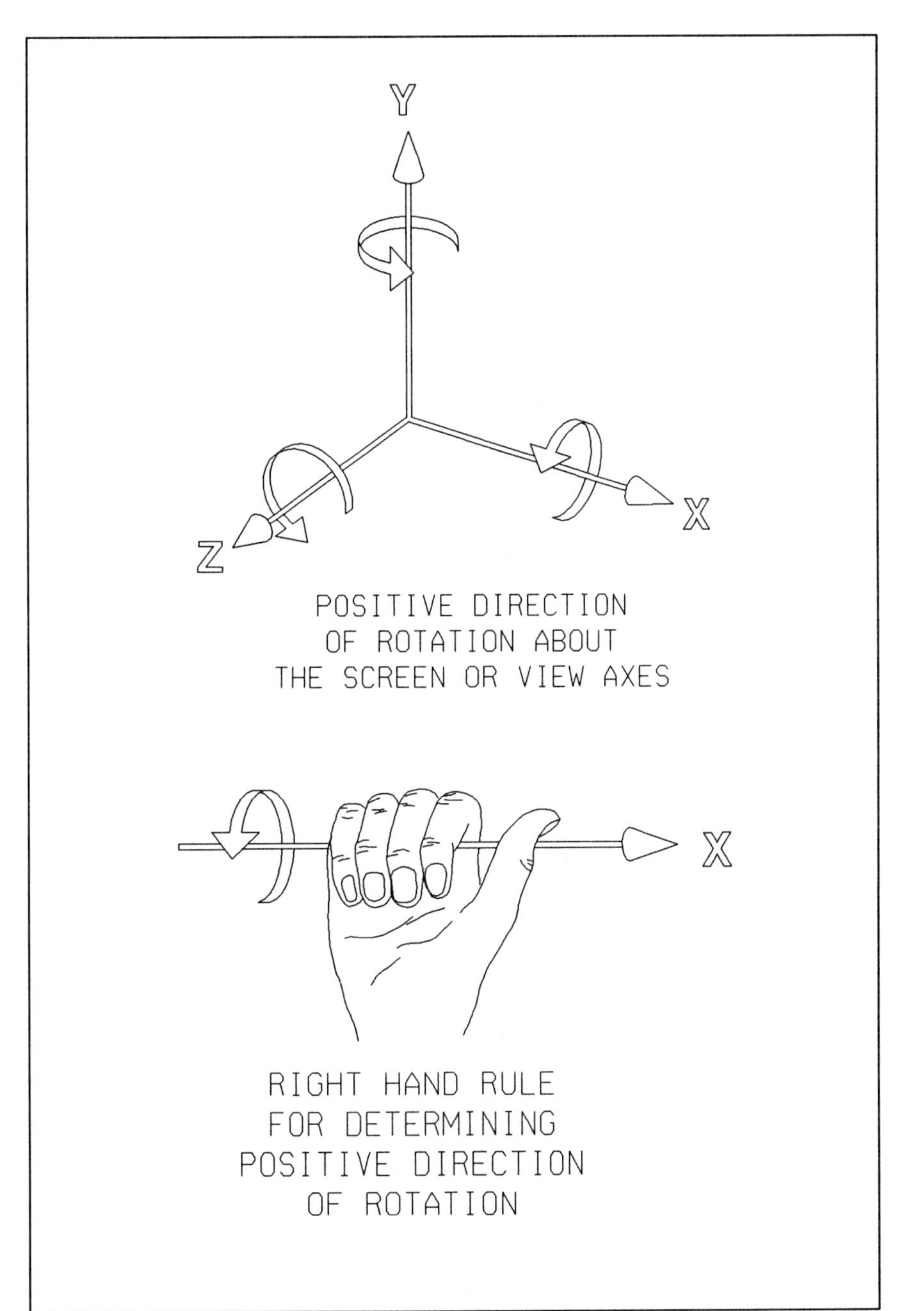

Figure 2.9 Positive Direction of Rotation

Syntax for the ROTATE VIEW RELATIVE key-in is:
RV = degrees about X axis, degrees about Y axis, degrees about Z axis.

Set up your file, with the model box, as we had it initially. That is, views 1-4 are ISO, TOP, FRONT and RIGHT respectively. Fit each view.

Rotate the TOP view minus 90 degrees about the X axis as follows:

- Key in RV = -90
- Place a single data point in the TOP view (further data points would continue to rotate it another 90 degrees each time).
- FIT the view and compare it with the FRONT view. They should be identical.

Now, rotate the FRONT view minus 90 degrees about the Y axis:

- Key in RV = ,-90
- Place a data point in the FRONT view.
- FIT, if necessary, and compare it with the RIGHT view, which should look the same.

Try rotating the views by various amounts, and combinations of values (e.g., RV = 20,45,30 for a combined rotation about the 3 axes). Rotating a view about the Z axis is similar to rotating a view in a 2D design file.

When a combination of values is entered (e.g., RV = 45,-10,30), the view is rotated, first about the X axis, then about the Y axis, and thirdly about the Z axis. Generally, rotating about one axis at a time is easier to comprehend and control. If a view is rotated with the key-in RV = 45,-10,30 for example, then keying in the reverse (i.e., RV = -45,10,-30) would not return the view to its original state. In each case the rotations would be about the axes in order of X, Y and Z. To retrieve the original view we would have to rotate the view -30 degrees about the Z axis, then 10 degrees about the Y axis and finally -45 degrees about the X axis. In other words, rotating it about each axis, separately, in the reverse order to the original rotation. If the original view had been a standard view, or a saved view, then we could have recalled it by using the 'VI = viewname' key-in.

Another point to be aware of when rotating views is that sometimes elements, or parts of elements, will disappear from the screen. This is because they have rotated out of the viewing area, or outside the display depth. The view is a defined volume relative to the screen. The design cube can be rotated through this volume. As we rotate the design cube containing our model, through the display depth, various elements will appear and disappear from the display. Remember that this is a display parameter only. Nothing happens to the elements, they just don't display if they are not within the display depth of the view. Figure 2.10 should help make this idea easier to comprehend.

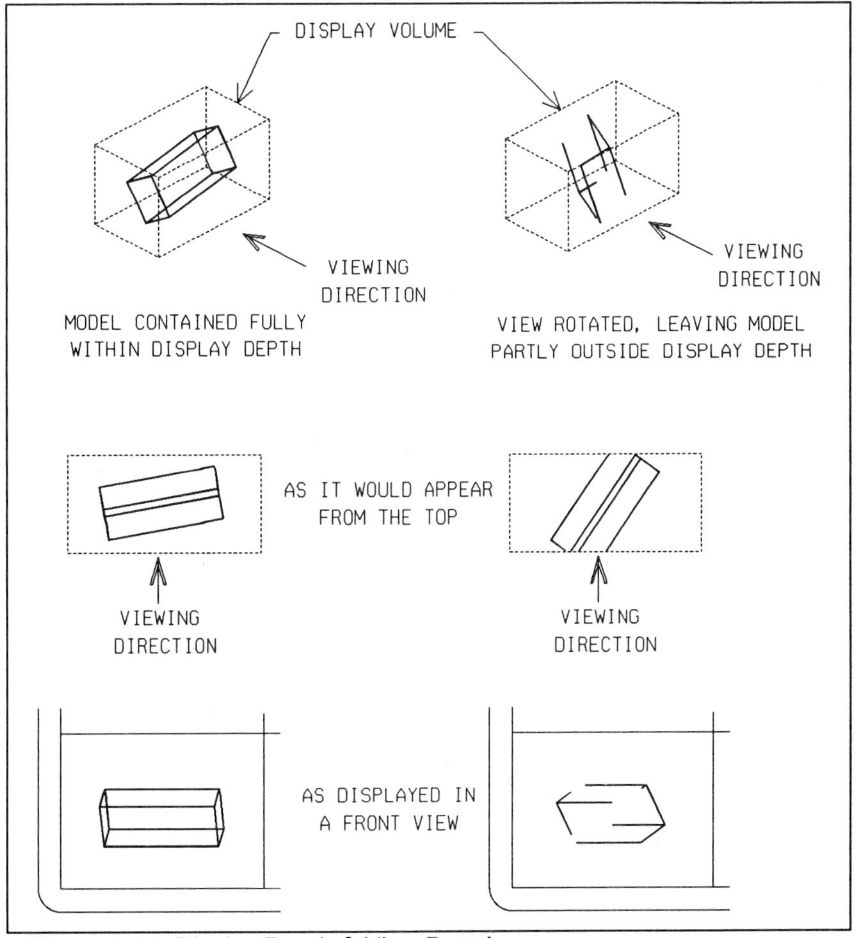

Figure 2.10 Display Depth & View Rotation

In figure 2.10 the dashed lines represent the display depth of the view. The rectangles at the bottom show the display as it would appear on the screen. Though parts of the model do not display in the right hand side rectangle, they are still there.

We can show this clearly with version 4, using the *Set Display Depth* tool. In figure 2.11 we see a model of a simple box that has been placed in the file slightly skewed. The *Set Display Depth* tool has been chosen and then the FRONT view (view 3) selected. In views 1,2 and 4 we can see the viewing volume for view 3, displayed as dashed lines, along with the active depth plane.

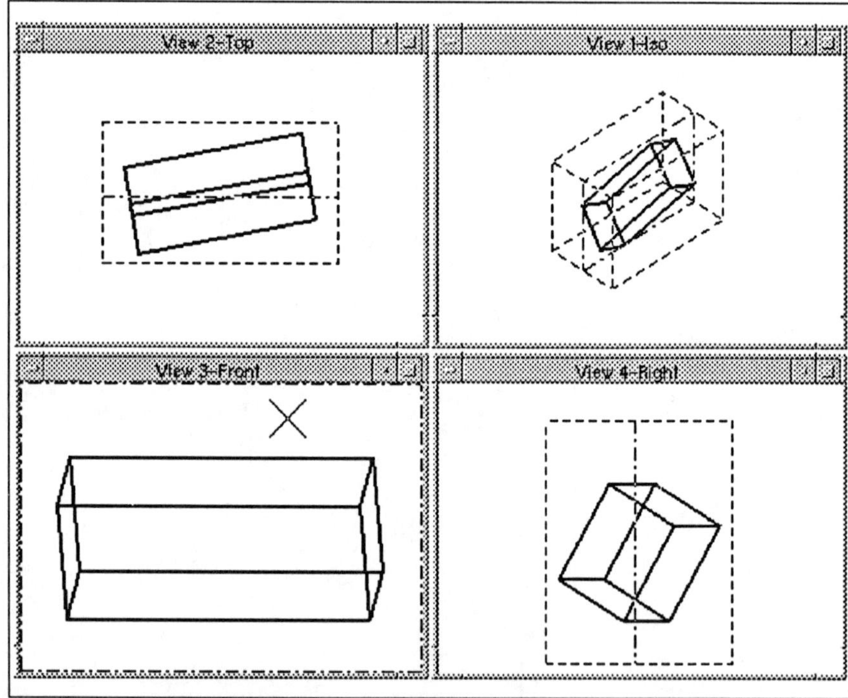

Figure 2.11 Display Depth & View Rotation ver.4

Figure 2.12 shows the same file except that view 3 has been rotated 30 degrees about the Y axis. Again the *Set Display Depth* tool was chosen (in the 3D palette), and view 3 selected as the one to be manipulated.

We can see clearly, in view 2, that the model now protrudes through the display depth of view 3. This is verified in view 3. Parts of the elements are missing, signifying that they are not contained in the display depth of the view.

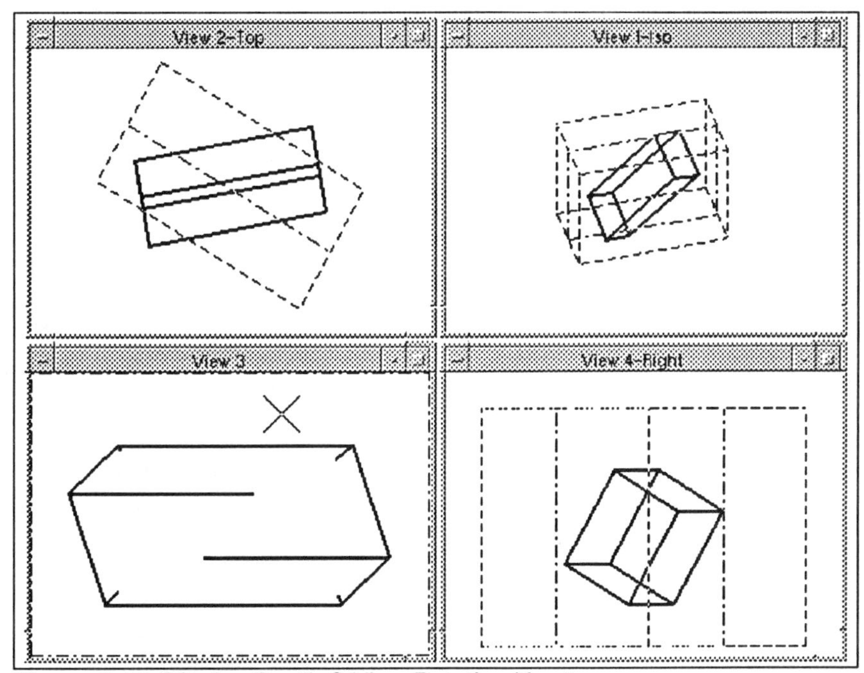

Figure 2.12 Display Depth & View Rotation Ver.4

When elements totally or partially disappear after a view rotation, we can re-define the display depth parameters or use FIT. Altering the display depth will not alter the window area parameter of the view, while a FIT may.

Having the standard views allows us freedom with view rotation during a design session. Often, it is convenient to rotate a view by a few degrees. This may be to see, more clearly, a particular part of the model. When we have finished, we can always return to a standard view. We don't have to remember how many degrees the view had been rotated. Standard views always return to the same orientation.

View Rotation Dialog Box (version 4)

With version 4 we can use also the View Rotation dialog box to specify view orientation. The view to be manipulated is chosen in the 'View' option menu. We have a choice of one of the standard views or a 'Custom' rotation. Standard views can be selected in the 'Std.' option menu as shown in figure 2.13.

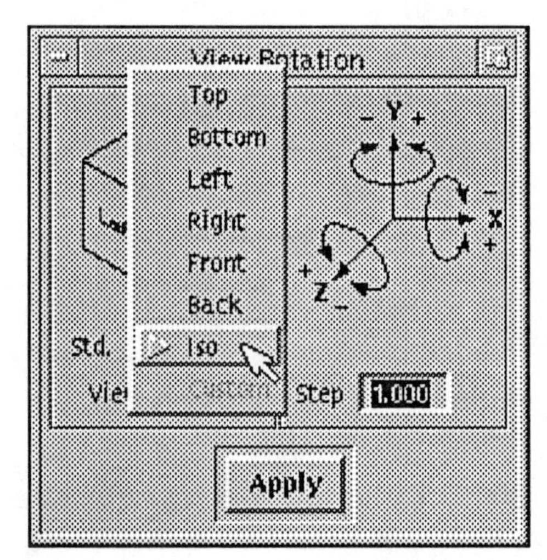

Figure 2.13 View Rotation Dialog Box

Custom rotations (relative) can be specified by clicking on the '+' or '-' controls of the axis system shown on the right side of the settings box. Each click on a '+' or '-' control rotates the view, about the corresponding axis. We control the amount of each relative rotation, by the amount indicated in the 'Step' field (in degrees).

As we rotate our view, the rotations are indicated, dynamically, by the cube on the left of the settings box (refer figure 2.14). When we are satisfied with the rotation, clicking on the 'Apply' button rotates the selected view to the specified orientation. To check the orientation of a view, we can pick a view from the 'View' menu in the dialog box.

Do the following to see how this works:

- Set view 1 as an ISO view.
- Select 'Rotation' in the 'View' pull-down menu.
- In the View Rotation dialog box select view 1 in the 'View' options.

Figure 2.14 shows the settings box after view 1 was selected. We can see in the 'Std.' field that the view is a standard ISO view. The 'View' field shows us that it is view 1 and the cube shows us how the view is rotated. This feature can be used to check the rotation of a view if we ever get 'lost' with rotations.

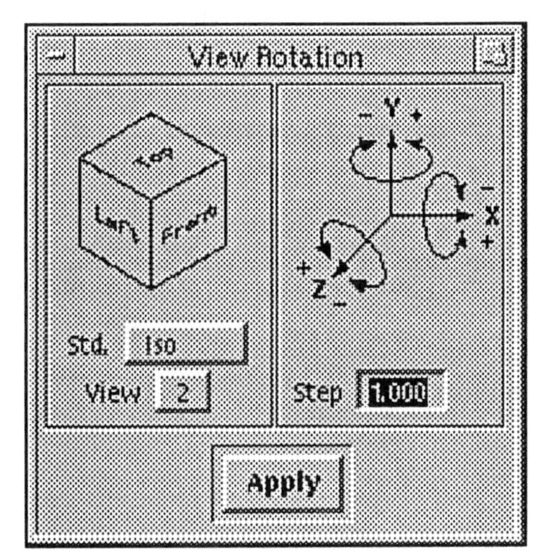

Figure 2.14 Setting ISO View

Rotate View Absolute (version 4)

With version 4, we have another key-in for rotating views to a specific orientation - ROTATE VIEW ABSOLUTE. Again, rotations are about the X, Y, and Z axes of the view or screen. The difference is that the rotations are absolute rather than relative. That is, the view is rotated by the specified angles from a known starting point. The starting point is where the view (or screen) and the design cube are exactly aligned - a TOP view. All ABSOLUTE view rotations are taken from a starting point of a TOP view. In a TOP view the absolute rotation angles are 0,0,0.

If we key in 'ROTATE VIEW ABSOLUTE 0,0,0' the resulting rotation will give us a TOP view, no matter how the view is presently oriented. ROTATE VIEW ABSOLUTE ignores the present orientation of the view.

If we use a rotation angle of '-90,0,0' we will get a FRONT view. The view will be rotated -90 degrees about the X axis from the starting point of a TOP view. This results in a FRONT view.

Rotate View by 3 Points

Sometimes, it is necessary to rotate a view to line up with a particular plane, or element in the design. The plane may not align with any of the standard views and the angle of rotation is not easily calculated. On these occasions we can rotate a view using three points, which allows a plane to be defined by 3 points in the design file. The view that we select is then rotated to make the specified plane parallel to the view or screen.

This tool does not appear in the pull-down menus or palettes with version 4. It must be selected from another menu (sidebar or digitizer), or the key-in 'ROTATE 3PTS' or 'ROTATE VIEW POINTS' used. With version 3.3 the key-in 'ROTATE 3PTS' may be used.

After selecting from the menu, or using the key-in, we are prompted for 3 points.

Point 1 defines the origin of the X axis.

Point 2 defines the positive direction for the X axis.

Point 3 defines the direction of the Y axis relative to the X axis.

The next prompt is to select a view to rotate to the required orientation.

To see how this works in practice, we will carry out a small exercise. In preparation, do the following:

- Set the screen up with the original views (in order from 1 to 4, ISO, TOP, FRONT and RIGHT) and FIT them.
- Make the active level 4 and the active color yellow.
- Window area in closer to the box in the ISO view.
- Using the *Place Shape* (not *Place Box*) tool, place a vertical rectangular shape. Locate it diagonally across the box from the front right corner to the back left corner as in figure 2.15.

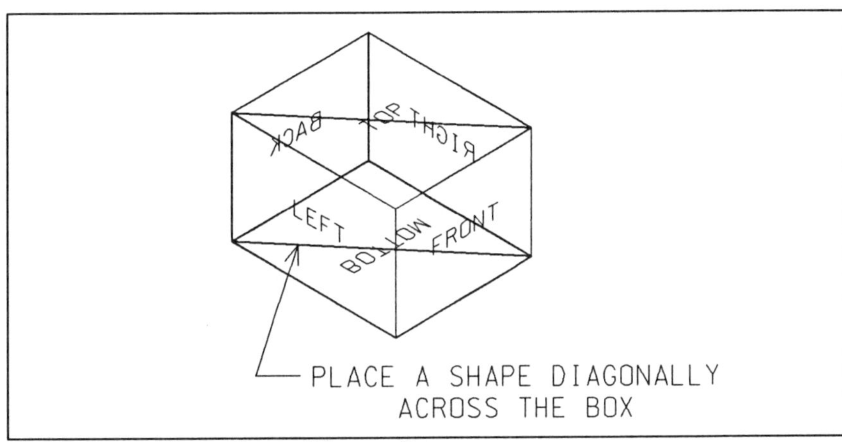

Figure 2.15 Place Shape Diagonally

We will place the word 'DIAGONAL' on the shape just created. Text is placed parallel to the screen, so we will need to rotate the view and make the diagonal shape parallel to our screen. Any view may be used for selecting the 3 points, but the ISO, here, is probably the simplest.

Referring to figure 2.16 for guidance:

- Select ROTATE VIEW BY 3 POINTS from the menu, or use the key-in 'ROTATE 3PTS'.
- Select a lower corner of the diagonal shape for the first point.
- Select the other lower corner for the second point. These two points define the X axis.
- Select an upper corner for the Y axis direction.
- Place a data point in the ISO view when prompted for a view to rotate.

We have rotated the view so that the diagonal shape is parallel to the screen as figure 2.16 shows. Its bottom edge is parallel to the X axis of the view or screen. Its vertical edge is parallel with the Y axis of the view or screen. The rotated view is a little crowded, however, with the text and the sides confusing the issue. We will restrict the display of the unwanted elements by setting the display depth to a small amount either side of the diagonal.

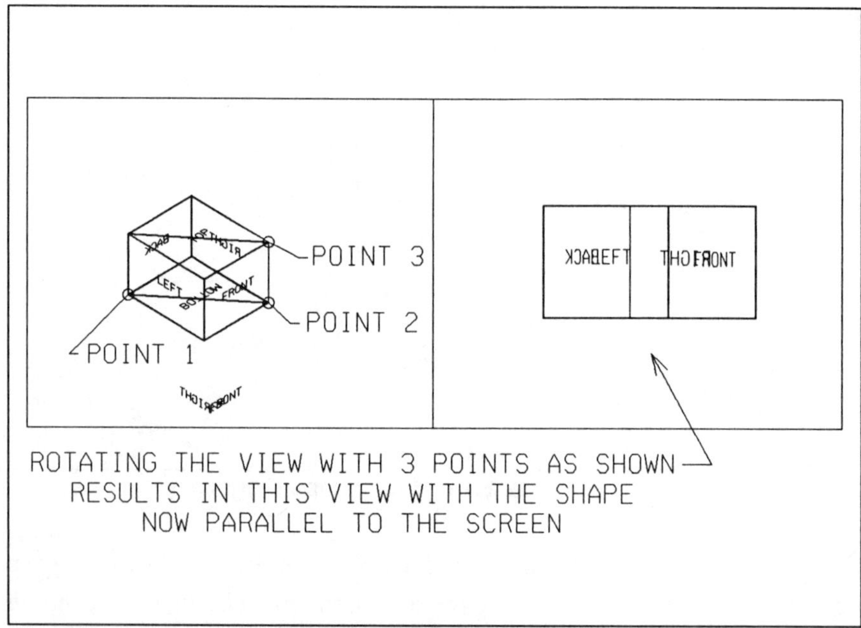

Figure 2.16 Rotating View by 3 Points

- Select *Set Display Depth* tool from the menu or key in 'DEPTH DISPLAY'.
- As shown in figure 2.17, set the DISPLAY DEPTH using a TOP view.

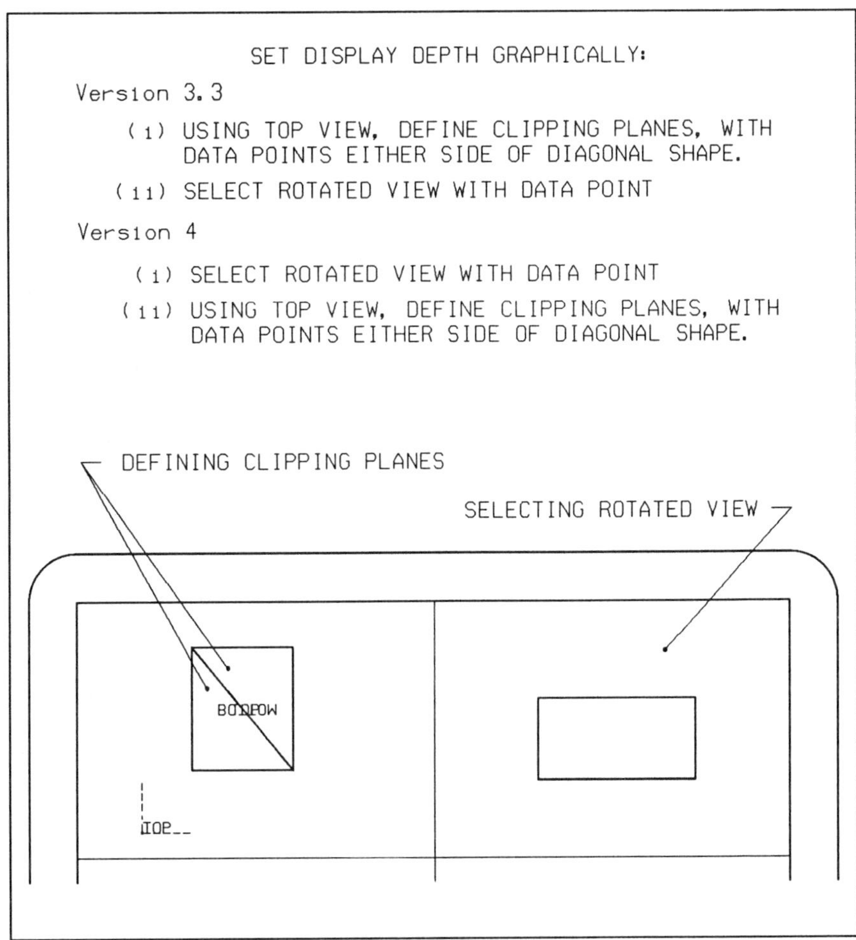

Figure 2.17 Clearing the View of Clutter

Setting the display to a small amount clears the screen of any elements in front or behind our area of interest. The elements are still there, we have just stopped them from displaying. We may now set the ACTIVE DEPTH to that of the diagonal shape and place the text.

Select *Set Active Depth* tool from the menu or key in 'DEPTH ACTIVE':

- As shown in figure 2.18, set the ACTIVE DEPTH using the TOP view to identify the diagonal shape.
- The rotated view is the view in which we set the ACTIVE DEPTH.
- Place the text and check in other views to see that it has been placed correctly, in the plane of the diagonal.

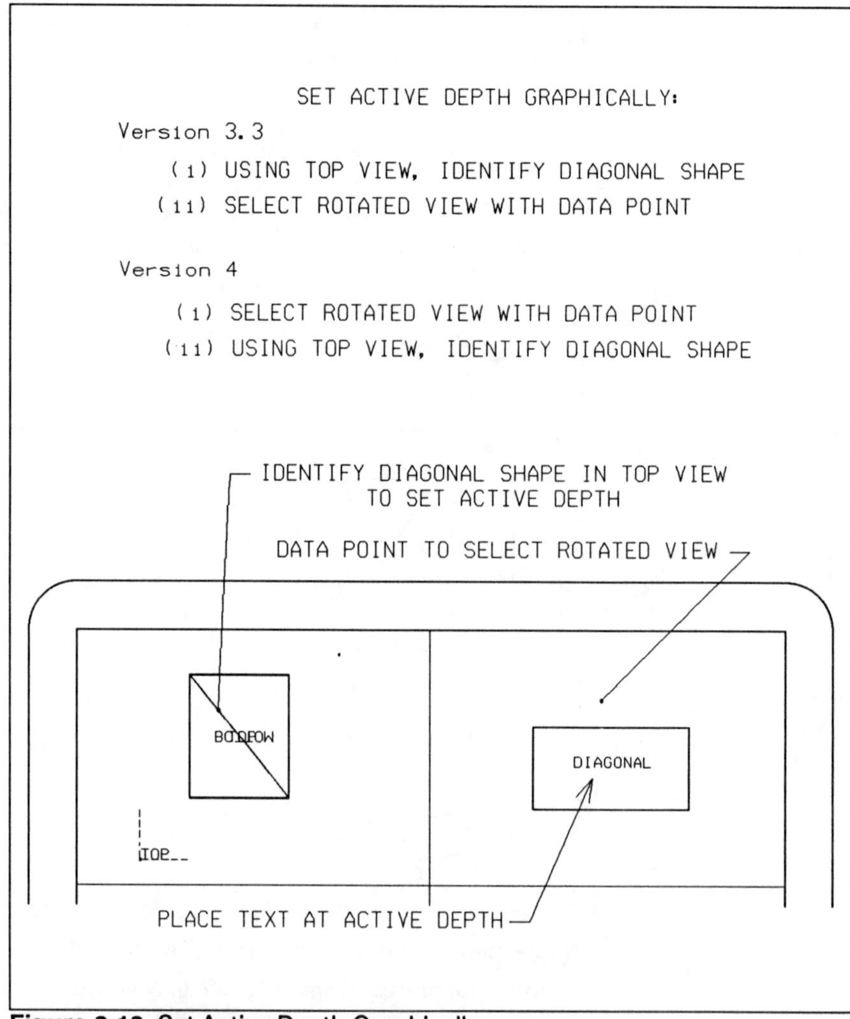

Figure 2.18 Set Active Depth Graphically

- Set the rotated view back to ISO (with VI=ISO). Your finished model should resemble figure 2.19.

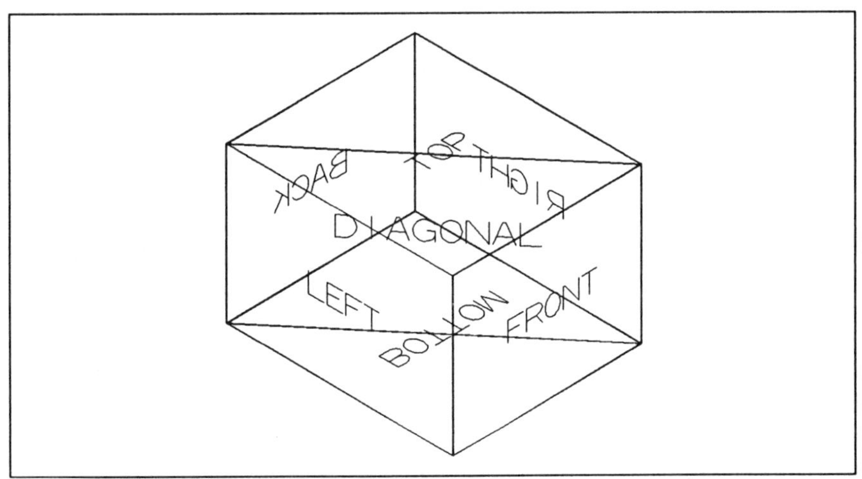

Figure 2.19 Iso View of Finished Model

All three procedures that we have just used - ROTATE VIEW BY 3 POINTS, SET DISPLAY DEPTH and SET ACTIVE DEPTH - are important tools for working in 3D. They allow us to manipulate, and work on our model as we wish. From our 2D work we know how to place elements and create shapes etc. It is the view manipulations that allow us to simply place these elements in any position in our three-dimensional design cube.

- We can rotate our design cube to align with any plane in our model.

- We can reduce any confusion in the wireframe model by looking at a discrete slice of it only.

- We can set the active depth to be anywhere within that slice for placing elements.

If you have a clear understanding of these tools and operations, then you are well on the way to mastering MicroStation 3D.

Rotate View Element (version 4)

We know how to rotate a view so that a plane, which we define with three points, will be parallel to our view or screen. With version 4 we have another key-in at our disposal - 'ROTATE VIEW ELEMENT'. This key-in requires only one data point to specify the view orientation, compared to three.

We can rotate a view/s so that a selected planar element is parallel to our view or screen, with the view Z-axis perpendicular to the element. The orientation of the view is taken from where we identify the element. This point specifies the part of the element that we want to align with the X-axis of our view. We then select a view/s to rotate and the rotation is performed.

In the previous example, we could have rotated the view by identifying the diagonal shape with a data point on its bottom edge as figure 2.20 shows.

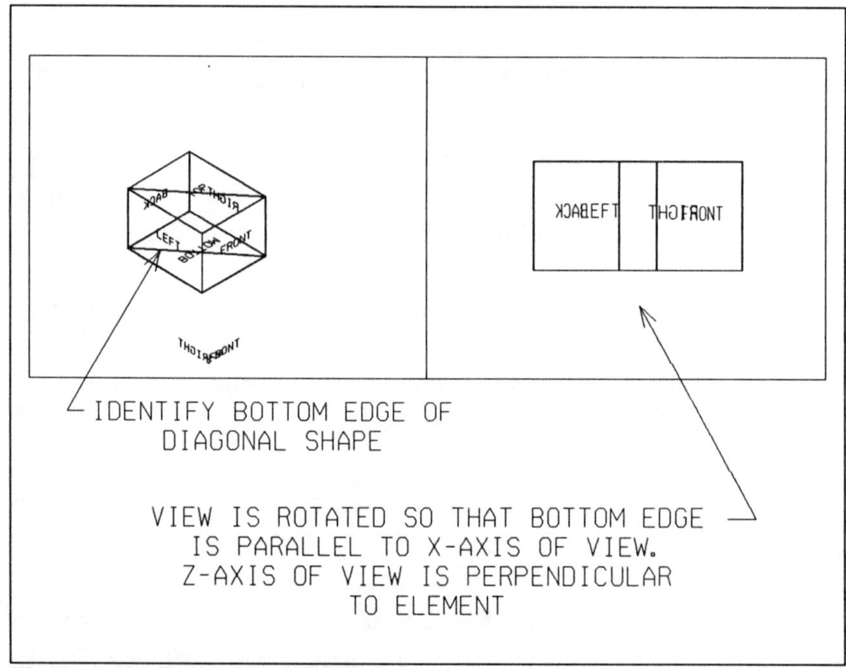

Figure 2.20 Rotate View Element

Align View

From your work in 2D you may be familiar with ALIGN VIEW. It allows us to align our views similar to how they would appear in a drawing. That is, identical elements in each view are lined up either vertically or horizontally, depending on the screen quadrants being used.

Turn off level 1 in each view, then FIT them. Now, zoom out different amounts in each view.

- Key in ALIGN VIEW or select from the menu.
- Choose the TOP view as the source view with a data point.
- Place data points in both the FRONT and RIGHT views when prompted for destination views.

Both the FRONT and RIGHT views are aligned with the TOP view. To make the RIGHT view aligned with the FRONT view, repeat the procedure, using the FRONT view as the source view and the RIGHT view as the destination view. Your screen should look something like figure 2.21 (without the dashed lines, which are only to indicate the alignment).

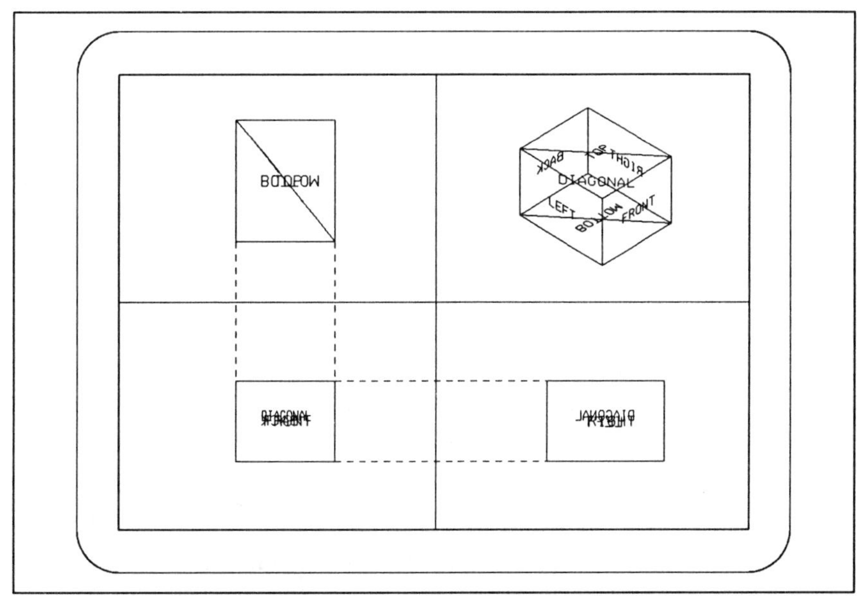

Figure 2.21 Aligning Views

Auxiliary Coordinate Systems (ACS)

Version 3.3 users can ignore this section as they do not have this facility (next section begins page 2-40). With version 4, we can define one or more auxiliary coordinate systems (ACS) to be different to the default design cube (world) system. One of these ACS systems can be active at any time, simultaneously with the default system. We have the choice of rectangular, cylindrical or spherical coordinate systems. You are already familiar with rectangular coordinate systems, as the default coordinate system of the design cube is rectangular. Locations are given as distances (x,y, and z) from the origin, along the three axes, .

ACS Precision Inputs

As with the default world system, we have precision inputs that relate to the active ACS. Having different key-ins allows us to work with the two coordinate systems simultaneously, the default world system and a defined ACS.

- *Where we use 'XY=' in the world system, we use 'AX=' with an ACS.*
- *The equivalent to 'DL=' key-in is 'AD=' with an ACS.*

It is not always possible to have a design or parts of a project align exactly with the three axes of our design cube. An ACS can be used where this situation arises. Parts of the design or project can be set up with their own ACS. Cylindrical and/or spherical systems can be used in work involving pressure vessels, for example. A rectangular ACS may be required with siteworks where the site 'North' varies from true North, or buildings do not align with one another. Often, in this situation, it is just as easy to rotate the view and use 'DX=' key-ins.

Cylindrical

We require three values to define a point in a cylindrical system. The three values are as shown for point 'P' in figure 2.22:

- Distance 'r' (along the X-axis).
- Angle 'A' relative to the X-axis (rotation about the Z-axis)
- Distance 'z' parallel to the Z-axis.

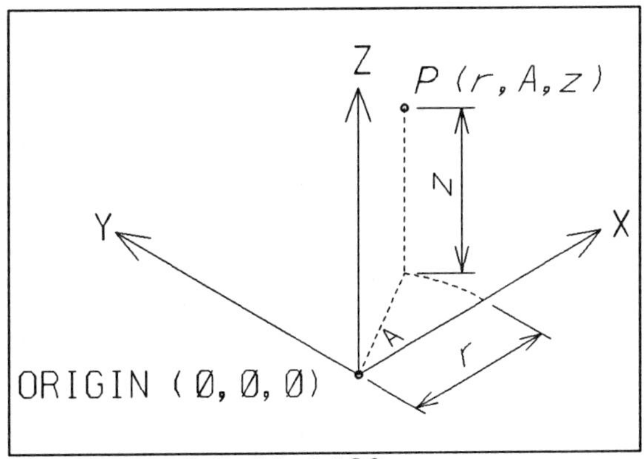

Figure 2.22 Cylindrical ACS

Spherical

With spherical auxiliary coordinate systems, points are defined with three values as shown for point 'P' in figure 2.23:

- Radius vector distance 'r' from the origin
- Angle 'A' which is the angle between the X-axis and the projection of the radius vector, on the XY plane.
- Angle 'B' which is the angle between the radius vector and the Z-axis.

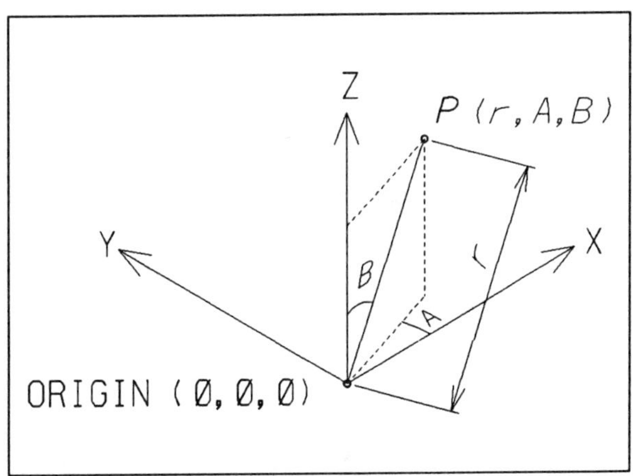

Figure 2.23 Spherical ACS

Rectangular

Like the default design cube system, rectangular coordinate systems define points by three distances as shown for point 'P' in figure 2.24. These three values are, in order:

– Distance 'x' along the X-axis.
– Distance 'y' along the Y-axis.
– Distance 'z' along the Z-axis.

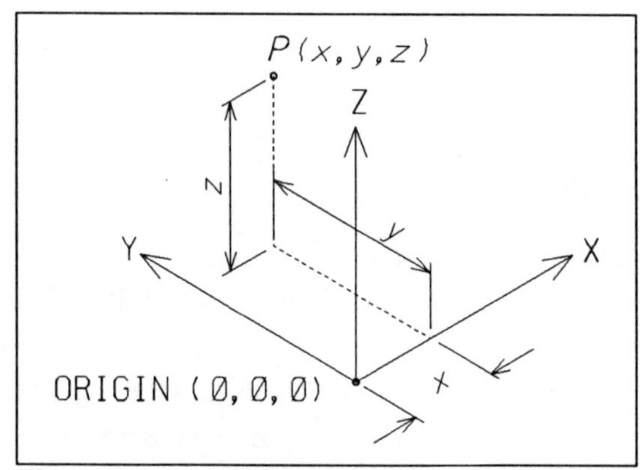

Figure 2.24 Rectangular ACS

Defining an ACS

We can define an ACS by:

1. Element

Where the XY plane of the ACS is parallel to the plane of the selected planar element. Here, the origin of the ACS is at the point of identification of the element.

2. Points

Defined by three points. Points one and two define origin and direction of the X-axis respectively. The third point defines the positive direction for the Y-axis. Positive direction for the Z-axis can be determined with the right hand rule as illustrated in figure 1.8, in the previous chapter.

3. View

Here the ACS takes the orientation of the selected view. That is, the ACS axes align exactly with those of the view selected.

Using an ACS

We work with the ACS as with 2D. We can select 'Auxiliary Coordinates' in the 'Palettes' pull-down menu to access the ACS tools palette. This contains the tools for defining, selecting, rotating and moving the ACS. We can also choose Rectangular, Cylindrical or Spherical in the 'Type' field (figure 2.25).

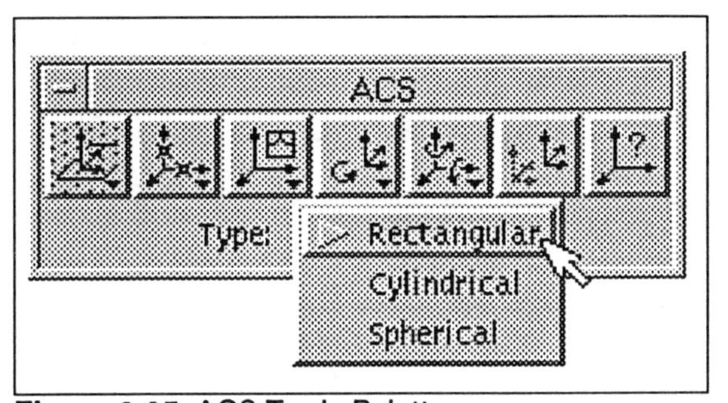

Figure 2.25 ACS Tools Palette

We will do an exercise to show how to define and use a rectangular ACS. We will use the model (rectangular box) that we created in the previous exercises.

Version 4

For this exercise we will work in the ISO view only, with the other views as a reference.

- Delete the diagonal shape and the word diagonal from the model of the rectangular box.
- Select the *Define ACS (By Points)* tool.
- Set the Type to Rectangular in the pop-down 'Type' field.
- Define the ACS, in the ISO view, using the three points as shown in figure 2.26.

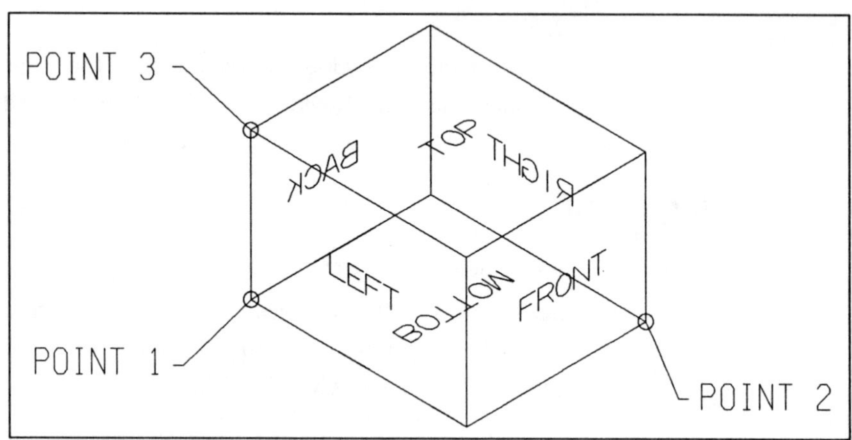

Figure 2.26 Defining ACS By Points (ver.4)

We have defined the ACS to be in the plane of the diagonal of the rectangular box. Previously, we placed a shape across the diagonal and then rotated the view to place the text. With an ACS there is no need to do this.

ACS Plane Lock

We can set the ACS plane lock on, which restricts all data points to the XY plane of the active ACS. This lock can be set in the lock settings box, or with the key-in 'LOCK ACS ON'. When we use the key-in, the system responds with the message 'ACS Plane Lock : ON'. If the lock is active, it appears as 'AP' in the list of active locks.

Do the following:

- Key in'LOCK ACS ON'.
- Select the *Place Block* tool.
- Place a block, in the ISO view, using the points as shown in figure 2.27.

Notice as you place the block that it aligns exactly with the plane of the diagonal of the rectangular box. This can be verified in the TOP view, where the block appears as a line (i.e., a block viewed edge-on). With an ACS defined, and the appropriate lock active, we can place elements at an orientation not aligned with any of the view or screen axes.

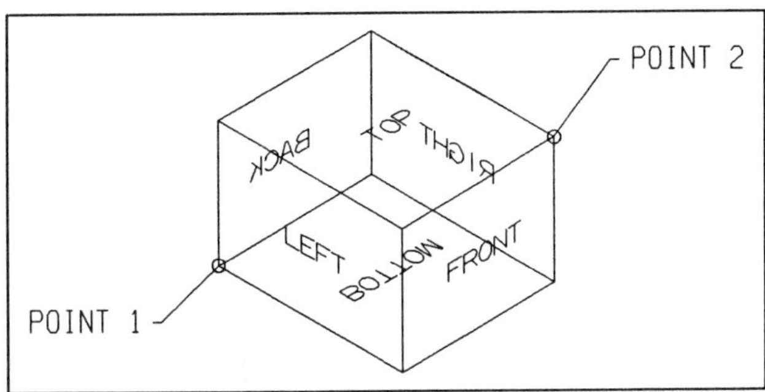

Figure 2.27 Using ACS Plane Lock

To finish the exercise, place the word 'DIAGONAL' in the middle of the block, again using the ISO view.

ACS Triad

If we wish, we can have a visual prompt as to the orientation of the ACS. This is in the form of a triad which displays in selected views. We can set this in the View Attributes settings box (figure 2.28) where it appears at the top of the left column.

Select 'Attributes' in the 'View' pull-down menu.

- Set the ACS Triad on and click on ALL.
- Fit each view

You will see a triad appear at the origin of the ACS in each view. Notice that the XY plane of the triad is directly aligned with the diagonal block that we placed in the ISO view. Figure 2.29 shows an ISO view with the ACS triad visible. It is located at the corner of the rectangular block that was selected as the origin of the X-axis.

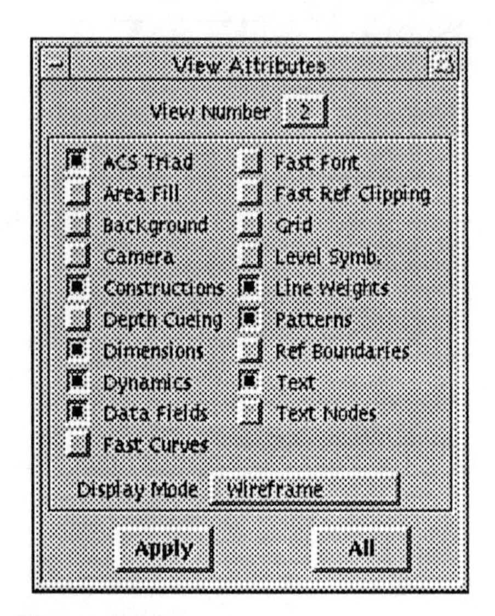

Figure 2.28

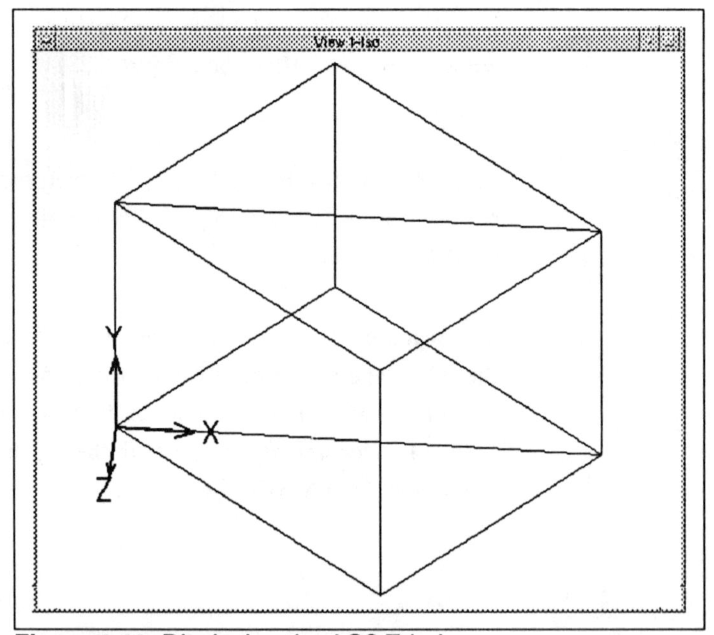

Figure 2.29 Displaying the ACS Triad

As was mentioned earlier, we can use an ACS in conjunction with the default or world system. This can be done by toggling the appropriate lock on and off as we will do now.

- Key in 'LOCK ACS OFF'.
- Place another block, again in the ISO view, using the same two points.

Checking the other views (e.g., TOP), notice that this block does not align with any of the views. Without an ACS and the associated lock active, element placement reverts to the default. That is, elements are placed parallel to the view or screen.

ACS Plane Snap Lock

Another facility that we have with auxiliary coordinate systems, is the 'ACS Plane Snap'. This is a snap lock that forces tentative points to fall on the XY plane of the active ACS. Again, this is a way of working in a plane that is 'skewed' to the view or screen.

This lock can be set in the lock settings box or with the key-in 'LOCK SNAP ACS ON'. When we use this key-in, the system responds with the message 'ACS Plane Snap Lock : ON'. If this lock is active, it appears as 'PS' in the list of active locks.

Saving ACS

We can have a number of auxiliary coordinate systems associated with a design file. Of these, only one can be active at any time. Once defined, we can save an ACS for future use. This is done via the Auxiliary Coordinates settings box, as with 2D files.

Once we have named the system and entered a description, clicking on 'Save' completes the procedure. This box can be used to recall an ACS when required. As with cells, saved views etc., it is important to give a description that is meaningful, for future reference.

3D Tentative Point

We learnt previously how our files have a plane, in each view, on which elements will be placed, by default. We can over-ride this default by using a tentative point to an existing element. With version 4 we can over-ride the default also, using an auxiliary co-ordinate system. We now will look at yet another option.

With the SET BUTTON key-in we can configure our cursor, or mouse buttons. We define buttons for DATA POINTS, RESET and TENTATIVE POINTS. We are then prompted for 3D TENTATIVE and 3D DATA POINTS. These can be configured to be key-board entries, if necessary.

When we place a tentative (or data) point in a vacant part of a file (i.e., without snapping to an existing element), the point falls on the active depth of the view. When we position the cursor on the screen, we are defining view X and Y coordinates. The Z value for the tentative/data point is determined by the system, and is set at the current ACTIVE DEPTH of the view.

With a 3D TENTATIVE point we require two inputs. The first determines the location of the point on the plane of the view. It locates the point in the view X and Y directions. The Z value could be anywhere along an imaginary line perpendicular to the screen. This imaginary line we call a bore line. A second input is then required to specify the view Z value (depth) of the plane. How this works in practice, is as follows:

When we place the first 3D TENTATIVE, bore lines appear in other non-parallel views, where possible (figure 2.30). The bore lines indicate where our tentative point is located, in two directions. We can place another 3D TENTATIVE in one of the other views to set the third value (i.e., Z value) of our point. Pressing the data button then accepts this point.

3D DATA points work in a similar fashion except that a data button is not required after the second input.

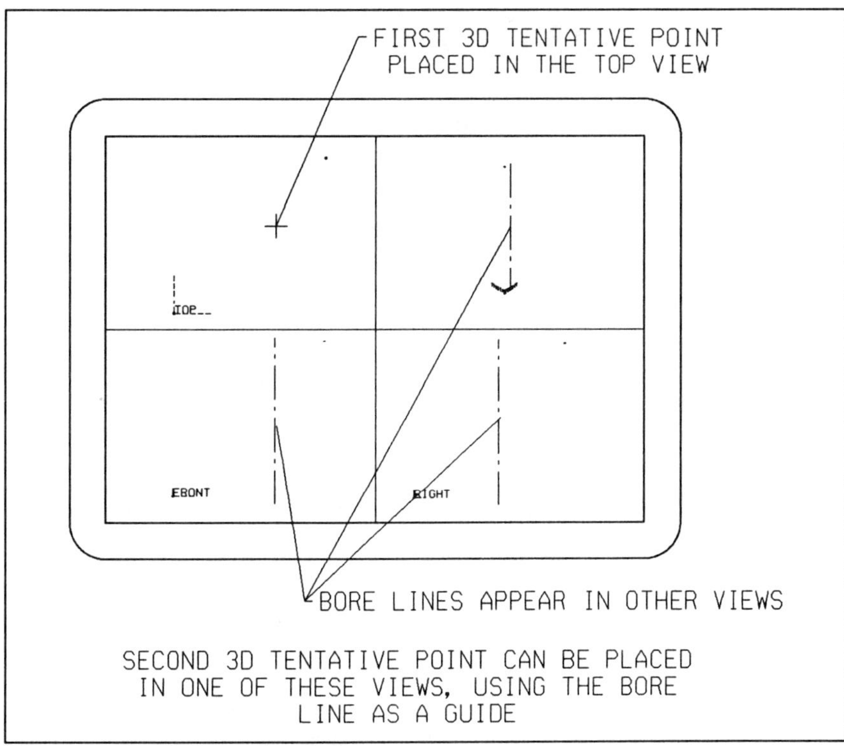

Figure 2.30 3D Tentative Point

Fence Manipulations

By now, you should be aware that each view in a 3D file displays part of the design cube volume. The boundaries of the volume are the area of the window along with the front and back limits (i.e., clipping planes) of the display depth. Elements, or parts of elements, not contained within the volume of the view, do not display.

Fence manipulations work within the same limits. Elements displayed in a view can be manipulated with a fence. Care should be taken with the use of OVERLAP LOCK. There may be parts of elements contained within the fence that, while they are displayed, they are not obvious. Two dots, for example, may be the opposite sides of a shape, viewed edge on, as we discussed previously (refer to figure 2.6).

To reinforce this point, do the following:

- Set the display depth in the FRONT view so that the rear face only (the BACK face) of the box displays.
- Set the active color to white.
- Place a fence around the back face, in the FRONT view. Make sure that the fence lock INSIDE is enabled.
- Change the color of the back face using the *Change Fence Contents to Active Color* tool.

Checking in the ISO view you should see that the rear face of the box, and its text, have changed color, but nothing else.

Now make the fence lock OVERLAP active and repeat the procedure.

This time you will notice that the left, right, diagonal, top and bottom faces of the box have changed color. Parts of each element were contained in the volume of the view. Although some of the elements were not obvious in a FRONT view, they were contained in the display depth of the view. They were included in the OVERLAP lock for the fence operation.

Figure 2.31 shows pictorially why the sides etc. changed color with overlap lock active.

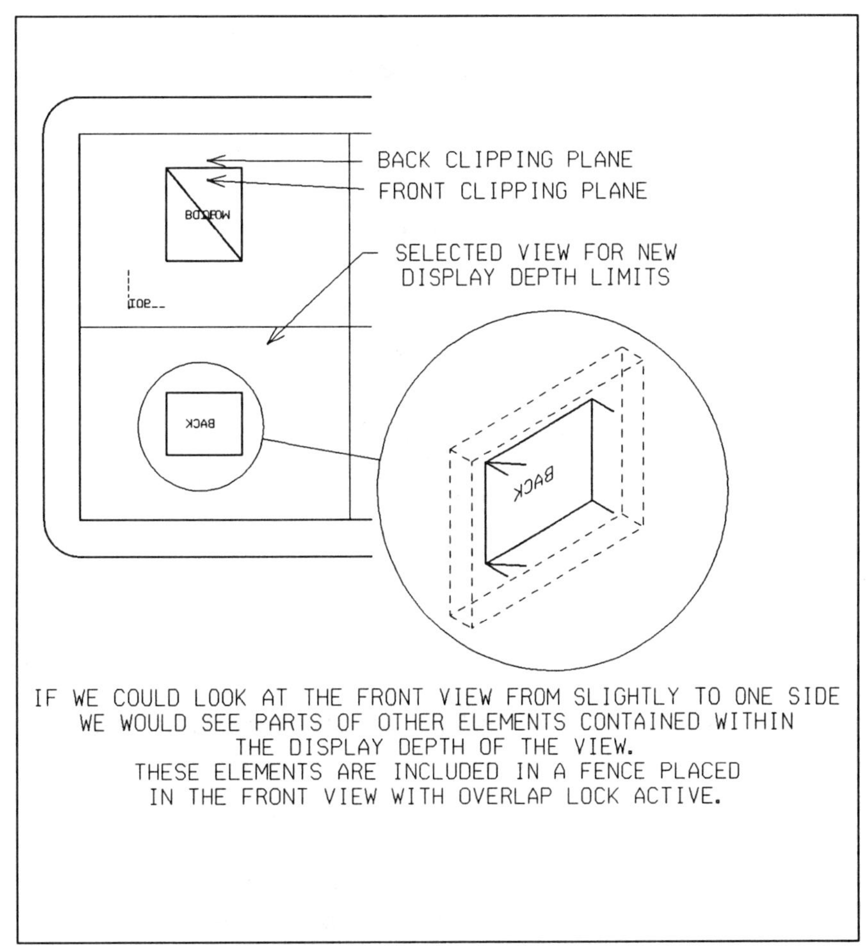

Figure 2.31 Using Fences with Overlap Lock

It is important when using fence operations to be sure that all the required elements are within the fence. Just as important, is to be sure that no unwanted elements are within the volume bounded by the fence and the display depth. This is particularly important where OVERLAP LOCK is active.

While we have the UNDO function if a mistake is made, this is not foolproof. Often, fence operations manipulate more elements than can be stored in the undo buffer. So, care has to be taken with selection of the LOCK to use and the placement of the fence. While in 2D we can see all the elements because they are on a flat plane, 3D is different. There may be elements, or parts of elements, contained in the view, that are not obvious. If you're not completely confident with a fence operation, create a BACKUP of the file prior to performing the task. This applies to any procedures that you are not sure about. Backing up will at least provide a means of getting back to where you were. Another option is to increase the undo buffer.

As with 2D files, we may use any view to specify distances to manipulate objects within a fence.

To see this, do the following:

- Fit each view and turn the text off in the TOP view (level 63). Place a fence around the box in this view.
- Select the *Copy Fence Contents* tool.
- For the origin, identify the lower left back corner in the ISO view (refer figure 2.32).
- Select the upper right front corner for the point to copy to.

When the fence operation is complete, fit each view then window area the two boxes in the ISO view as in figure 2.32.

In the previous exercise, the distance we specified had X, Y, and Z vectors. That is, we copied the box a horizontal distance in the X and Y directions, also vertically in the Z direction. We placed the fence in the TOP view. To specify the distance with a precision input was an option but required calculations. We simplified our task by using the ISO view to select the distance to copy the contents of the fence. We used a graphical precision input. The copy was of the box as displayed in the TOP view, with no text. The view we used for specifying the distance had the text displayed. With 3D, more so than 2D, it is important to understand how to use the views to help with construction and manipulations.

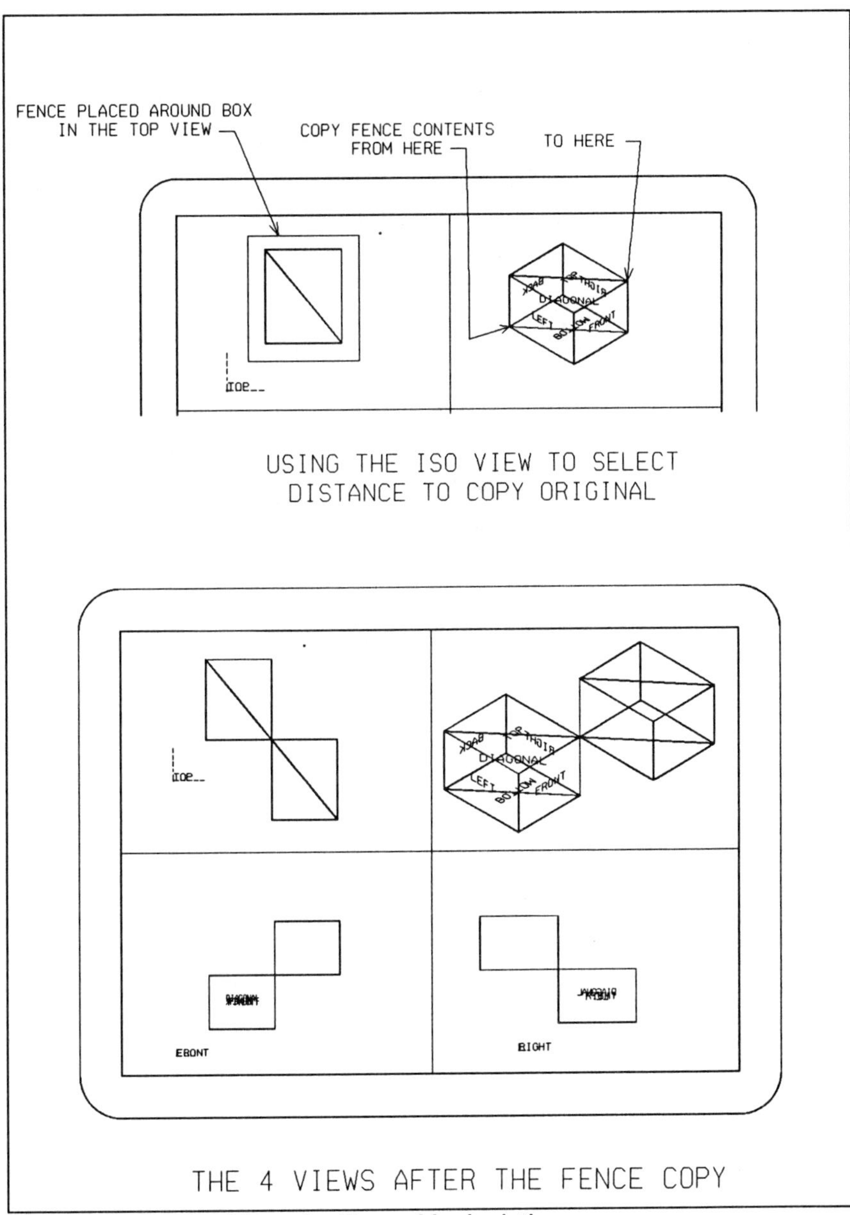

Figure 2.32 Using Views for Fence Manipulations

Summary of Chapter 2:

In this chapter we discussed the tools we use to move in and around the design cube. We learnt how to create a simple model and how to use precision inputs. In the following chapters we will be using these tools and procedures. Be sure that you feel comfortable with the topics we have covered. They are the basis of most of our work in 3D.

Important subjects covered in chapter 2 included:

- Precision Inputs
- Using Active Depth and Display Depth
- View Rotation
- Auxiliary Coordinate Systems (ACS)
- ACS locks
- 3D Tentative/Data Points
- Fence Manipulations

3 : Project & Surface of Revolution

In this chapter we will be using some of the more powerful 3D tools. Till now, we have been using two-dimensional elements only, to create a model. In some instances, this may be the only way an object can be constructed. Now we will look at how to create PROJECTED surfaces and SURFACES OF REVOLUTION. Before we do that, however, we will learn the rudiments of running the HIDDEN LINE REMOVAL utilities. This will help us in checking our models as we construct them.

Basic Hidden Line Removal

We will be looking at rendering and hidden line removal, in detail, in chapters 9,10 and 11. Here we will learn how to run the basic hidden line removal. With hidden line removal we can see our model in a more realistic fashion. The system displays only those elements that would be seen in real life. That is , any elements or lines that would be obscured by other parts of the model are removed from the screen display.

Again the methods used differ between version 3.3 and version 4 of MicroStation, so we will look at each individually.

Version 3.3

To run the RENDERING and HIDDEN LINE REMOVAL utilities, with version 3.3, they first have to be installed. This is usually done at the time of installing the rest of the MicroStation software. If you have problems getting HIDDEN LINE REMOVAL to operate, check first that the required program files have been installed.

The relevant files are located in the MicroStation RENDER directory, and should include:

HIDDEN.EXE
RENDERU.EXE
HLINE.UCM
SHADES.UCM
EDGES.UCM

Here, we will be using the HIDDEN LINE REMOVAL utility. To initiate the process, we select UTILS - HLINE from the sidebar menu. If you are not using the sidebar menu, key in 'UC=MS_RENDER:HLINE'. You will be prompted to 'Enter options or select a view'. For now, we won't worry about the options - we will just select a view to have the hidden lines removed.

Version 3.3

We will use the ISO view of your training file, which should have the two boxes after the fence copy in Chapter 2.

° Select HLINE from the menu or use the key-in.
° Select the ISO view.

The resulting hidden line removal should be similar to figure 3.1.

The process is a raster hidden line removal. That is, it only shows us, on the screen, how the model would look with hidden lines removed. If we update the screen, we lose the hidden line removal view and return to wireframe mode. Any view may be selected for hidden line removal.

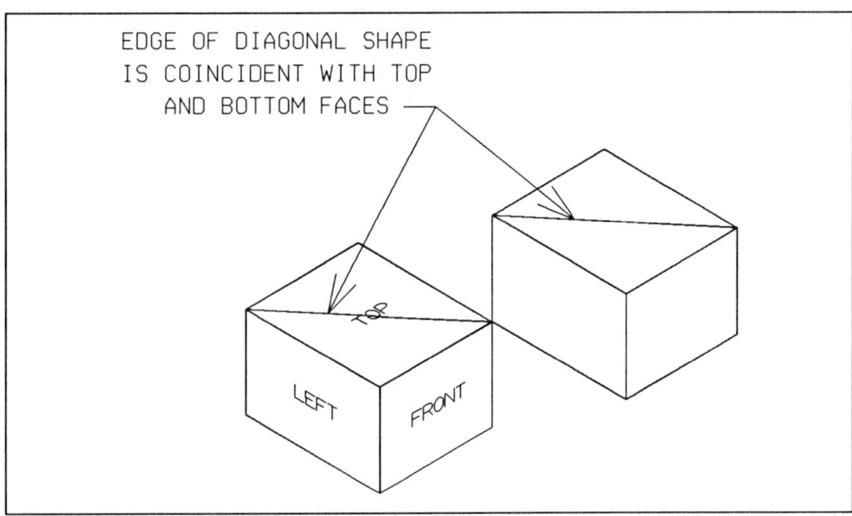

Figure 3.1 Hidden Line Removal

Version 4

To perform a hidden line removal with version 4, we can select the tool in the pull-down menu (refer figure 3.2). Alternatively, we can use the key-in 'RENDER VIEW HIDDEN'.

We are then prompted to select the view to render.

Version 4

Try this with the ISO view of your training file, displaying the two boxes:

- Choose View - Render - Hidden Line in the pull-down menus (figure 3.2), or use the key-in 'RENDER VIEW HIDDEN'.
- Select the ISO view.

The result of the hidden line removal should be similar to figure 3.1. As with version 3.3 this is a raster hidden line removal. It is a screen display only. Updating the view will return it to wireframe.

You will notice that there is a diagonal line across the top surface of both boxes. This is the diagonal shape. Its edge is coincident with the top (and bottom) surface of the box.

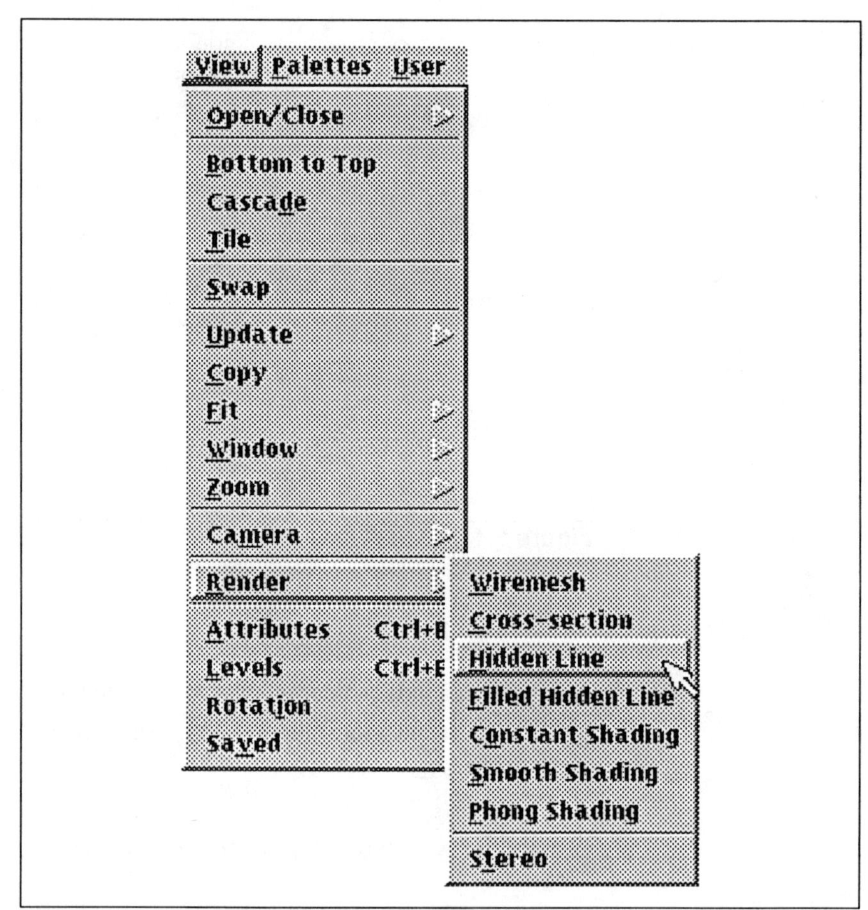

Figure 3.2 Choosing Hidden Line Removal (ver.4)

Projected Surfaces

With the *Construct Surface/Solid of Projection* tool we can create three-dimensional objects from two-dimensional elements. Lines, linestrings, complex strings, arcs, circles, blocks, shapes and complex shapes (including cells) may be projected. Version 4 also allows for B-spline curves to be projected.

Elements are projected, or copied, a distance and direction that we specify. Surfaces are created between the original location of the element and the projected location. In wireframe mode projected surfaces are shown as lines joining the key points of the original and projected elements. *In creating projected surfaces, the original element is not moved or altered in the process.* Figure 3.3 shows projected surfaces of a block and a circle in an ISO view.

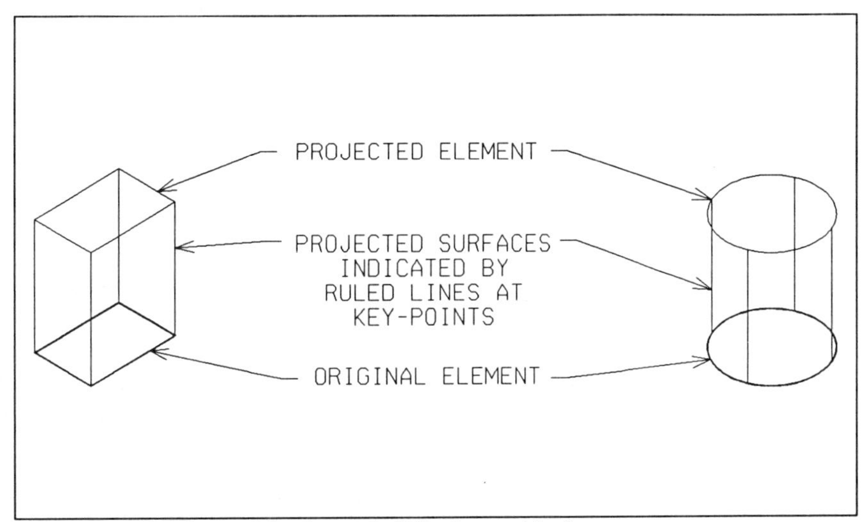

Figure 3.3 Projected Surfaces in Wireframe Mode

When we use the *Construct Surface/Solid of Projection* tool, we are first prompted to identify the element to be projected. This is followed by a prompt for the projection distance. The projection is constructed from the point where the originating element was identified, in a straight line, to the second point.

When using data points, the selection of the identifying point for the element is important as figure 3.4 illustrates. The solid lines in this diagram show the projection if we identify the element at I.D. point 1 and project it to point X. The dashed lines shows the projection if we start from I.D. point 2 with the same final target projection point X.

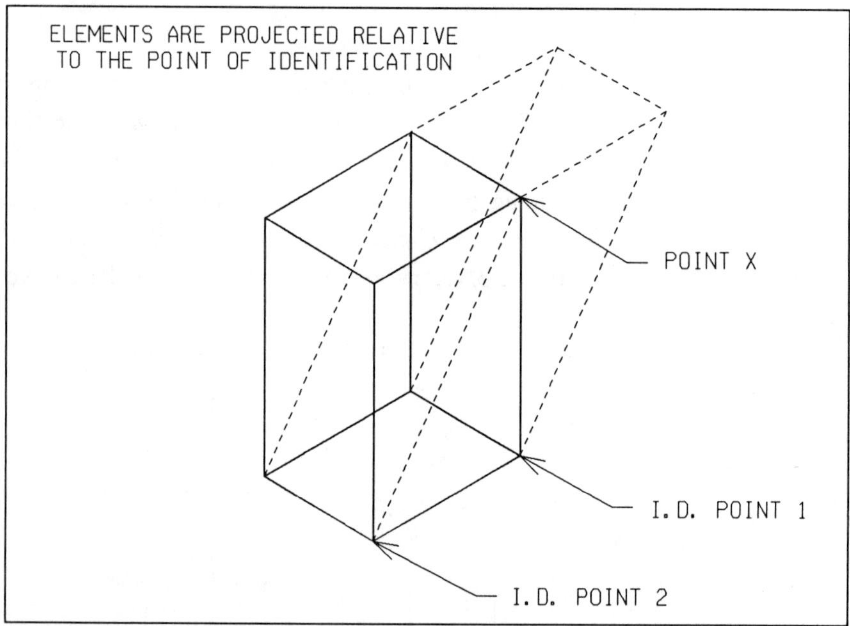

Figure 3.4 Projecting an Element

ACTIVE ANGLE and ACTIVE SCALE both affect projected surfaces. There are times when it is necessary to create a projected surface with an active angle, active scale, or both. For most cases the requirement will be for straight projections with an active scale of 1 and active angle 0. Either way, it is good practice to check both prior to creating the projected surface.

With version 4, when we select the *Construct Surface/Solid of Projection* tool in the 3D palette (bottom left corner), the resulting dialog box shows us the current status of active angle and active scale (figure 3.5). If required, we can change them in this dialog box.

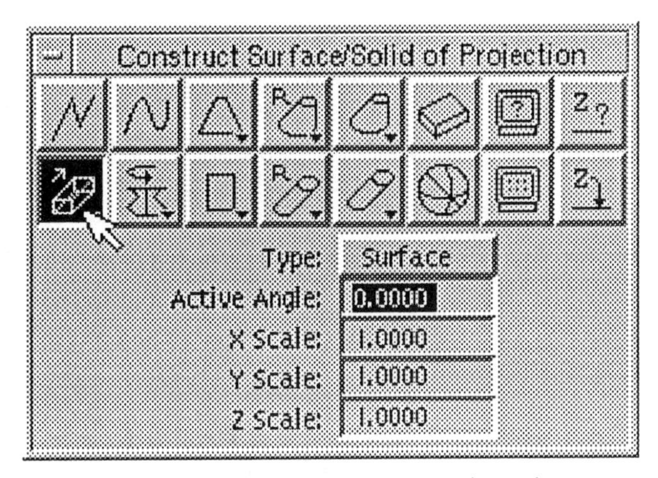

Figure 3.5 Selecting the Project Tool (ver. 4)

Construct a projected surface as follows:

° Copy the base of your box model to level 10.
° Set the views up with only levels 1 and 10 displaying. Fit each view if necessary.
° Make sure that Active Scale is 1 and Active Angle is 0. With version 4 this can be checked after selecting the *Construct Surface/Solid of Projection* tool in the 3D palette (as shown in figure 3.5).
° Select *Construct Surface/Solid of Projection* in the palette or from the menu.
° Identify the element to be projected.
° Key in DL=,,400 to specify the distance and direction of the projection (i.e., 400 units in the Z direction of the model).

The result is a box like your original model. The difference is that this one is produced from a single element. The original box consisted of 6 separate blocks, one for each face. Projected elements are more efficient, in that they take up less disk space. While it is not always appropriate to use projected elements to create a model, they should be used where possible. A projected element can be manipulated and modified as a single element.

Experiment with placing various elements and projecting them. Use both precision inputs and data points. Check in the other views to see what is happening in each case. If parts of the elements disappear, they may have projected through the DISPLAY DEPTH. Use tools or key-ins such as FIT, DEPTH DISPLAY etc., to display the whole element.

Elements may be projected in any direction and in any view. A circle placed in a front or right view and projected horizontally could be a pipe-line. The same circle placed in a top view and projected vertically could represent a tank.

Capped and Uncapped Projections

Two forms of projection of closed shapes are available, CAPPED and UNCAPPED. These are referred to also as SOLIDS and SURFACES respectively. Shapes projected as capped surfaces (solids) have both ends closed. Uncapped (surface) projections produce surfaces only, without closed ends.

In wireframe mode it is not apparent whether an object is a capped or uncapped surface. When we perform a hidden line removal, however, it becomes obvious. Figure 3.6 shows examples of capped and uncapped projections after hidden line removal.

We have other ways of finding out whether a projection is capped or not. It isn't always convenient to run hidden line removal, particularly if the element is one of the last to be processed. Two other methods are:

1. Select the *Copy Element* tool and identify the projection. MicroStation displays information on the identified element in the message area. Capped surfaces register as SOLID while uncapped surfaces register as SURFACE. Once the object has been identified, we can reset without altering anything. Other tools could be used in this manner also.

2. We can use ANALYZE and identify the required element. Again, the element registers as a SOLID if it is capped, and as a SURFACE if it is uncapped.

With version 4 we can select 'Info' from the 'Element' pull-down menu.

Note:

While the capped surfaces are referred to as SOLID by MicroStation, they are not solids in the true sense. They are fully enclosed, but hollow.

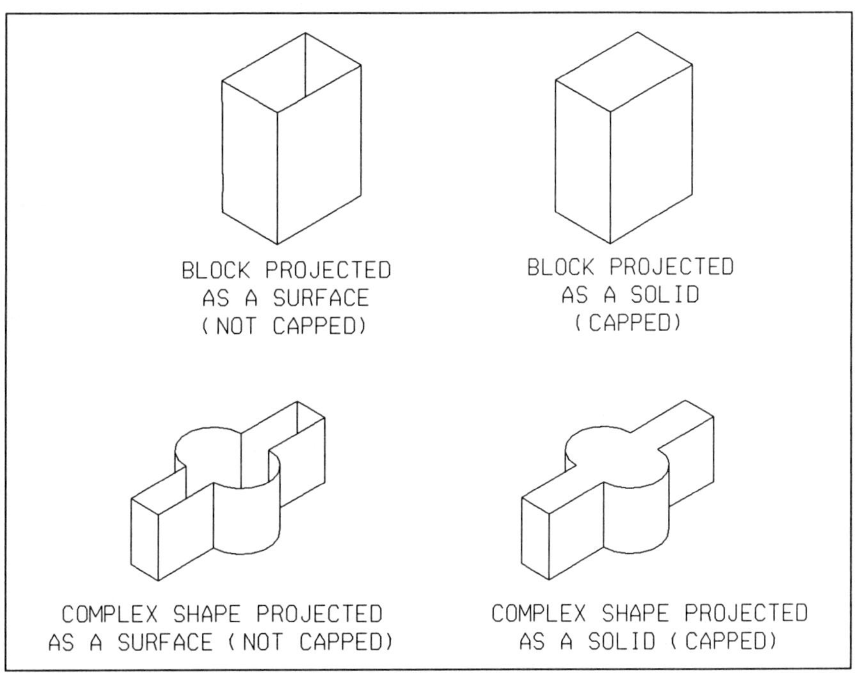

Figure 3.6 Capped and Uncapped Projections

Keying in ACTIVE CAPMODE ON or OFF allows us to project closed shapes either as solids or as surfaces. With ACTIVE CAPMODE ON the shape is projected with a top and bottom. ACTIVE CAPMODE OFF causes the shapes to be projected as surfaces only, without a top or bottom (apart from the original element used to create the surface).

We can select active capmode via the sidebar or tablet menus also. Figures 1.3 and 1.4 in chapter 1 show the locations on these menus.

With version 4, when we select the *Construct Surface/Solid of Projection* tool in the 3D palette, we are given the opportunity to select the Type as either SOLID (CAPPED) or SURFACE (UNCAPPED), as shown in figure 3.7.

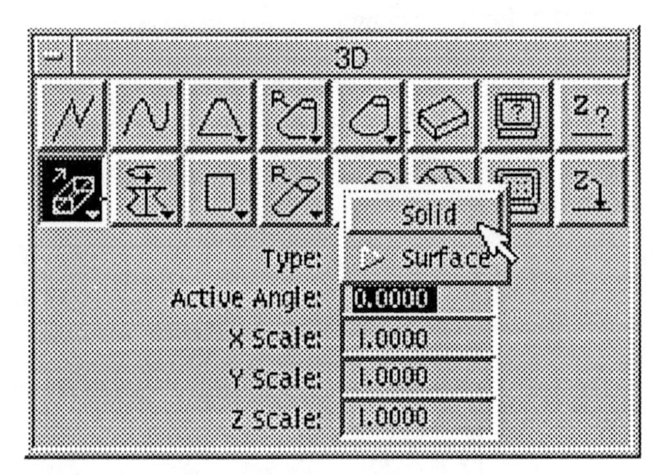

Figure 3.7 Selecting Solid or Surface (ver.4)

To check which mode is enabled (default is ON) we can key in ACTIVE CAPMODE. The system will respond with 'Capped Surface Placement : ON' if it is enabled (solid mode) and 'OFF' if it is disabled (surface mode).

Delete or undo the elements and projected surfaces created in the previous exercise and place two 200 x 100 blocks in the TOP view as follows:

- Place one at XY=300,300,200 (followed by DL=200,100) and the other at XY=600,300,200 (followed by DL=200,100). Remember to make the TOP view current before using the precision inputs.
- Set ACTIVE CAPMODE ON (alternatively, with version 4, type can be set to SOLID when the *Construct Surface/Solid of Projection* tool is selected).

- Select the *Construct Surface/Solid of Projection* tool (version 4 users can set Type to SOLID, here, if the 3D palette is being used).
- PROJECT the first block 400 vertically (e.g., DL=,,400)
- Set ACTIVE CAPMODE OFF (set Type to SURFACE - version 4 using the 3D palette).
- PROJECT the second block 400 vertically.

Your ISO view should look like part (A) in figure 3.8. Both boxes look identical in wireframe mode. To see the difference, use the hidden line removal utility. The result is as shown in part (B) of figure 3.8.

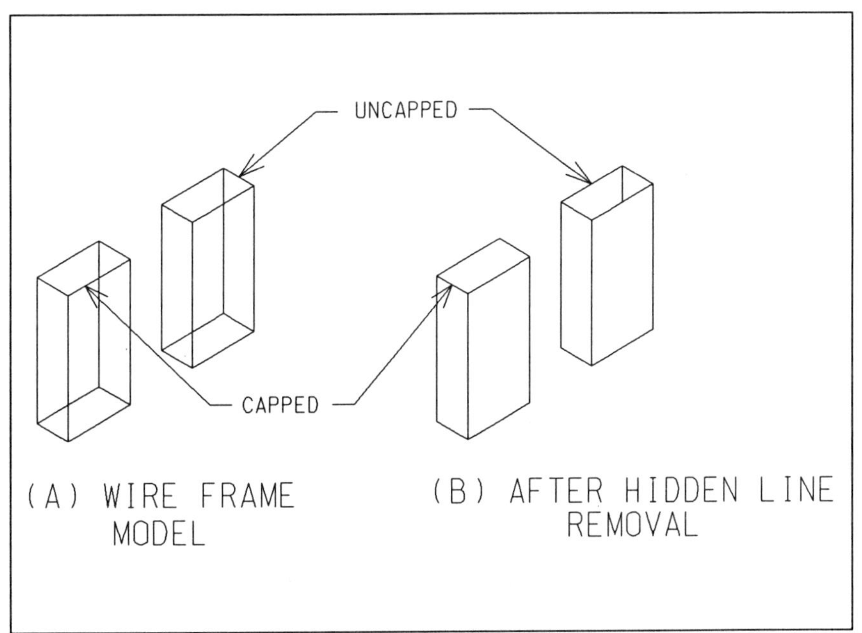

Figure 3.8 Capped and Uncapped Surface

Dropping Projected Surfaces

Like shapes, projected surfaces (or solids) can be dropped. When dropped, they revert to their separate components. The separate components are the original element, the copied or projected element and the ruled lines joining them. Figure 3.9 illustrates, after hidden line removal, the results of dropping various projected surfaces.

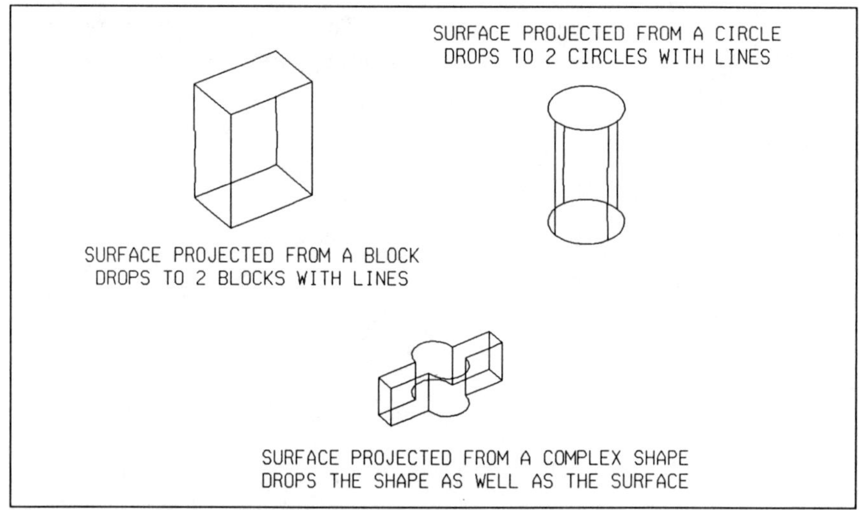

Figure 3.9 Dropped Surfaces

Projecting with Active Scale

Projecting with an ACTIVE SCALE specified causes the projected element to be scaled up or down by the scale factor. Surfaces are created between the original and projected element as for normal projections. Scaling of the element is performed about the point where it was identified.

Scale factors may be specified for the X, Y or Z direction or any combination of the three. Scaling factors are view dependent. They relate to the axes of the screen in the view being used. This is the view in which we specify the projection distance - the current view, when using precision inputs.

Delete the surfaces and the blocks created for the previous exercise, then:

- Place a 200 x 100 block in the FRONT view. Make the FRONT view current. Place the block with its origin at XY=200,200,300 followed by DX=200,100 or DL=200,,100.
- Make the active scale 2:

  Keying in AS=2 makes the scaling factor 2 for all three axes. If we wanted the scaling to be in the X direction only we would key in XS=2 as for 2D files. We use similar key-ins (YS= and ZS=) for setting the scale factor, individually, for the other axes.

  With version 4 we can set the scale factors in the dialog box when we select the *Construct Surface/Solid of Projection* tool.

- Check that the active angle is 0.
- Select SURFACE from the menu or key in ACTIVE CAPMODE OFF to select a non-capped surface.

  Again this can be set in the dialog box with version 4.

- Select the *Construct Surface/Solid of Projection* tool in the palette or from the menu.
- Identify the block at the lower left hand corner in the FRONT view.
- Key in DX=,,-400 for the projection distance (400 into the screen from the FRONT view).

Your screen should look similar to figure 3.10. You will notice that the projected element has been scaled up. The scaling is about the point that we used to identify the element - the lower left hand corner in the FRONT view.

Figure 3.10 Projecting with Active Scale

Projecting with Active Angle

Similar to projecting with active scale, elements projected with an ACTIVE ANGLE are rotated about the point at which they were identified. Rotation is always relative to the screen Z axis, of the view where we specify the projection distance.

Delete or undo the surface created in the previous exercise. We will use the same block for the projection.

Now:

- Set the ACTIVE SCALE to 1 (AS=1).
- Set the ACTIVE ANGLE TO 30 degrees (AA=30). With version 4 these can be set after we select the *Construct Surface/Solid of Projection* tool in the 3D palette.
- Select the *Construct Surface/Solid of Projection* tool in the palette or menu.
- Identify the block at the lower left end in the FRONT view.
- Key in DX=,,-400 for the projection distance.

This time the element has rotated as it was projected as in fig 3.11. The rotation is about the point that we used to identify the element.

Note:

The surface created in figure 3.11 should be a warped surface. That is, it should have a twist in it. Because we created the surface in one operation MicroStation projects the copy of the element, rotating it the specified angle. Ruled lines are then shown between the original element and the projected element. This creates surfaces that are non-planar. That is, the four points defining the surfaces are not co-planar as we can see in the FRONT and RIGHT views. In chapter 7 (Advanced Techniques), we will look at this problem and a work-around. We will also see an alternative way to create the same construction with version 4.

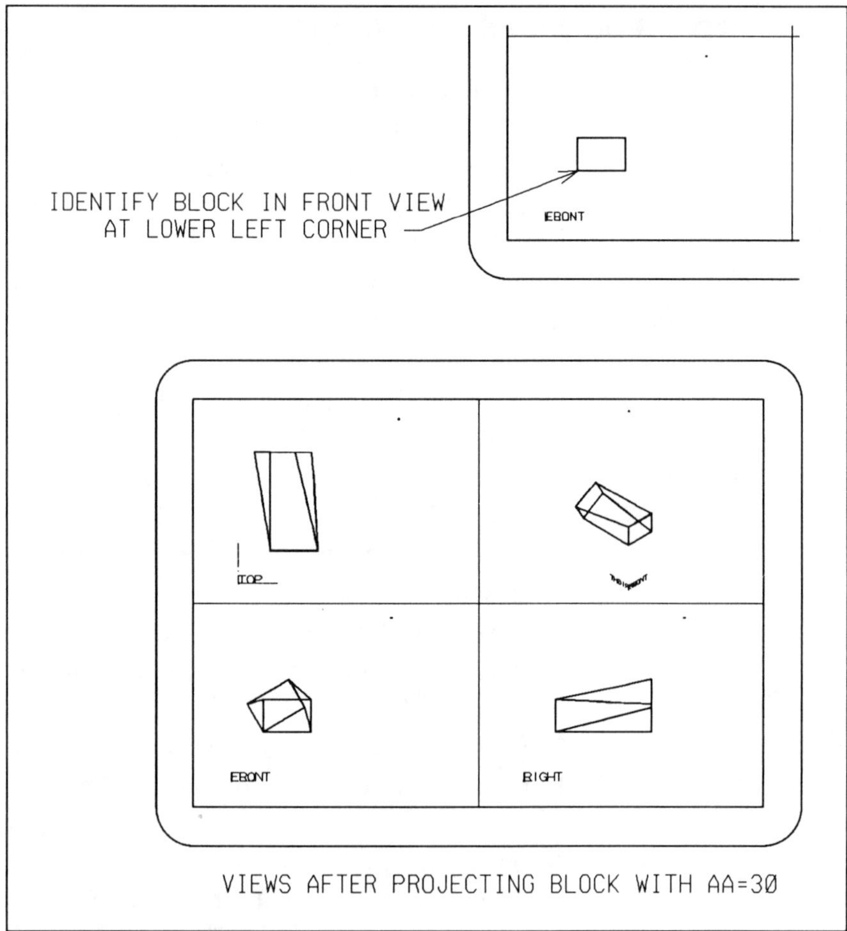

Figure 3.11 Projecting with Active Angle

Using both together

Projecting with ACTIVE ANGLE and ACTIVE SCALE together can be used for creating objects such as mitred corners and the like. These construction methods have more significance for version 3.3 users. With version 4 we have other surface construction tools at our disposal, such as those for b-spline surfaces. Again, all of these topics are discussed in chapter 7. The exercise included with that chapter has examples using both methods. For now, it is more important to grasp fully, the fundamentals.

Projecting Fence Contents (version 4)

We can project the contents of a fence with version 4. We use the key-in: 'FENCE SURFACE PROJECTION' to initiate the process. With fence projections we can project a number of objects simultaneously. For example, the columns in a building could be projected inside a fence.

Our projections can be as capped (solids) or uncapped (surfaces) elements. These can be set with the ACTIVE CAPMODE ON/OFF key-in. Fence projections are affected by ACTIVE SCALE and ACTIVE ANGLE settings.

Another consideration is the fence lock setting. Figure 3.12 shows a TOP view of three elements with a fence that partly overlaps each. Also shown are ISO views of the results of projecting the fence contents with various fence lock settings, as indicated. Where one of the CLIP fence lock settings is used, *the original element is altered,* just as with any fence clip operation.

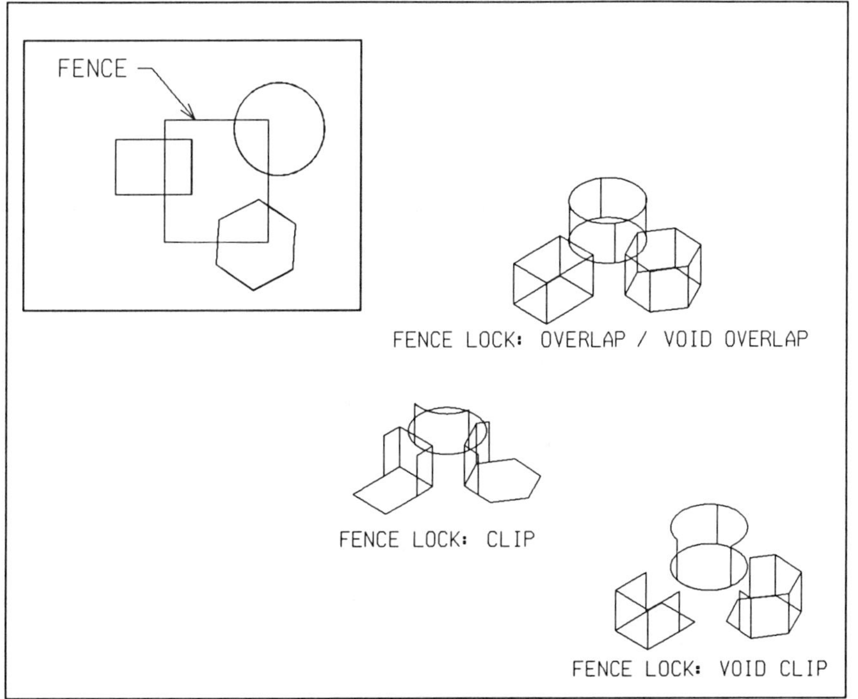

Figure 3.12 Projecting Contents of Fence (ver.4)

Surface of Revolution

Surfaces of revolution are similar to projected surfaces. A surface of revolution is formed by projecting an element in a circular path about an axis or pivot point, the AXIS OF REVOLUTION. In figure 3.13 we have a circle which is to be used to form the surface of revolution in figure 3.14.

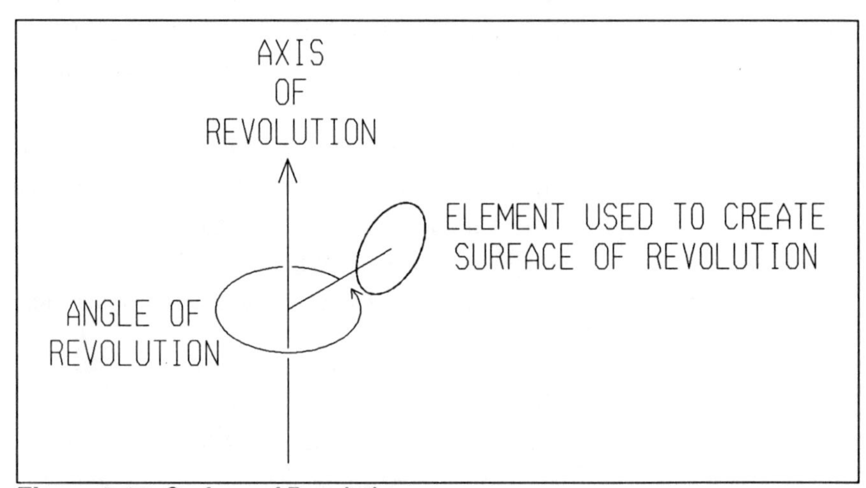

Figure 3.13 Surface of Revolution

To make the surface of revolution, the circle is projected in a circular path about the AXIS OF REVOLUTION. This is similar to rotating and copying the element. Curved surfaces are formed between the original and projected position. The surfaces formed are shown, in wireframe mode, as curved lines from the key-points of the originating element (the circle, in this instance). Figure 3.14 shows the surface of revolution of the circle in both wireframe form and after running hidden line removal. The shape of the surface is quite evident, after hidden line removal.

While ACTIVE ANGLE and ACTIVE SCALE have no effect on surfaces of revolution, ACTIVE CAPMODE does. We can create capped (solid) or uncapped (surface) surfaces of revolution.

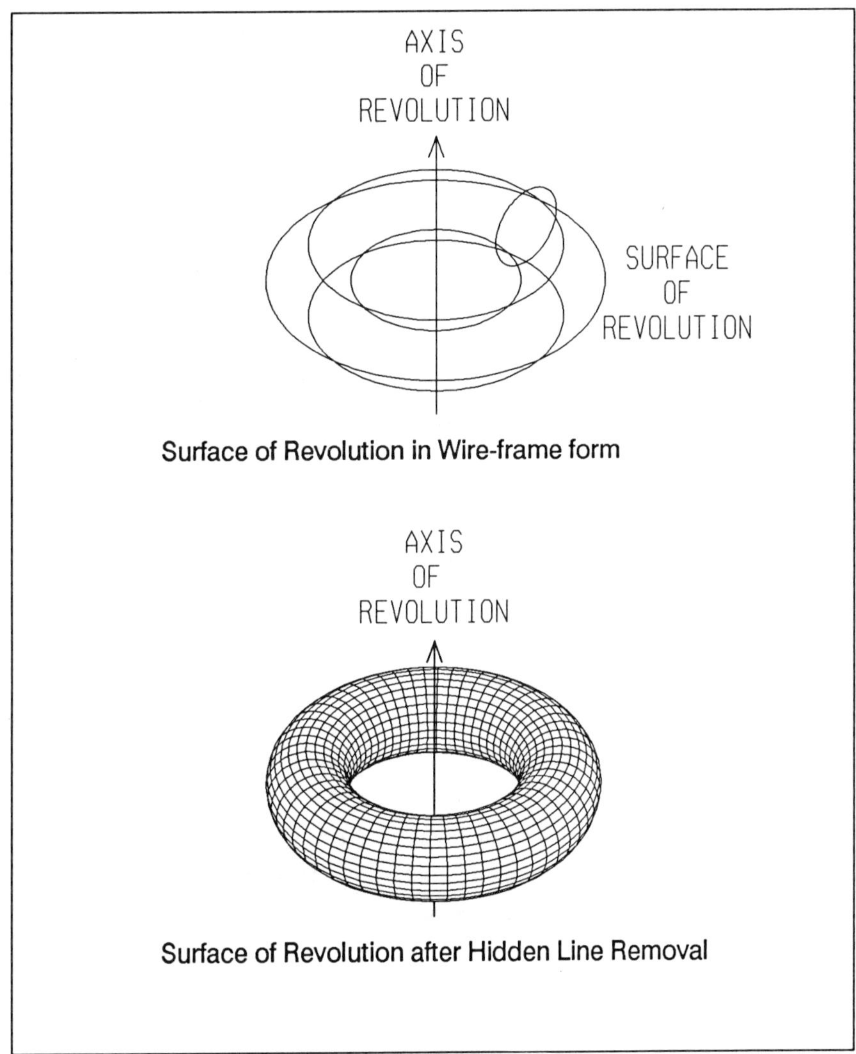

Figure 3.14 Displaying Surfaces of Revolution

Selecting the correct view for defining the AXIS OF REVOLUTION is the most important consideration for creating a surface of revolution. As with all rotations, the angle specified is about an axis parallel to the Z axis of the view.

Note!

The AXIS OF REVOLUTION is always perpendicular to the screen for the view in which we select it. It always aligns with the Z axis of the view or screen.

With version 4 we can select the *Construct Surface/Solid of Revolution* tool in the 3D palette (second from the left, bottom row). We then have the opportunity to set our ANGLE OF REVOLUTION and the type of construction to SOLID or SURFACE (figure 3.15).

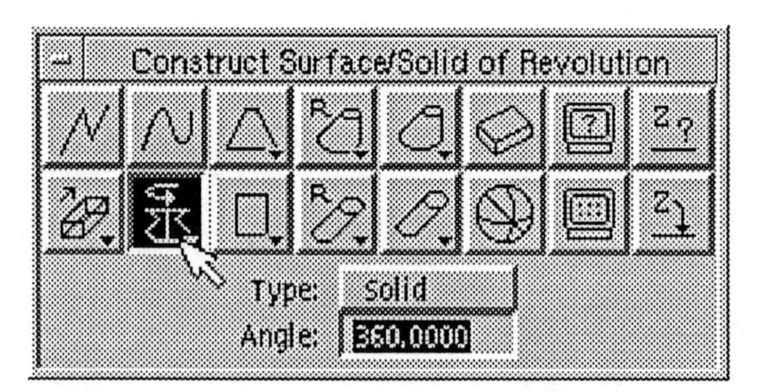

Figure 3.15 Choosing Surface of Revolution (v.4)

When we select SURFACE OF REVOLUTION from the sidebar or digitizer menus, both with version 3.3 and version 4, we are first prompted for the revolution angle in degrees. Positive angles are anti-clockwise (as per the right hand rule shown in figure 2.9). We are then asked to identify the element. This may be in any view. Next, we have to define the AXIS OF REVOLUTION. It is about this data point that the SURFACE OF REVOLUTION is created. The correct view must be used for defining the AXIS OF REVOLUTION as will be seen in the following exercises.

For the exercise, we will use the symmetrical shape shown in figure 3.16. Placing the bottom of the 'vee' at XY=500,500,500 will ensure that the shape is contained within our 1000x1000x1000 imaginary cube.

Make the active level 20 and construct the shape in a FRONT view:

- Use lines, and then chain them together to form the finished shape.
- Check the other views to make sure that all the lines are co-planar before creating the shape.

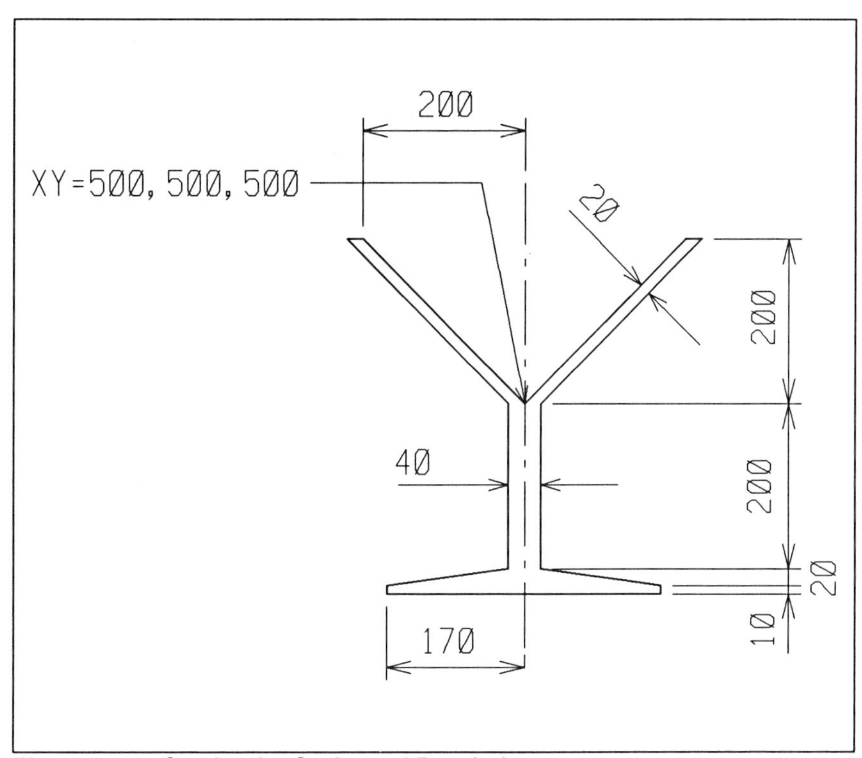

Figure 3.16 Section for Surface of Revolution

Set the views up as they were for the earlier tutorials. That is, views 1 to 4 set to be ISO, TOP, FRONT and RIGHT respectively.

We are going to create two surfaces of revolution from the shape that we have just created. This exercise will demonstrate the importance of selecting the correct view for the AXIS OF REVOLUTION.

For the first one we will rotate our shape 180 degrees, about its center, in a TOP view. Our second version will be rotated 360 degrees about an axis of revolution specified in a RIGHT view.

With figure 3.17 as a guide, for selecting the AXIS OF REVOLUTION, do the following:

- Make the active level 20 (same level as the shape is drawn).
- Check that KEYPOINT SNAP is enabled
- Select *Construct Surface/Solid of Revolution* in 3D palette or menu.
- Enter 180 for the angle of revolution.
- Identify the shape, in any view.
- To specify the AXIS OF REVOLUTION, tentative to the center of the shape in the TOP view. Don't accept until one of the lines on the inside of the 'vee' highlights. This can be seen in the FRONT view.
- When the correct element highlights an accept (data button) will create the surface of revolution.

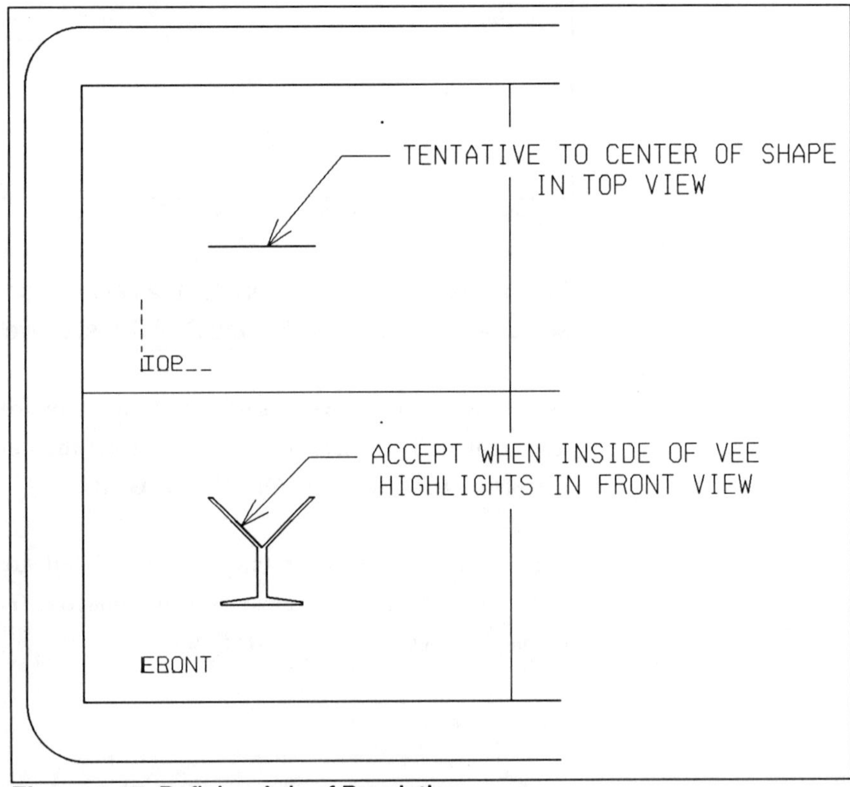

Figure 3.17 Defining Axis of Revolution

Running hidden line removal on the ISO view will show that the object we have created is like a cocktail glass. Figure 3.18 shows both the glass both in wireframe form and after hidden line removal has been performed.

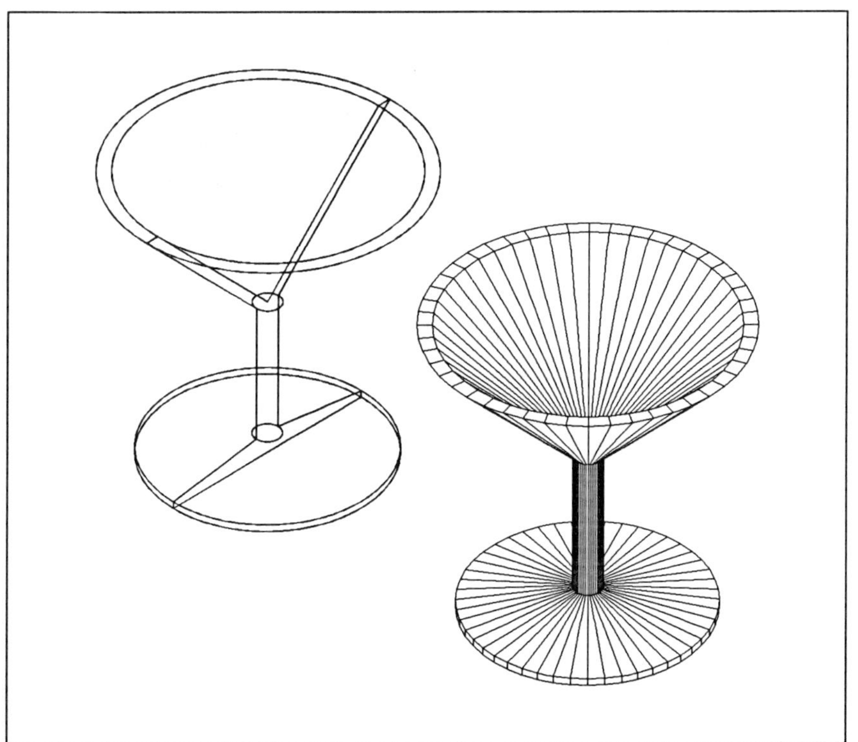

Figure 3.18 Cocktail Glass created with a Surface of Revolution

We rotated the shape 180 degrees about its mid-point to create the cocktail glass. The same model could have been created by drawing a half-section of the glass and rotating it by 360 degrees.

We don't have to select a point on the original element to specify the AXIS OF REVOLUTION. It can be a point offset from the element as our next exercise will show.

Copy the original shape to level 21, then turn off level 20 in all views to clear the previous model from the screen.

Now, with the active level 21, do the following, using figure 3.19 as a guide where necessary:

- Select *Construct Surface/Solid of Revolution* in the 3D palette or menu.
- Enter 360 for the angle of revolution.
- Identify the shape, in any view.
- To specify the AXIS OF REVOLUTION, tentative to the bottom of the shape in the RIGHT view. Check in the FRONT view that the bottom line of the shape highlights. If not, keep placing tentative points in the RIGHT view until the bottom line does highlight.
- When the correct element highlights, key in DX=,-100 to initiate the creation of the surface of revolution.

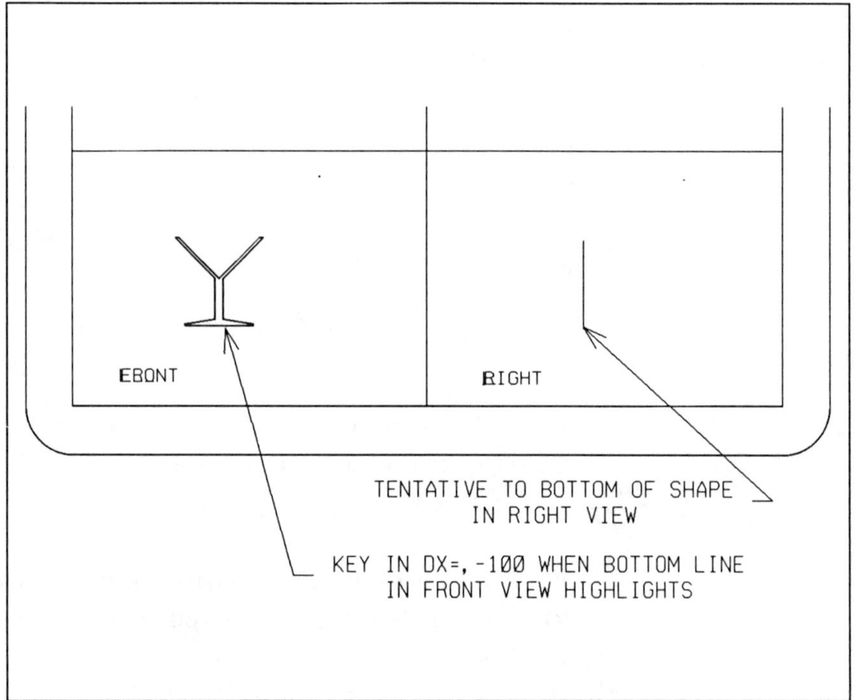

Figure 3.19 Creating the Second Surface of Revolution

You may have to FIT your views to see the whole model, because of its size. We have used the same initial shape as before, but the AXIS OF REVOLUTION was specified in a different view. The resulting model, in this instance, is a pulley wheel. The hole in the center has been created by off-setting the AXIS OF REVOLUTION from the shape.

Figure 3.20 shows an ISO view of the pulley wheel in both wireframe mode and after hidden line removal.

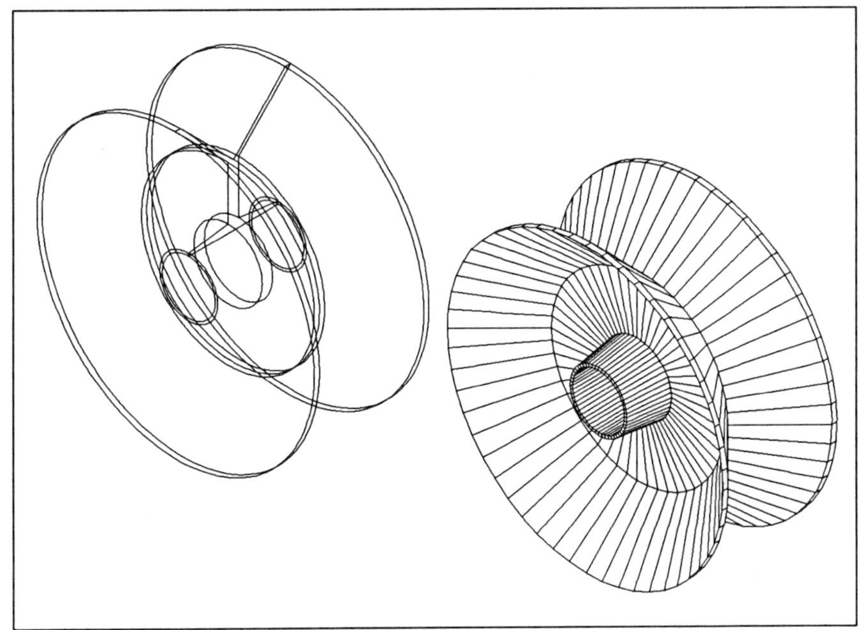

Figure 3.20 Pulley Wheel Formed from Surface of Revolution

For the above exercise we used the same shape to create both the cocktail glass and the pulley. The only difference was the specification for the axis of revolution. The view used to specify the axis of revolution in each case, had its Z axis aligned, or parallel, to the required axis of revolution.

For example, to form a SURFACE OF REVOLUTION about an axis parallel to the Y axis of the model, the axis of revolution has to be selected in a FRONT or BACK view. In both these views we know that the Y axis of the model aligns with the Z axis of the view or screen. To select an axis of revolution parallel to the X axis of the model, we use the RIGHT or LEFT views. For an axis of revolution parallel to the Z axis of the model a TOP or BOTTOM view is used for its selection.

Producing surfaces of revolution is easy with the standard orthogonal views. There is always another standard view available at right angles to the one we are using. At times we have to create a surface of revolution from an element that is not aligned with any of the standard views.

Surfaces of revolution may be created at any angle or direction in the design file. The only restriction is the view in which we select the AXIS OF REVOLUTION. The view must be rotated such that the axis of revolution, for the surface, is parallel with the Z axis of the view or screen.

In chapter 7 (Advanced Techniques), we will go through the creation of a surface of revolution from an element that is not aligned with any of the standard views.

Wireframe Display of Surfaces of Revolution

Probably you will have noticed that in wireframe mode, a surface of revolution is not clear always in each view. The glass that we created, for example, is difficult to decipher in the RIGHT view (refer figure 3.21).

With version 4, the view or views can be set to WIREMESH mode with the SET VIEW WIREMESH key-in or tool. This gives a much clearer idea of the surfaces, in all views, but slows the system. This, and other viewing modes for version 4 are covered in chapter 11 (Rendering Version 4).

Another way to make the display more clear, with both version 3.3 and 4, is to select a smaller angle of revolution. In our example, for the glass, we used an angle of 180 degrees. Breaking this into 4 lots of 45 degrees improves the display in wireframe mode.

Breaking the surface of revolution into smaller sections gives a display, in wireframe mode, that is similar to that of the WIREMESH display available with version 4.

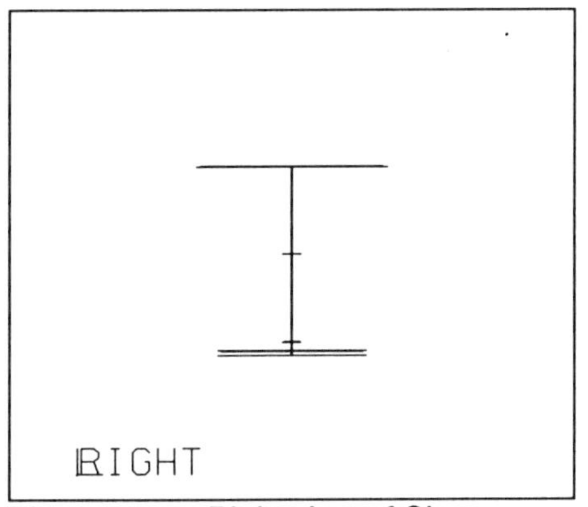

Figure 3.21 Right view of Glass

Try this for yourself:

- Copy the original shape to level 22, which will be the new active level. Turn off levels 20 and 21.
- Select SURFACE OF REVOLUTION in the 3D palette or from a menu.
- Key in 45 for Angle of Revolution.
- Identify element in any view.
- Define the Axis of Revolution in the TOP view, as before, with a tentative and accept to the center of the shape.

This creates a 45 degree surface of revolution. For the other 3 quarters, key in DL=0 each time, to specify that the Axis of Revolution is the same for each.

DOS users can use DL=0|3 to perform the above in one operation, where '|' is the DOS pipe symbol, usually found on the backslash key (\) of the key-board.

The RIGHT view now should look like figure 3.22.

Our glass, in wireframe mode, now is more clearly defined in the RIGHT view. Breaking the revolution angle into smaller sections makes it easier to see the design in all views.

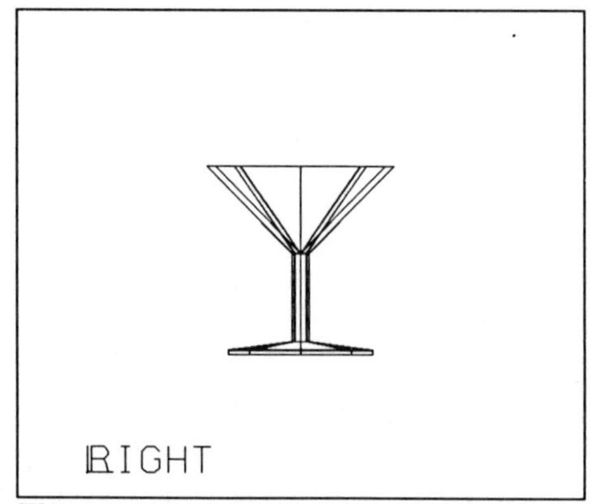

Figure 3.22 New Version of Glass

Although we have created the surface of revolution in smaller sections, it is still treated as a single element for any manipulations.

Group Holes

With the *Group Holes* tool we can define a surface with holes in it. This tool creates an 'orphan' cell containing the elements making up the required surface. An orphan cell is merely a cell, having no name, created without requiring a cell library. The orphan cell may be used to create three-dimensional elements with holes in them. This is done with the PROJECTION or SURFACE OF REVOLUTION tools. We cannot put a hole in a projected surface. The element must be a closed element. Entities used to create the surface with holes must all be co-planar closed elements such as shapes, blocks, circles, complex shapes etc. Elements used to create the solid and hole entities need not be on the same level.

We can select the *Group Holes* tool in the Chain sub-palette (version 4). This is located at the right end of the palette (figure 3.23). Alternatively, we can select it from the sidebar or digitizer menus. Keying in 'GROUP HOLE' is another option.

We are first prompted, with the *Group Holes* tool, to identify the solid element. After that, we are asked to identify the holes within the solid element. The resulting element is an orphan cell which, in this case, is a surface containing the hole elements.

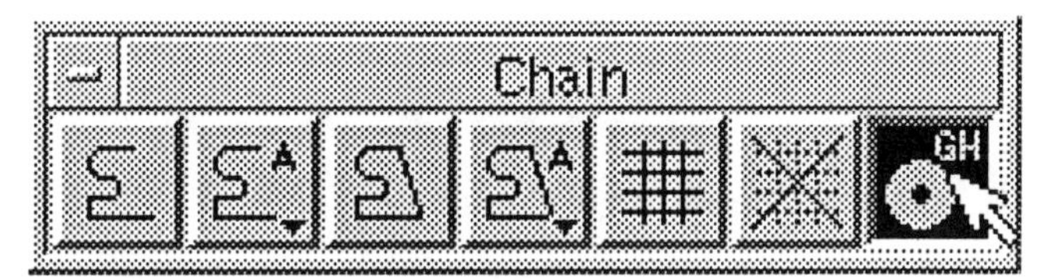

Figure 3.23 Chain Sub-palette

We will use the *Group Holes* tool to create a pipe section. From this we will create a 90 degree pipe bend.

- Make the active level 30 and turn off the other levels (except level 1, if you need it).
- In the TOP view, place a 125 radius circle at XY=600,400,400. Remember to make the TOP view current prior to placing the circle.
- Place another circle of 100 radius with the same center.
- Key in GROUP HOLE or select the tool in the CHAIN palette or menu.
- Identify the outer circle as the solid element.
- Identify the inner circle as the hole.

This gives us the cross-section of our pipe.

- In the TOP view, place an active point 300 to the left of the center of the pipe. Refer figure 3.24.

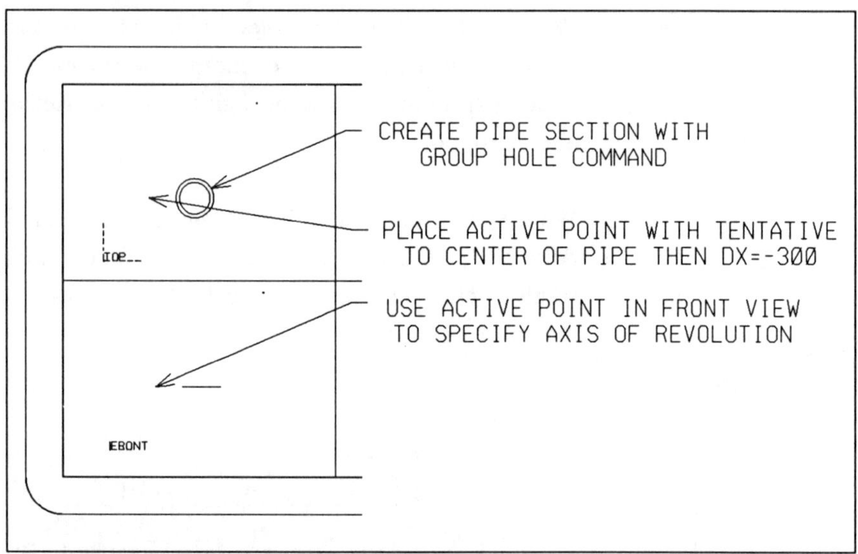

Figure 3.24 Creating a Pipe Bend

We can use this point in the FRONT view to specify the axis of revolution when we create the pipe bend. This is an alternative method to placing a tentative point at the center of the circle, in the FRONT view, followed by a key-in.

Finding the active point is easier than finding the center of the circle in the FRONT view. It is a simple construction to make life easier for us. Using active points in this fashion is convenient, but remember that they take up space in the file. They should be deleted when no longer needed.

To finish the pipe bend:

- Set ACTIVE CAPMODE ON by key-in or menu selection. Version 4 users choosing the *Construct Surface/Solid of Revolution* tool in the 3D palette can set this in the Type field of the dialog box.
- Select the *Construct Surface/Solid of Revolution* tool.
- Key in 90 degrees for the angle.
- Identify the cell. If a tentative followed by a data point doesn't work, use a straight data point on the cell.
- For the axis of revolution, identify the active point in the FRONT view (figure 3.24).

Figure 3.25 shows how your screen should look now. Running hidden line removal will prove that the model is a pipe bend.

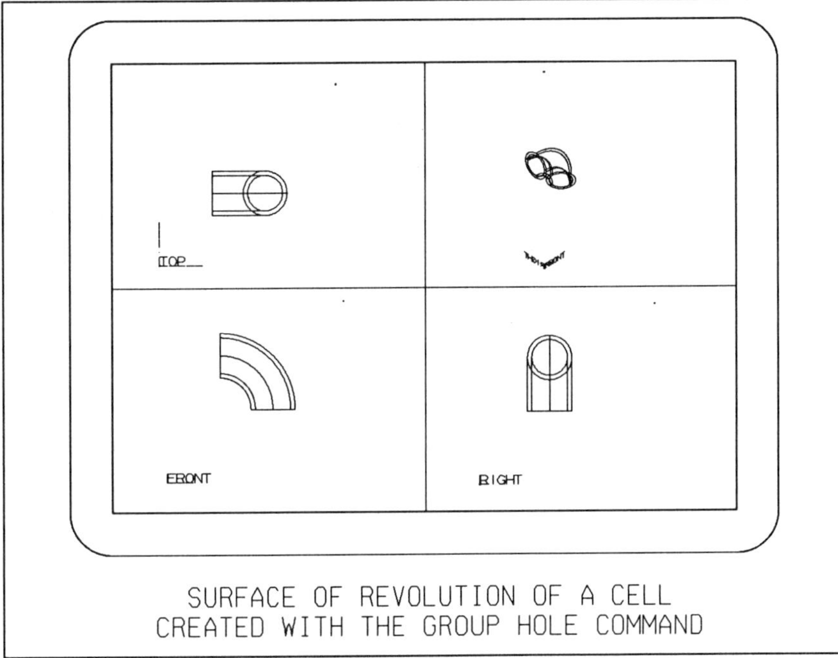

Figure 3.25 Creating a Pipe Bend

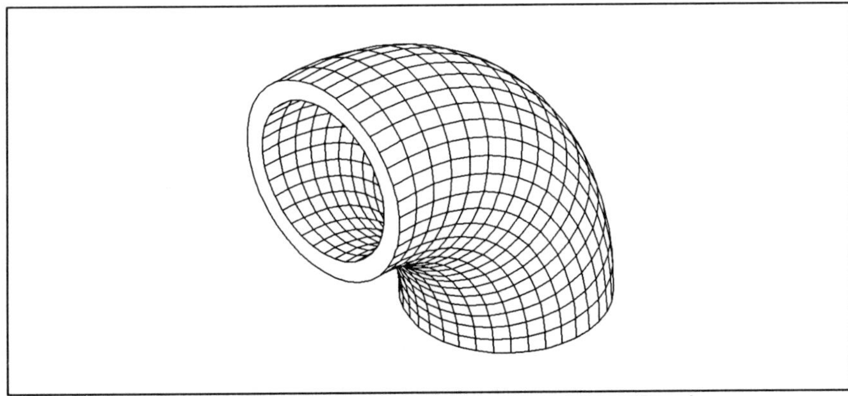

Figure 3.26 Hidden Line Removal of ISO view of Pipe Bend.

Another option we have with the *Group Holes* tool is the ability to create 'cutaway' sections in our models. We can place the 'hole' elements on a different level to the 'solid' elements. Using the *Group Holes* tool we then create the orphan cell. We can enable or disable the cutaway sections by enabling or disabling the display of the levels where they are located.

This allows us, for example, to place holes in the exterior walls of a model. We could then produce hidden line removals, or shaded images, which show inside the model, through the walls. If we then re-processed the view, with the levels containing the cutaway sections disabled, we would no longer be able to see through the walls.

Chapter 3 Summary

We have looked at, in this chapter, two of our more powerful 3D tools, *Construct Surface/Solid of Projection* and *Construct Surface/Solid of Revolution*. As well, we looked at the *Group Holes* tool for creating surfaces with holes in them.

Special note should be made of the following points:

- Use the *Construct Surfaces/Solid of Projection* tool, where possible, in preference to constructing a model from separate elements. Projected elements take up less disk space than the same model constructed with individual elements.
- Elements are projected in a straight line.
- Projected elements are affected by active angle and active scale. All scaling and rotating is about the point by which the element was identified. Scaling and rotating is relative to the view in which the projection distance is specified.
- With version 4 we can project contents of a fence. As well as active angle and active scale, the fence lock setting affects the results.
- With the *Construct Surface/Solid of Revolution* tool, the surface is created about an axis perpendicular to the view, or screen. The view is that used to specify the AXIS OF REVOLUTION.

Further exercises in projecting and creating surfaces of revolution can be found in chapter 14 (Practice Examples).

4 : Cells & Reference Files

From your 2D work you should be familiar with cells and reference files. All the same features are available with 3D, the only difference being that of the depth factor. Cells in 3D are contained in a volume of space, and reference files also occupy a volume of space.

Cells

While we can attach 2D cell libraries to 3D files and use the 2D cells, we cannot create 2D cells from a 3D file. Also, 3D cell libraries cannot be used with 2D files.

Creating 3D Cells

We use a similar procedure to that used in 2D. There must be a 3D cell library attached to the design file (RC=cell library). We create the cells by placing a fence around the elements, defining an origin, then keying in CC=cellname to initiate the cell creation. When we create a cell in a 3D file, we are dealing with a volume of space in the design cube.

We know from earlier exercises that a fence in a 3D file has depth - the display depth of the view in which it is placed. Our 3D cells consist of the elements contained within the confines of the fence and the display depth of the view. The display depth should be the minimum required to include all the wanted elements. Just as the fence sets the area being used for the cell, the display depth sets the depth occupied by the cell.

It is good practice when creating cells to set the display depth to include only the volume required to contain the elements of the cell. The fence also should be placed to just include all the required elements.

As an example, we will create a cell of our 90 degree pipe-bend.

- Attach your training cell library to your file (RC=cell library name).
- Set the display depth in the FRONT view to just include the bend.
- Place a fence around the bend in the FRONT view.
- Define the origin with a tentative to the center of the circles. The TOP view is the most convenient to use for this.
- Create the cell with the CC=cellname key-in. In the description, it is a good idea to include the view in which the cell was created (e.g., TOP, FRONT etc.).

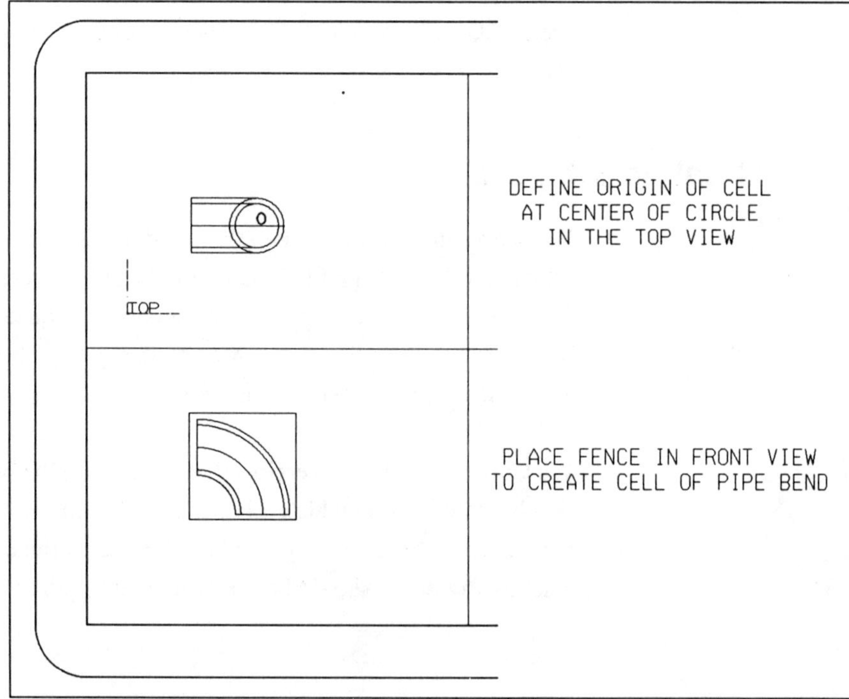

Figure 4.1 Creating a 3D Cell

With version 4 we can use the Cells dialog box to create a new cell. Once the fence and the cell origin have been placed, we select 'Cells' in the 'Settings' pull-down menu.

We then click the 'Create' box and the 'Create New Cell' dialog box pops up. This contains fields for the cell name and its description (figure 4.2). Once these have been completed, clicking on 'Create' completes the task.

Our new cell is placed in the library and we are returned to the initial dialog box.

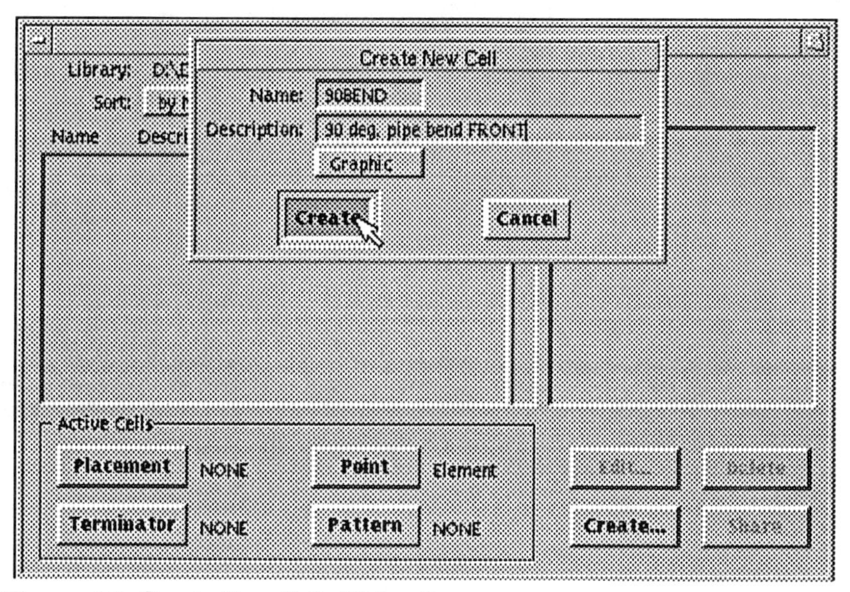

Figure 4.2 Create New Cells Dialog Box

We are now given a chance to review our newly created cell. The cell is highlighted and on the right of the dialog box we see four views of it. Unlike 2D where cells are flat (two-dimensional) objects, in 3D our cells have depth. The four views are set as we had them for the training file. That is, views 1 to 4 are ISO, TOP, FRONT and RIGHT respectively. The cell is displayed as though it was placed in the TOP view (top left quadrant).

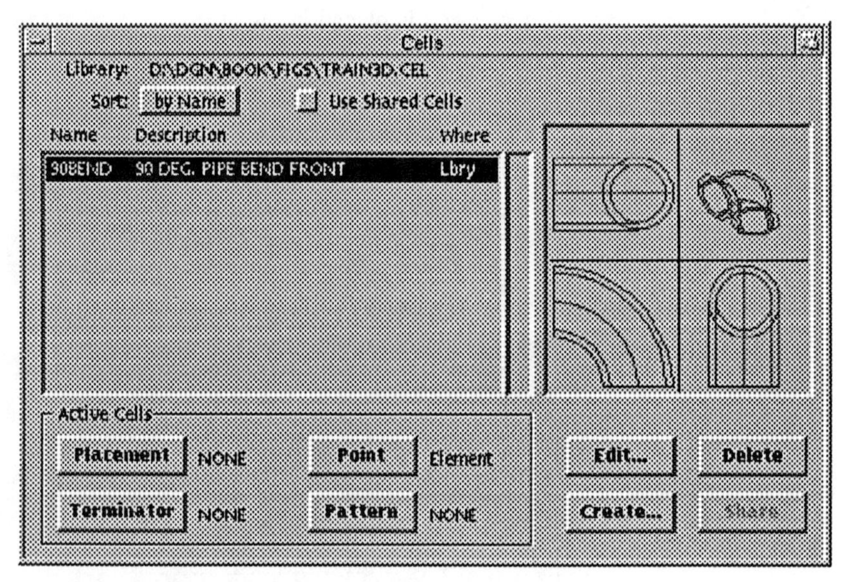

Figure 4.3 Reviewing the New Cell

Placing Cells in 3D Files

In 3D files, cells are placed with the same orientation relative to the view or screen, as when they were created. A cell created in a TOP view, will be placed as though it is in a TOP view, no matter which view we use to place it.

2D cell libraries may be attached to 3D files and 2D cells may be used in 3D files but not vice versa. A 2D cell when used in a 3D file is placed like other elements. That is, parallel to the view or screen and at the ACTIVE DEPTH.

Try placing the cell you created in the previous exercise.

- Place it in each view and see how it appears. Use the 'AC=cellname' as in 2D to recall the cell.
- Notice that no matter which view you use when placing it, the cell appears in that view as it did when it was created (i.e., as if in a FRONT view).

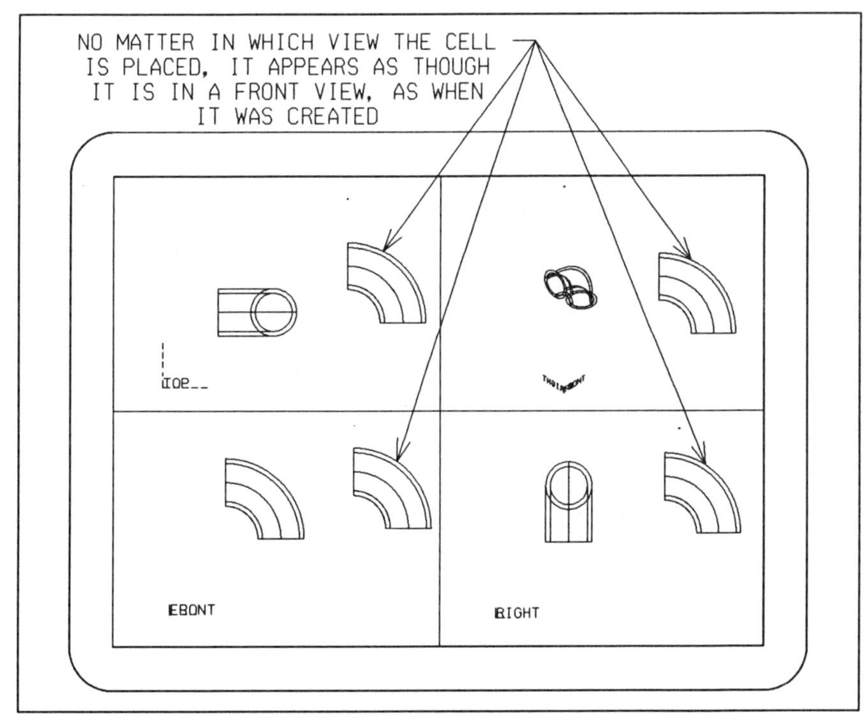

Figure 4.4 Placing Cells in 3D

Reference Files

We can use reference files in 3D similar to how they are used in 2D. A three-dimensional model can be created by a number of people, working on various parts of it, simultaneously. The different parts can be referenced together to create the finished model. In a networked environment, we can reference other files, as they are being used.

Reference file information is read into the system when the active file is first accessed. To see any current changes to reference files, with version 3.3, we have to retrieve design (RD=filename) back into the active file. Keying in 'NEW' automatically performs an 'RD' back into the active file.

With version 4 we can use the *Reload Reference File* tool or the alternative key-in: REFERENCE RELOAD 'filename/logical name'. This reloads the specified reference file information without leaving the active file, or re-reading the other reference files.

Until the release of version 4 of MicroStation we have not been able to reference 2D files to 3D or vice-versa. A feature with version 4 is that we can reference 2D files into our 3D designs. With the earlier versions, the 2D files have to be converted to 3D first. This is done with the 2D to 3D conversion program that comes with MicroStation. It can be accessed via the MCE menu environment.

Using reference files is a very powerful feature of MicroStation. With 3D there are a few extra points to be aware of compared to working in 2D files.

Depth Considerations

While in 2D the only concern is the area in which the reference file appears, in 3D we also have to consider the depth factor. If a reference file doesn't 'appear' when attached to 2D files, zooming out will generally overcome the problem. With 3D this does not always apply. Elements in a 3D reference file may be outside the display depth of the view in the active file. In this case zooming out won't work, because the display depth is not changed. ZOOM OUT increases the area of the file that can be seen in a view, but not the depth.

Just as a WINDOW AREA limits what we see in the X and Y directions (horizontal and vertical), DISPLAY DEPTH limits what we see in the Z direction (depth) of a view. We can change the window area of a view with the ZOOM and WINDOW AREA tools. To change the display depth we have to use the DISPLAY DEPTH or FIT tools.

Version 3.3

FIT operates on the active file only. Reference file elements outside the volume containing the elements in the active file may not display. We have to reset display depth to include the volume containing the reference file data. We can place active points at diagonally opposite corners of the volume containing the elements. FIT will then include them and any other data enclosed by the volume that they define.

Version 4

Version 4 has a feature that allows us to FIT any file. The relevant key-ins are:

FIT - Fit the active file.

FIT REFERENCE x - Where 'x' is the number of the reference file

FIT ALL - Fit all files (the active file and all reference files).

For example, if we key-in FIT REFERENCE 1-3, we are prompted to 'Select a View'. When selected a FIT is performed on reference files 1, 2, and 3.

Scaling, Rotating, Clipping and Moving

Operations such as scaling, rotating, clipping and moving can be used as with 2D reference files. There are some differences, however, due to the third dimension. Reference files, in 3D, may be rotated about the X,Y or Z axis of the view or screen. The key-in to stipulate the angle of rotation is similar to the way view rotations are specified.

For instance, when prompted for the angle to rotate the reference file:

- keying in '30' means we want to rotate the Reference file 30 deg. about the X axis.
- keying in ',30' indicates 30 deg. about the Y axis.
- keying in ',,30' indicates 30 deg. about the Z axis.

The directions of rotations are as for view rotations and rotating in general. As with view rotations, when a reference file is rotated, it may rotate through the display depth. This can cause some elements to disappear from view.

Normally, we would rotate a reference file only when setting up a file to create a drawing. This is demonstrated in chapter 12 (Drawing Production). All the parts of a model should be located in their correct place, and orientation in space, without needing to be rotated. An exception is when we use a file containing an item that we want to reference a number of times in different orientations. It may be too large to efficiently create a cell from.

One of the main advantages with reference files, over cells, has been that the information is on disk once only. Placing large cells, a number of times quickly increases the size of files, because when we place cells, the information is repeated with each placement. Version 4 overcomes this problem, to some degree, with shared cells.

Clip bounding a reference file works as it does in 2D. We place a fence block around the section of the reference file that we want to see and select the *Define Reference File Clipping Boundary* tool. With version 4 we have other tools also, for masking part of a reference file, and to mirror a reference file about a horizontal or vertical axis. These operate as for 2D, remembering that they are view dependent.

There are also tools for clip bounding the front and back planes of a reference file. This is similar to setting the display depth for reference files separately. A similar effect can be achieved with SAVED VIEWS.

Version 3.3 supports fence blocks only, for reference file clipping. If we use a fence shape it defaults to a block. The block completely contains the fence shape. With version 4 we can use fence shapes as well as blocks for reference file clipping.

Using Saved Views

We can reference files using saved views. Saved views in 3D files define a volume of the file. A saved view is bounded by the window area and the display depth of the view. When a file is referenced to another with a saved view, the saved view is placed in the active file similar to how a cell is placed. That is, we may create a saved view in a FRONT view. We can reference the file, with the saved view, and place it in the TOP view. It still appears to be a FRONT view.

Using saved views is similar to attaching a reference file that is already rotated and clip-bound. We could attach the reference file normally, rotate it and clip-bound to achieve the same result as the saved view.

5 : Placing & Manipulating Elements

With 3D MicroStation we work in a three-dimensional 'world' in the design file. Most of the basic elements we use are, in themselves, two-dimensional. Many of these elements form surfaces (or holes, when needed), but have no depth. Such elements include SHAPES, COMPLEX SHAPES, BLOCKS, CIRCLES and ELLIPSES.

We also have some three-dimensional elements. These are SPACE LINESTRINGS, SPACE CURVES, B-SPLINE CURVES, CONES and CYLINDERS. Version 4 also has HELIXES, SLABS and SPHERES.

Placing Elements Generally

In 2D we place all elements on a flat plane, similar to drawing on a sheet of paper. In 3D also, many of the elements we place are flat, or two-dimensional. We have to specify where, in the design cube, we want to place them. We also have to specify the orientation. That is, whether we want the element horizontal, vertical, or at some other angle.

By default, elements are placed at the active depth of the view we are using. The active depth is a plane, parallel to the view being used. It is easier to remember that the active depth is a plane parallel to our monitor screen. No matter which view is being used, the active depth is parallel to the monitor screen.

Any two-dimensional elements (such as blocks, circles etc.) that we place with data points - without snapping to existing elements - will be placed at the active depth of the view. Also, they will be parallel to our screen. Most of our work in 3D starts with these two-dimensional elements. They are then put together, or projected, to form a three-dimensional object.

Planar Elements

We rotate views so that our screen is parallel to the plane in which we want to place particular elements. In this way we can place them with the correct orientation. We need view manipulations, such as this, only for planar elements that require less than three points to place them.

As a 'rule of thumb':

IF AN ELEMENT REQUIRES A MINIMUM OF 3 POINTS TO DEFINE IT, WE MAY PLACE IT USING ANY VIEW.

With these elements we define their plane with the first three points. Where an element requires less than three points to define, it takes the orientation of the current view, by default.

Blocks, Circles and Polygons

In order to place a BLOCK (orthogonal), POLYGON or CIRCLE (by center,radius or diameter), we first have to rotate the view. This is to align the design cube with our screen. We want our screen to be parallel with the plane in which we wish to place the element. These elements require two points only, to place them. They take, by default, the orientation of the view being used - the current view.

Alternatively, we can place BLOCKS, POLYGONS (by edge) and CIRCLES, so that they slope into or out of the view or screen. Here, the width/height component of the block or polygon, or a diameter of the circle will be aligned with the Y axis of the view or screen.

Figure 5.1 shows examples of elements placed obliquely in a view. In each instance, the first point was a data point and the second point was specified with a 'DX=' key-in as indicated. Each element has its second point 300 into the screen, caused by the Z value of -300 in the 'DX=' key-in. In the lower half of the figure we see the same elements in a view rotated -90 degrees about the X axis. That is, we are looking along the Y axis of the view in which the elements were placed. It is obvious that the elements are aligned with the Y axis of that view, because they appear as lines in this rotated view.

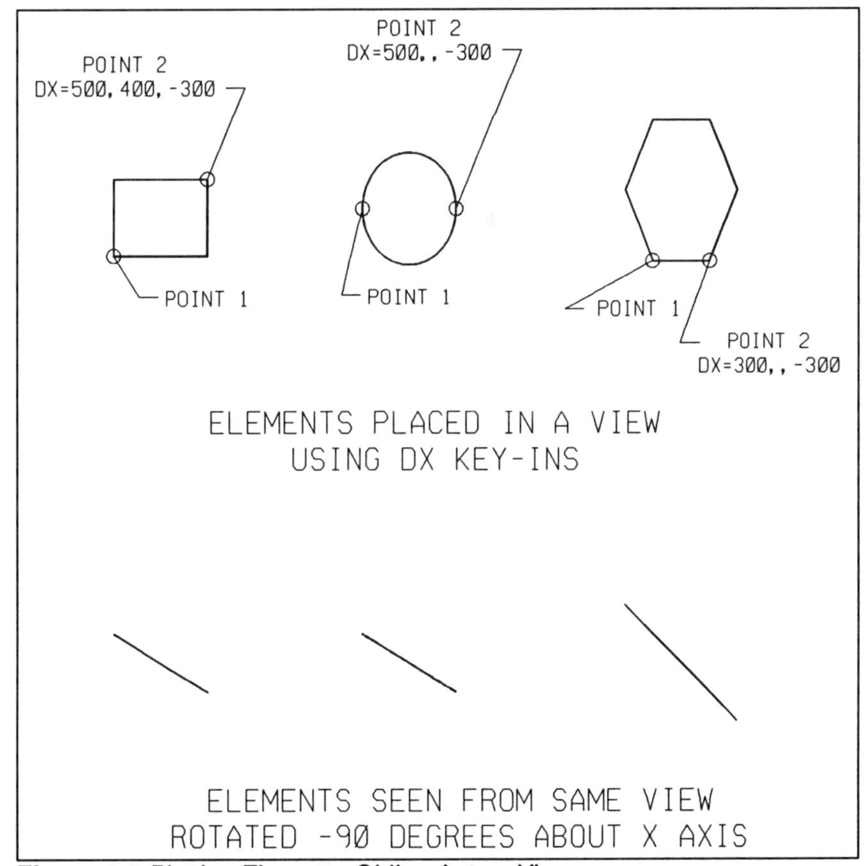

Figure 5.1 Placing Elements Obliquely to a View

Shapes

We can define the orientation of a plane with a minimum of three points. Placing a shape, other than a triangle, requires more than three points. Once the first three points have been defined, all other points must fall on the same plane. The orientation of a shape is determined by the first three points placed.

Often, we use the *Place Shape* tool to create surfaces between existing elements, or known points. Once the first three points have been placed, defining the orientation of the plane of the shape, we are restricted to that plane. This can cause some confusion, when the line seemingly will not go where we put the cursor. If we tentative to an element that is in the required plane, however, the line re-attaches itself to the end of the cursor. The explanation is quite simple.

Our cursor, by default, remains at the active depth of the view, until we tentative to an existing element. The system shows us the line from the previous vertex to the next location. Often the active depth is in a different plane to the defined plane of the shape. Here, the system shows the next point as a normal to the plane, at the cursor location.

Figure 5.2 shows two examples of placing a shape. The shape was placed in a TOP view with the first three points defining the plane as sloping relative to the view.

In example 1 the active depth plane in the TOP view is in front of the next point, causing the cursor to separate from the line indicating the next point. This is illustrated in the view looking from the right of the top view. We can see that the cursor is at a point normal to the plane of the shape.

Example 2 shows a similar situation, but with the active depth plane behind the next point of the shape. Again the view looking from the right of the top view shows that the cursor is at a point normal to the plane of the shape.

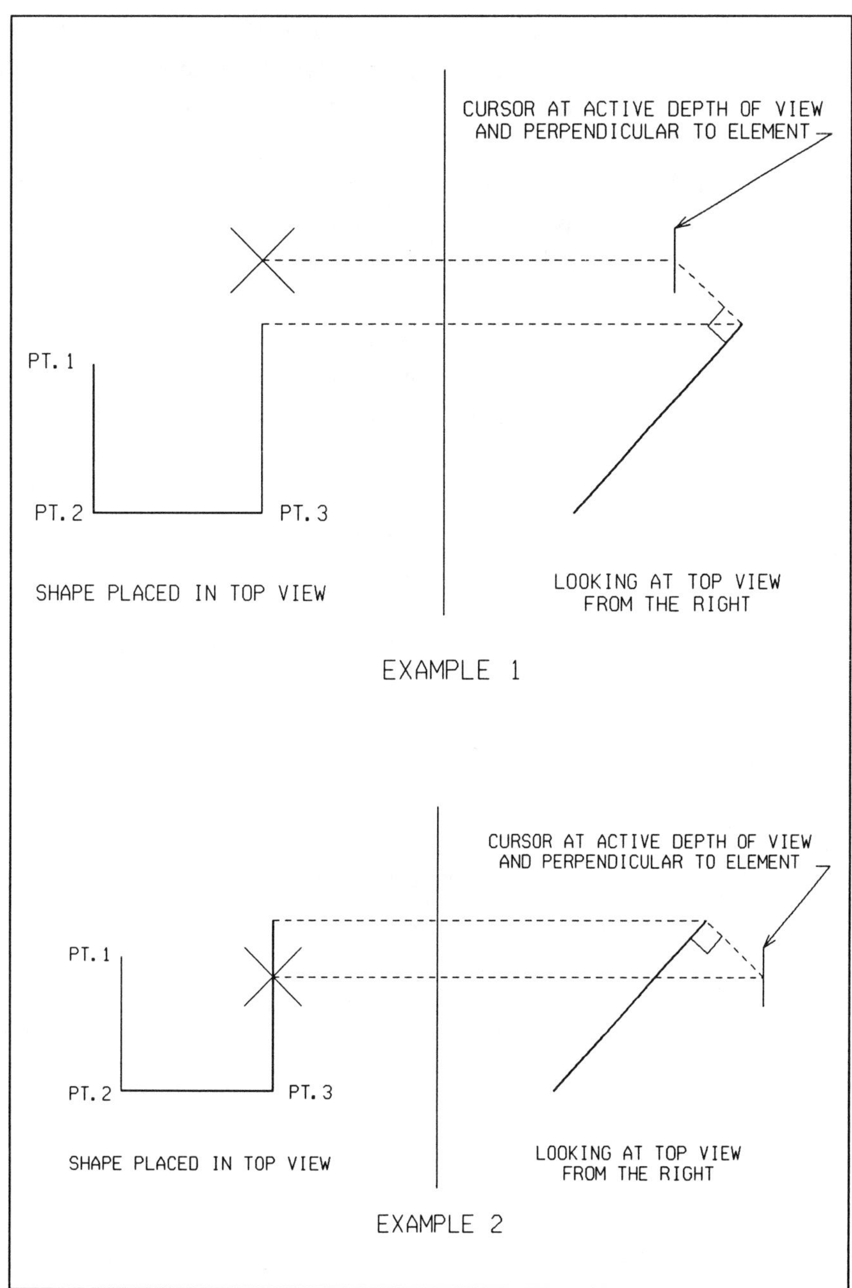

Figure 5.2 Placing a Shape in 3D

Another situation, resulting in a similar display to example 1 of figure 5.2, is when the next point is outside the display depth of the view. This is discussed, later in this chapter, in the section on extending lines.

From the previous example we can see that even where the plane of the shape is not aligned with any of the design cube's axes, the same rule for the orientation of its plane applies. After the first three points, the plane is defined. All other points must lie on that plane.

Being able to place elements in any view is very useful. It allows us to work in a rotated view, for example, where it is easier to 'see' the model. There are times where it is quicker to place a block using the *Place Shape* tool, rather than the *Place Block* tool. We can place a shape in any view. To place the same rectangular shape, with the *Place Block* tool, we first have to rotate the view to the correct orientation.

Summing up, we can separate planar elements into 2 categories:

1. View Dependent

Elements that take the orientation of the view. They are view dependent. They require less than three points to define them because they take their orientation from the view being used. The points only determine their dimensions not their orientation.

These elements include - blocks, circles with radius, circles by diameter/center and polygons (figure 5.3). Version 4 also has multilines which are view dependent.

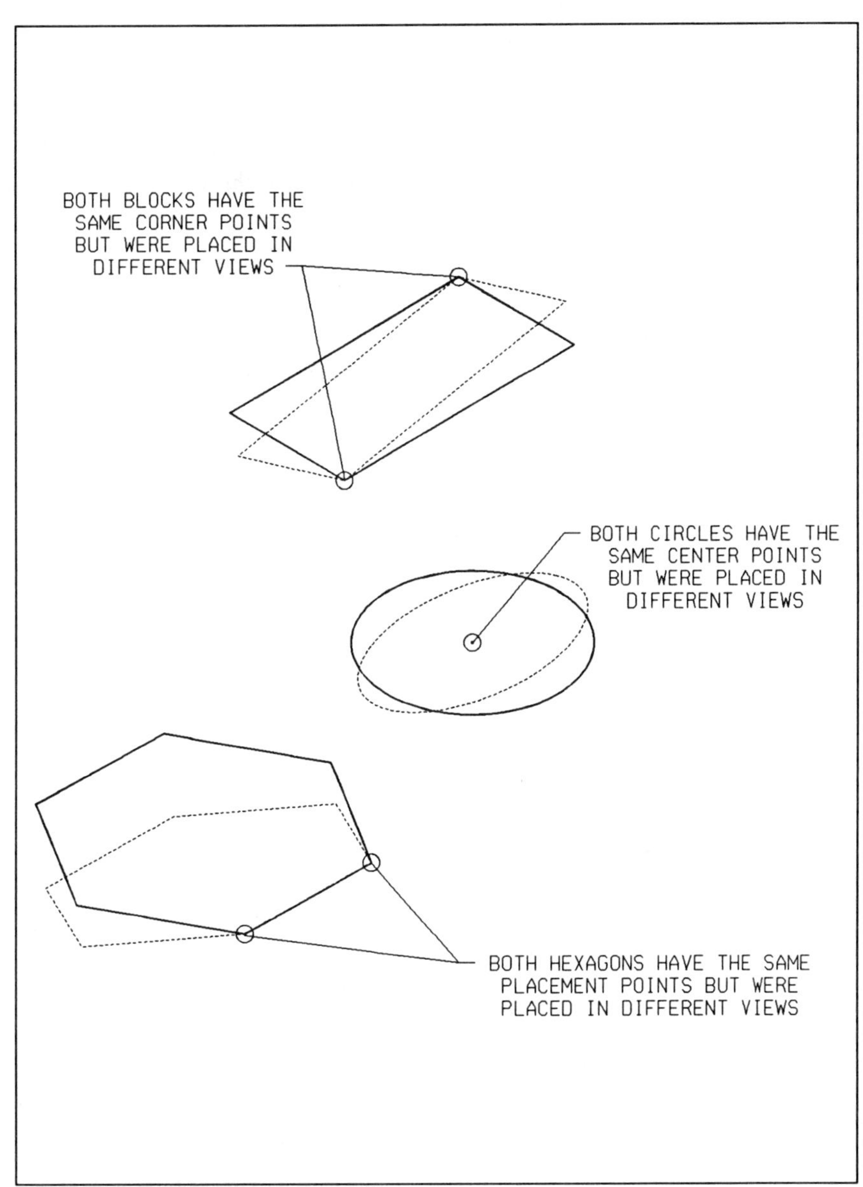

Figure 5.3 View Dependent Elements

2. View Independent

Elements that we can place in any view. They are not view dependent. The minimum of three points required to describe them also provides their planar orientation. Once the first three points have been specified, any further points will fall on the same plane. This plane is not necessarily aligned with our view or screen.

Elements in this category include shapes, circle by edge, ellipses and rotated blocks. Figure 5.4 shows two such elements which were placed in an ISO view. The three points required to describe them, for placement, also fixed their orientation.

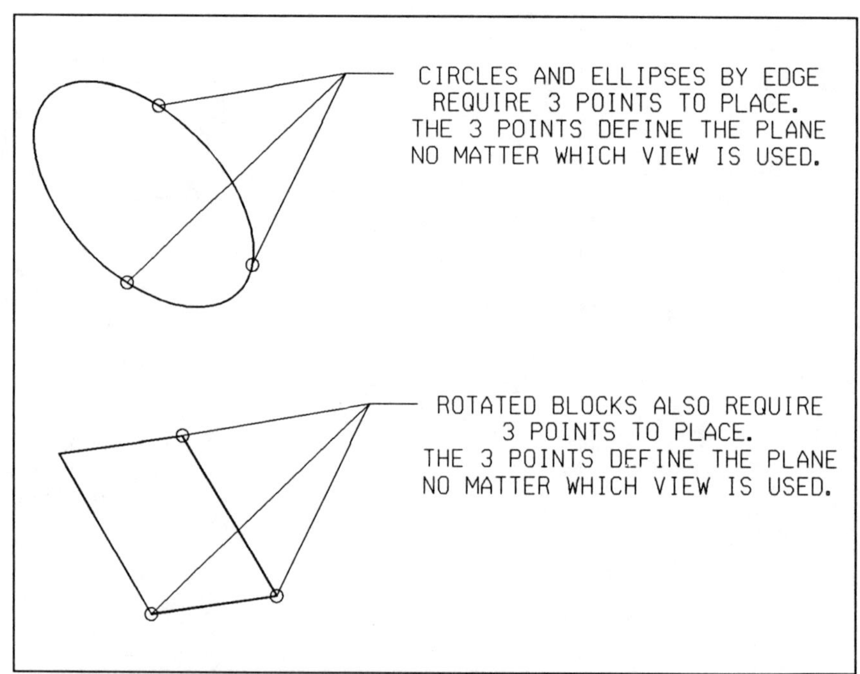

Figure 5.4 Placing Elements with 3 Points

2D Linear Elements

Most linear elements such as LINESTRINGS, ARCS, PARTIAL ELLIPSES and CURVES are two-dimensional. All their vertices (or key-points in relation to arcs and ellipses) are co-planar. As with shapes, circles etc., the same rule applies. The first three points used to define the element, also fixes the orientation of the plane in which the element lies.

Standard Linestrings and Curves

When we place elements like standard LINESTRINGS and CURVES, the first three points define the plane. For example, if we place a linestring with precision inputs thus:

XY=250,250,50 XY=500,300,50 XY=300,350,50 XY=200,450,75

then the first three points would establish that the plane of the linestring is at a Z value of 50 (design file units).

When we key in the fourth value, the vertex will have the specified X and Y co-ordinates, but the Z value will remain at 50. The plane of the linestring has been defined, by the first three points, as being at a Z value of 50.

Any further inputs for the linestring would allow variations in the X and the Y values only. The Z value would remain constant because a linestring is a two-dimensional element. All its vertices must be co-planar. If necessary, we can modify it after it has been placed, and change the Z values of vertices.

Curves are similar to linestrings. We place them as two-dimensional objects with the option of modifying them later.

Arcs and Partial Ellipses

Elements such as arcs and partial ellipses are strictly two-dimensional or planar elements. They require three points to define them. These points also define the orientation of the element.

3D Linear Elements

Elements that we have discussed so far should be familiar to you. They are standard elements that can be used in 2D files. We will now look at some new elements. Elements that are truly three-dimensional. That is, they are not planar, nor do all the points describing them need to be co-planar. These elements are placed independent to the rotation of the view.

Space Linestrings/Curves

SPACE LINESTRINGS and SPACE CURVES are three-dimensional alternatives to the standard linestrings and curves. Unlike the standard types, SPACE LINESTRINGS and SPACE CURVES in 3D files allow for variations in all three dimensions (X, Y, and Z). The first three points in both cases don't define a plane on which all other points must fall.

As well as from the sidebar and digitizer menus, both these tools can be selected in the 3D tool palette. Figure 5.5 shows the 3D palette where the *Place Space Linestring* tool has been selected. Immediately to the right of this is the *Place Space Curve* tool.

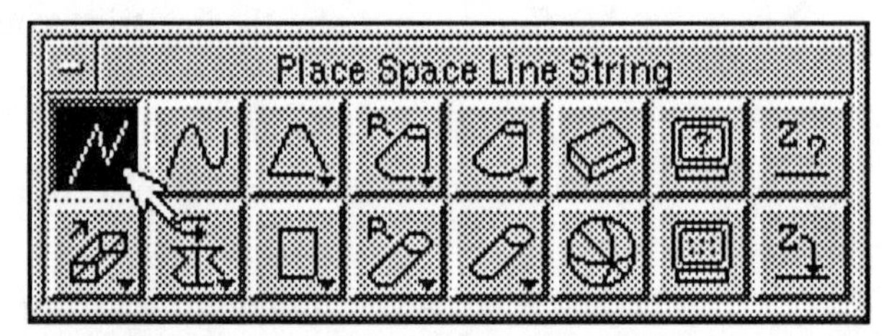

Figure 5.5 Place Space Linestring Tool

With both these elements, we can freely specify values of X,Y, and Z for each vertex. Figure 5.6 shows examples of both. In each case the element was constructed using points A,B,C,D and E on the corners of a rectangular box as shown. It can be seen that the elements are non-planar with parts of each lying in three different planes of the box.

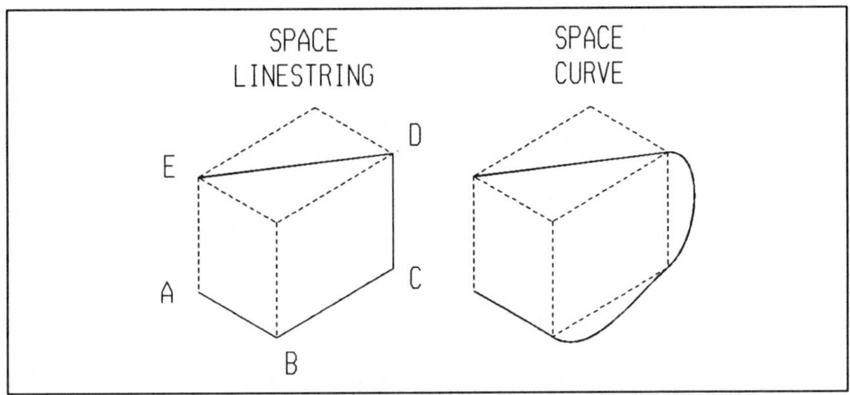

Figure 5.6 Space Linestrings and Curves

B-spline Curves

B-SPLINE curves also, are three-dimensional elements in 3D files. They too have no restrictions on input values for X, Y, and Z values. With version 4 we have tools to place three variations of b-spline curves. They can be placed by 'poles', 'points' or 'least squares'. Version 3.3 has the first only (by poles). In figure 5.7 we have an example of each, constructed using the same 5 points (A,B,C, D, and E) on a rectangular box. In each case, the control polygon is displayed. It can be seen that the curves are not limited to a single plane.

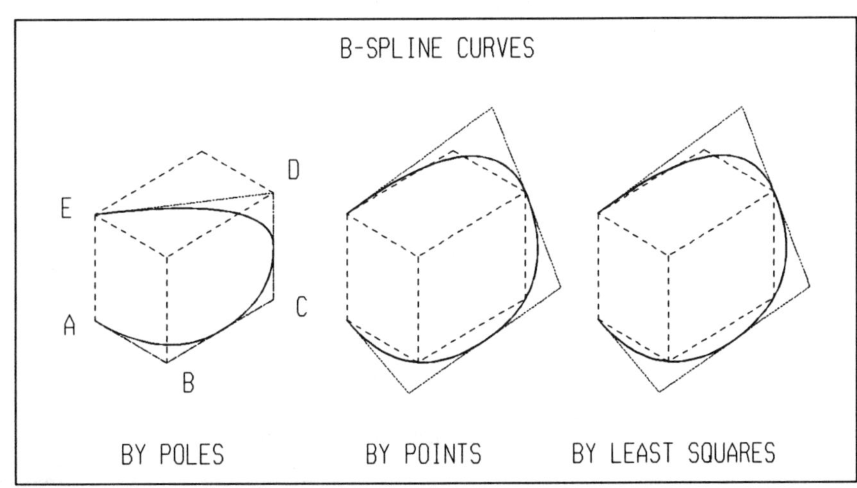

Figure 5.7 B-spline Curves

Lines and Multi-lines

Lines, in a fashion, are three-dimensional elements. They can be placed anywhere in a 3D file, using any view.

Multi-lines (version 4) are two-dimensional linear elements, and view dependent. In 3D files they rely on the view for their orientation.

Helixes (version 4)

Version 4 has a construction for helixes. These are three-dimensional b-spline curves that may be placed in any view. The *Place Helix* tool is found in the 'Space Curves' palette which is opened from the '3D B-Splines' option of the 'Palettes' pull-down menu. Pop-down fields are provided for entering the 'Pitch' and the 'Thread' of the helix.

'Vertical' orientation of a helix is in the plane of the first three points used to define it. That is, the points defining the centers of the base, top, and the radius of the base. A helix may be modified just as a b-spline curve can be modified. When we select the *Place Helix* tool we are prompted for four points. They are, in order (refer figure 5.8):

- Center of the base (Pt.1)
- Radius of the base (Pt.2)
- Center of the top (Pt.3)
- Radius of the top (Pt.4)

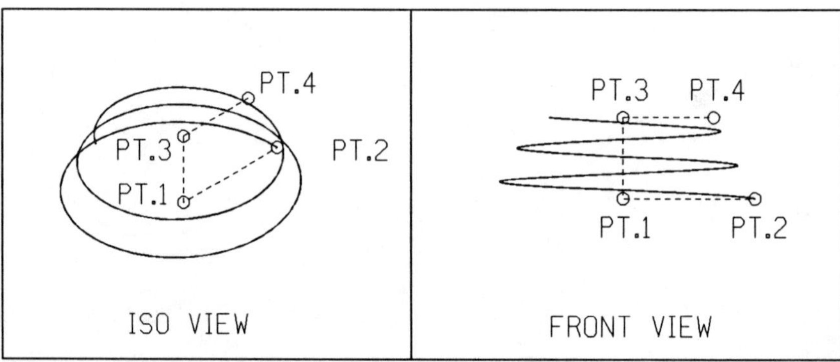

Figure 5.8 Placing a Helix (ver.4)

Surfaces & Solids

We form most surfaces and solids by projecting shapes, complex shapes and the like. MicroStation also has tools for creating CONES and CYLINDERS. In addition, version 4 has SLABS and SPHERES.

Cones and Cylinders

We have three types of cones and cylinders. They are :

RIGHT CYLINDERS/CONES
CYLINDERS/CONES BY KEYED-IN RADII
SKEWED CYLINDERS/CONES

With the sidebar menu these options are given to us after we select PLACE CYLINDER/CONE. Both the digitizer menu and the 3D tool palette have individual icons for the respective types.

Right Cylinders/Cones

RIGHT CYLINDERS/CONES take their orientation from an imaginary line joining the centers of their base and top. With both, the planes of their top and bottom faces are normal to this line (figure 5.9). No matter which view is used to place a RIGHT CYLINDER/CONE the orientation of the element is determined by the placement of the center points of its top and bottom face.

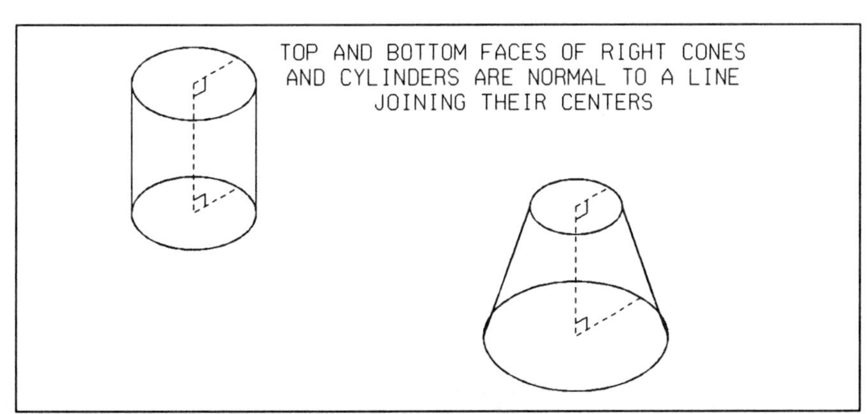

Figure 5.9 Orientation of Right Cones & Cylinders

When we use the *Place Right Cylinder* or *Place Right Cone* tool we are prompted for data points to define the element. For both constructions the orientation of the element is defined by the first and third points - the centers of the base and top respectively.

Cylinders and cones can be solids or surfaces. We can specify this with the ACTIVE CAPMODE key-in, or select from the menu. Alternatively, with version 4 we can specify the type (solid or surface) in a pop-down field, when we select the tool from the 3D palette (figure 5.10).

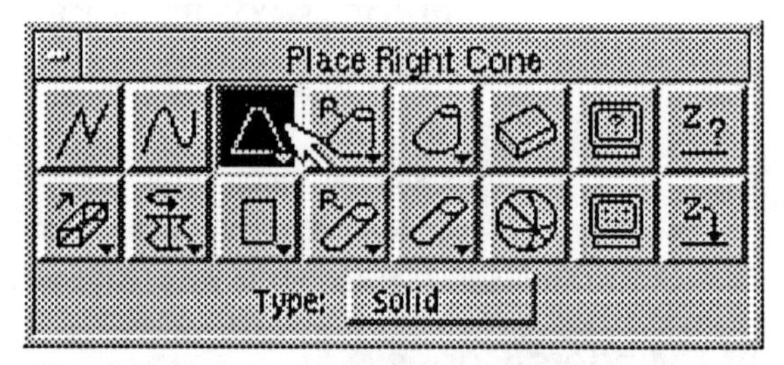

Figure 5.10 Place Right Cone Solid

Figure 5.11 shows examples of both right cylinders and right cones. For the two cylinders the same two starting points were used. Likewise, the two cones have the same starting points. Each was placed, in a TOP view, with data points. If the same points were then used in a rotated view, as shown, the elements created in the rotated would be identical to the original elements. This is because their orientation is taken from the center points of the base and top faces, as explained previously.

For RIGHT CYLINDERS we are asked for:

- The center of the base (Pt. 1)
- A point on the surface (Pt. 2)
- The center of the top (Pt. 3)

RIGHT CONES require similar inputs, except that we have to place points defining the radii at both the base (point 2) and top (point 4).

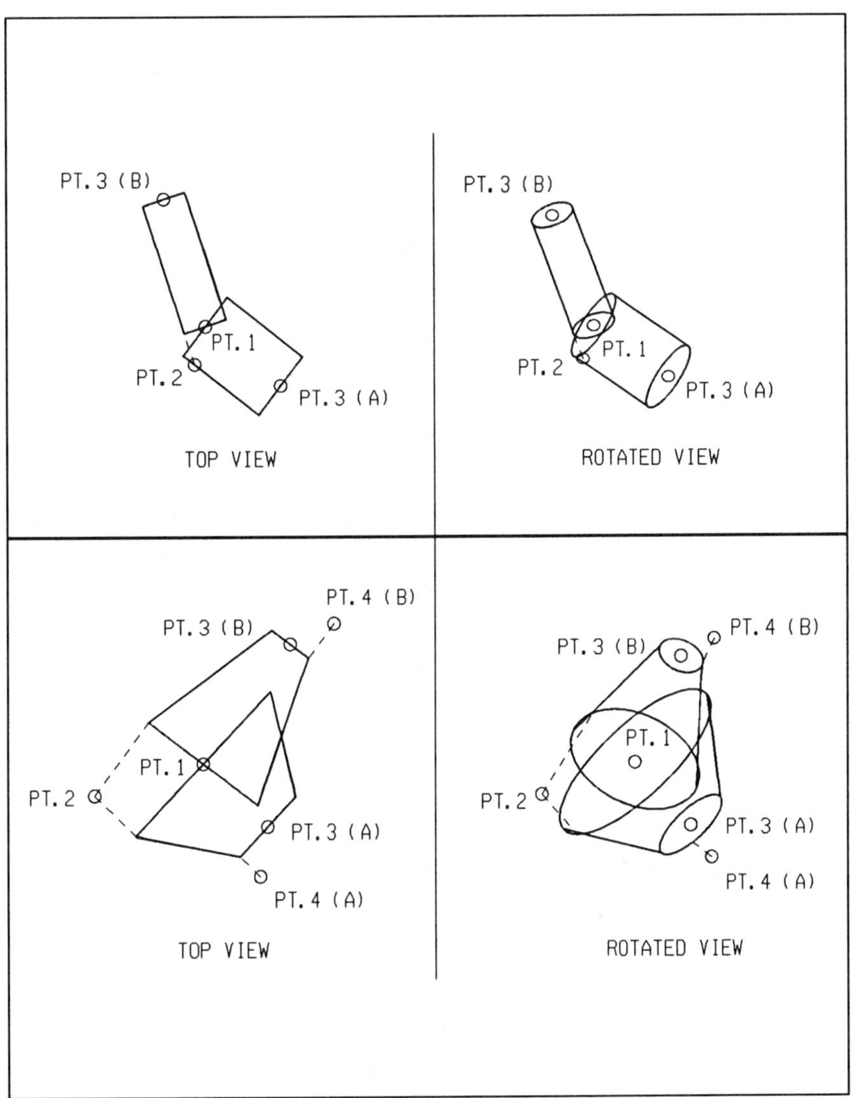

Figure 5.11 Placing Right Cylinders & Cones

Right Cylinder by Keyed-in Radius

Where we know the radius of the cylinder we can use the *Place Right Cylinder by Keyed-in Radius* tool. With version 3.3 (or, if you are not using the 3D palette with version 4) we are prompted to key in the radius. With version 4, we can select the tool in the 3D palette. As figure 5.12 shows, the radius is entered in a pop-down field. The type (solid/surface) is also selected here. After setting the radius, we then place the center points of the base and top of the cylinder. The orientation of the cylinder is taken from these points.

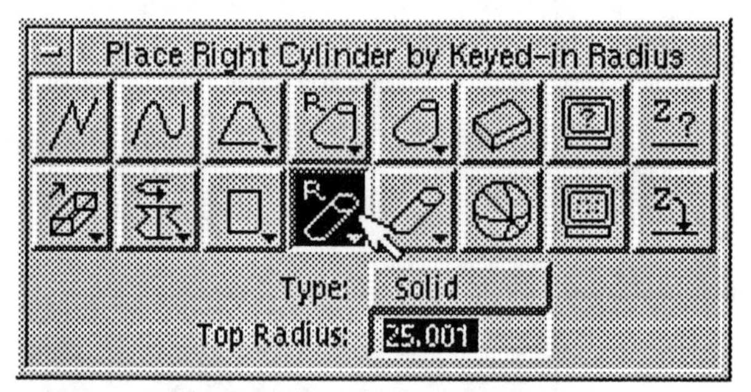

Figure 5.12 Entering the Top Radius

Right Cone by Keyed-in Radii

With the *Place Right Cone by Keyed-in Radii* tool we are prompted for two radii, those of the base and the top of the cone. If the tool is selected in the 3D palette, with version 4, we have pop-down fields in which to enter these values.

Placing cones with keyed-in radii operates similarly to that for cylinders. That is, the orientation of the cone is taken from the centers of the base and top, irrespective of the view being used.

Skewed Cylinders/Cones

As well as the right cones and cylinders, we have 'skewed' versions of both. SKEWED CYLINDERS/CONES take their orientation from a combination of the first two points placed, and the view being used. The two points define the center and radius of the base. With a cylinder, this is the radius of the cylinder. Two points are not sufficient to describe a plane, so the orientation of the view is used, along with these first two points. The planes of the base and top are normal to the view used. That is, they align with the Z axis of the view.

To place a skewed cone we use the *Place Skewed Cone* tool. On the sidebar menu, this is the 'Pnts' option, after you select PLACE CONE.

We determine the radius of the base and, along with the view, its orientation with the first two points. The next point specifies the center of the top. The fourth point determines the radius of the top. This is the minimum distance between the fourth point and a line joining the center of the base to the center of the top (or its extension).

Figure 5.13 shows two skewed cones which were placed using the same two starting points. In each case the orientation of the base and top has been determined by these two points. Also indicated is the line joining the centers of the base and top along with the dimension for the radius of the top.

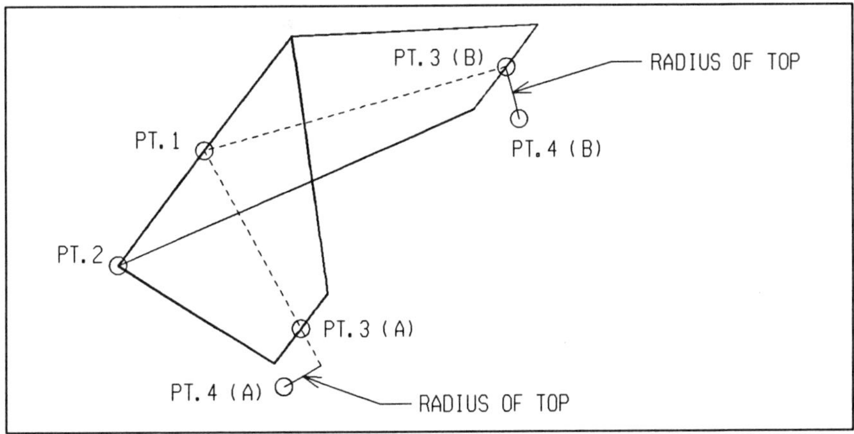

Figure 5.13 Placing Skewed Cones

To place a skewed cylinder, we use the *Place Skewed Cylinder* tool. On the sidebar menu, this is the 'Pnts' option after you select PLACE CYLINDER. Orientation of the cylinder is taken from the first two points placed, and the view being used. The plane of the top and bottom of the cylinder is normal to the view. Also, the radius of the cylinder is determined by the two points. We place a third point to specify the top of the cylinder. Figure 5.14 shows two skewed cylinders. Both were constructed using the same two starting points. We can see that these points, along with the view, have set the orientation of the cylinder.

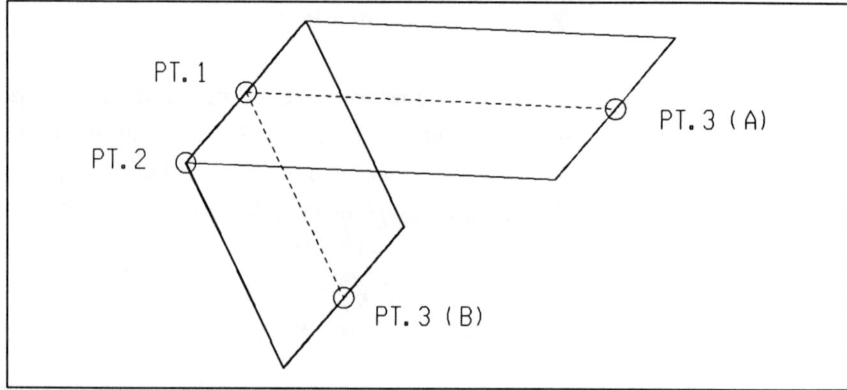

Figure 5.14 Placing Skewed Cylinders

Typical Uses of Cylinders/Cones

Placing cylinders by radius is particularly useful for straight sections of piping. Center lines for the piping layout can be placed first. We can then place cylinders with radius to create the pipes, by snapping to the ends of the center lines. This can be done in any view.

Figure 5.15 shows a simple example of how cylinders and cones can be used in piping. The pipes were constructed using the *Place Right Cylinder by Keyed-in Radius* tool. They were placed by snapping to the ends of the center lines. Cones were used for the reducers. One reducer is a right cone which was placed in the ISO view by snapping to the center lines and the outer edges of the pipes.

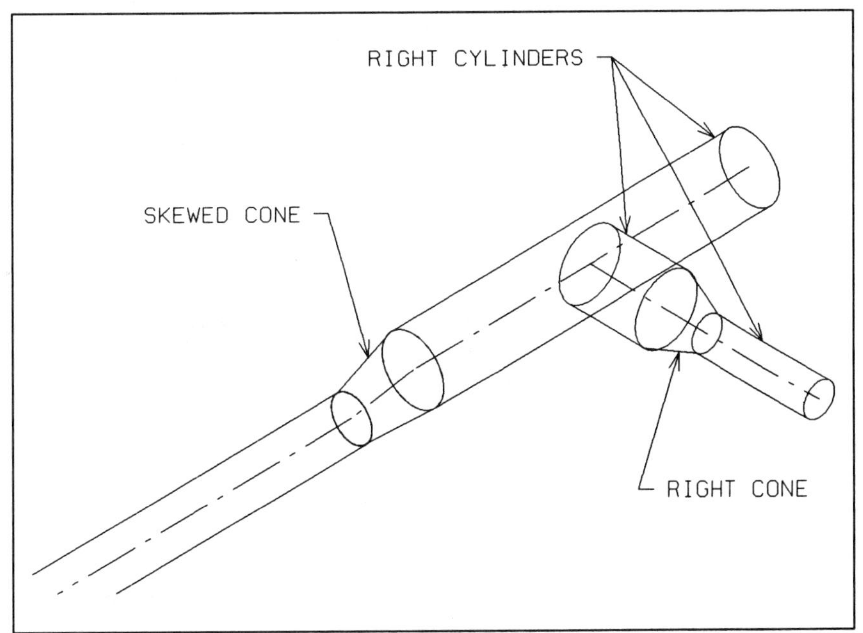

Figure 5.15 Using Cylinders and Cones in Piping

The second reducer is a skewed cone. It was placed in a view that was rotated to be parallel to the center line of the pipe as shown in figure 5.16.

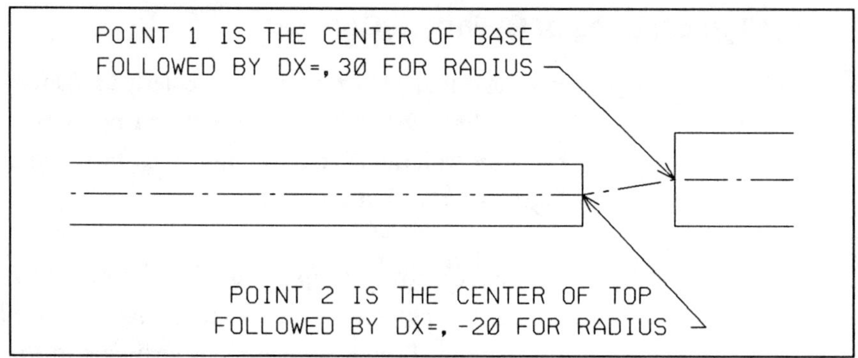

Figure 5.16 Placing a Cone by Points

Modifying Cylinders/Cones (version 4)

While cylinders and cones cannot be modified in version 3.3, version 4 does have this facility. We can modify the location of their end points by snapping to the center of the base or top (figure 5.17) and moving the cursor to the new location. The orientation of the base and top faces (or cylinder ends) does not alter, but the location can be modified. For example, a right cone may be modified, resulting in it becoming a skewed cone.

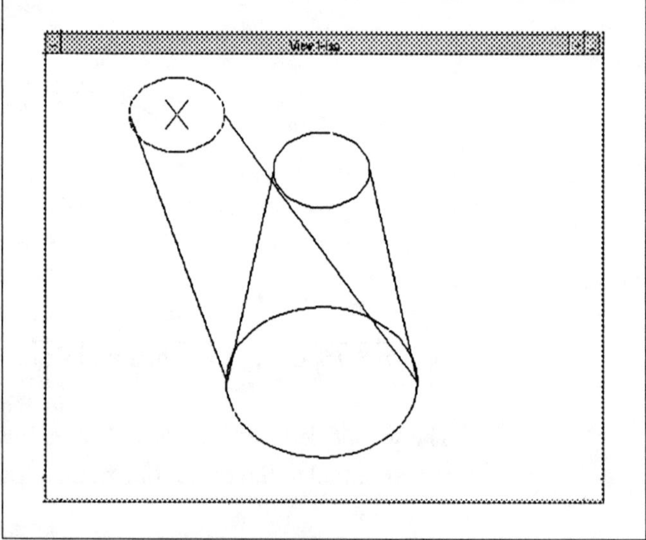

Figure 5.17 Modifying Cones

Spheres and Slabs (version 4)

A SPHERE is defined with two points, its center and a point on its surface. It may be defined in any view.

A SLAB is defined by 3 points, but it is an exception to our 'rule of thumb'. The first two points define its area. This is similar to placing a block, therefore it is view dependent. The third point defines the thickness of the slab. This may be defined graphically in any view, or with a key-in (e.g., DX=,,200) in the view used for the first points.

Element Manipulation

The same element manipulation tools available in 2D may be used in 3D design files. There is, of course, the third dimension to consider when scaling, rotating, copying, placing fillets and chamfers etc.

Scaling

ALL SCALING IS RELATIVE TO THE AXES OF THE VIEW OR SCREEN.

That is, the scaling is in the X, Y and Z directions of the view or screen. When scaling equally in all directions (i.e., X, Y and Z axes), we can use any view to identify the point about which we want the scaling to occur.

Where there are different scale factors involved, for one or more axes, the view used for selecting the scaling origin is very important. We must ensure that the item being scaled is positioned correctly, relative to our screen. It must be located so that the direction/s in which we want to scale it are aligned with the X, Y, and Z axes of our screen. We can use 'ROTATE 3PTS' to align the object with the axes of our screen.

Rotating

ALL ELEMENT ROTATIONS ARE ABOUT THE Z AXIS OF THE VIEW OR SCREEN.

This means that rotations are about an axis perpendicular to the screen. The positive direction for the active angle is anti-clockwise.

Rotations that we are discussing here are for manipulating elements only. We can rotate reference files or views about any of the X, Y, or Z axes. Whenever we rotate elements - whether singly or within a fence, working set or graphic group - the rotation is about the Z axis of the view or screen.

Fillets and Chamfers

We construct fillets and chamfers as for 2D files. The main criterion is that the two elements, or their extensions, must physically intersect at some point. For example, the two lines shown in figure 5.18 appear to intersect when viewed from the TOP. When we see them from another direction, it is obvious that they do not intersect. They are separated in the Z direction of the design cube. The construction would not be completed. The error message 'illegal definition' displays in these cases.

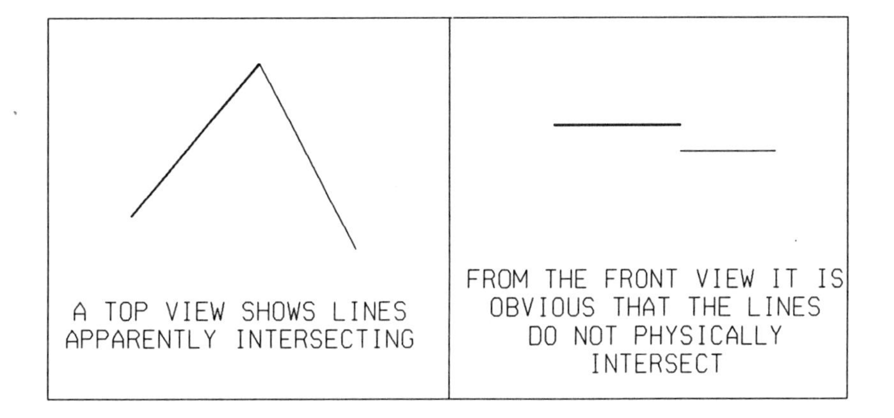

Figure 5.18 Viewing Lines in 3D

Extending Lines

Extending lines works as for 2D with some exceptions. When we use the cursor to make the extension it sometimes cannot be made to extend the line far enough. This occurs though there is enough 'screen' room. The reason for this anomaly can be that the line is at an angle into, or out from, the screen. Our cursor is always at right angles to the identified line, so it appears to move away from the line in particular views. Figure 5.19 illustrates the point.

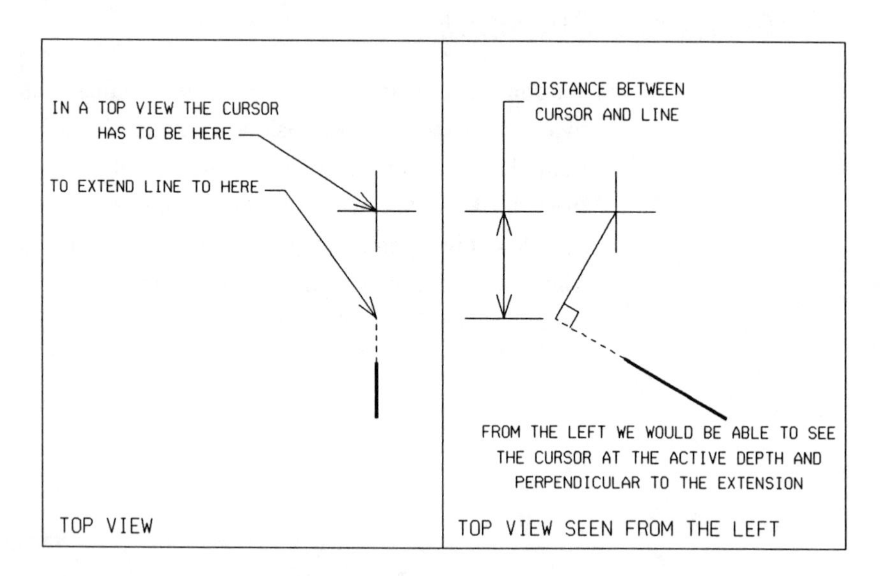

Figure 5.19 Extending Lines in 3D

Another apparent anomaly is when we extend a line and it won't extend past a certain point on the screen. This occurs when the extension protrudes through the display depth of the view.

Elements or parts of elements that are outside the display depth of a particular view do not appear in that view. They are still in the design file, they merely do not display in the view. This same law applies to elements that we are manipulating. If we attempt to move or modify them so that they are outside the display depth of the view then they disappear from view.

Figure 5.20 shows a line that is being modified in a TOP view. An attempt has been made to extend it to a point that is outside the display depth of the view. The right side of the figure shows the same view, if we could see it from the left. It shows clearly that the line is being extended beyond the display depth of the TOP view.

In these situations, the line would be extended but we would not see the complete element unless we change the display depth parameters.

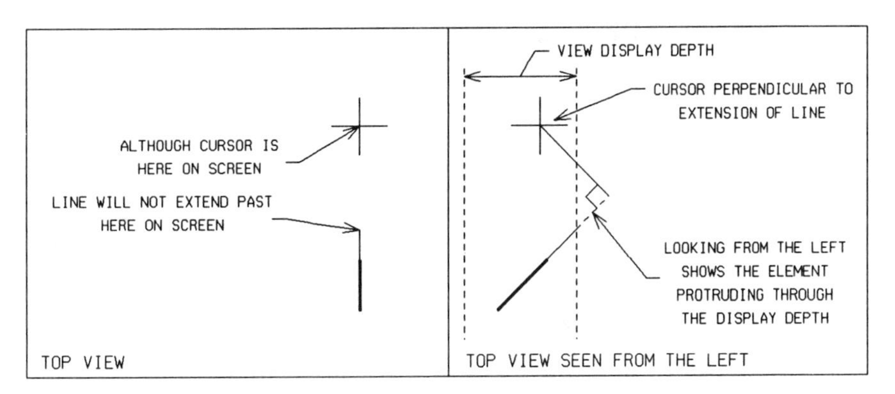

Figure 5.20 Modifying Lines in the TOP View

Extending to Intersection

Extending elements to intersection is a view dependent operation. The intersection point is along a line perpendicular to the view or screen. Where two elements physically intersect this means nothing. Where they are separated in the Z direction of the view or screen then the point of intersection depends on the angle from which we are viewing them, as figure 5.21 shows.

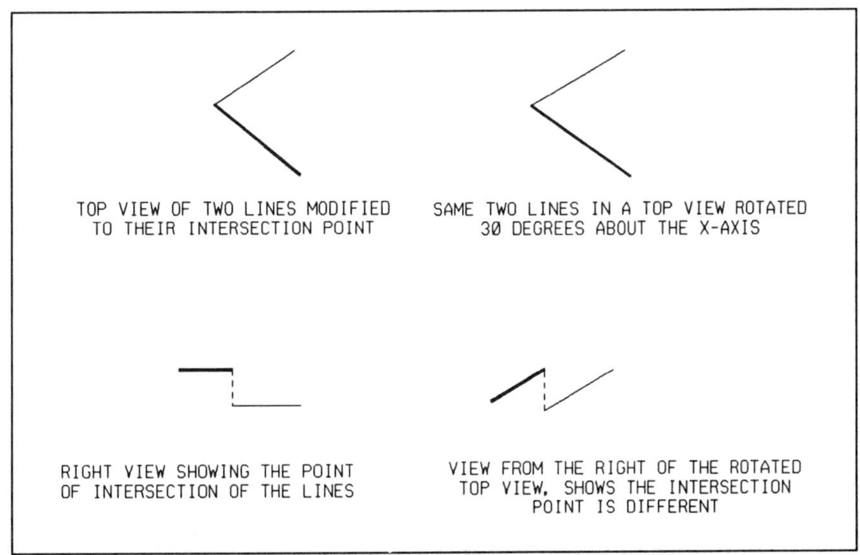

Figure 5.21 Extending to Intersection in 3D

Intersect Snap

Intersect snap is available in 3D, but it too is view dependent. As we saw when discussing the *Extend Element to Intersection* tool, lines that are separated in the Z direction of the view or screen will appear to intersect at different places. This depends on the angle from which we view them. The intersection point is along a line perpendicular to the view or screen.

With version 3.3, the second element identified, when using intersect snap, is the element to which the cursor 'snaps'.

We may have to draw a line, for example, from the intersection of two other lines. With version 3.3, the new line will be 'attached' to the second line identified with the intersect snap.

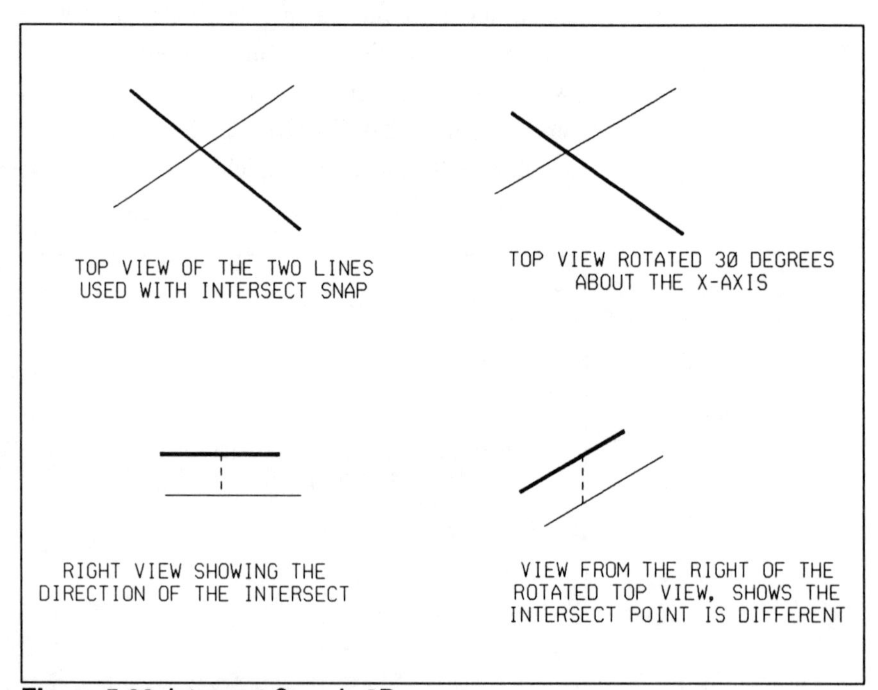

Figure 5.22 Intersect Snap in 3D

Copying and Moving

When we copy and/or move elements, they retain their relative orientation. That is, vertical elements stay vertical, horizontal elements stay horizontal etc. Copying and moving may be performed in any view with precision inputs or using tentative points to existing elements. We can copy or move elements in any direction. That is, they can be copied or moved in the X, the Y, or the Z direction, or any combination of the three.

When we use the DX key-in the elements are copied/moved relative to the view or screen coordinates. Screen coordinates are horizontal, vertical or perpendicular to the view or screen (X, Y, Z respectively).

When we use the DL key-in the elements are copied/moved relative to the design cube axes. This applies no matter in which view we are working.

Rotated copies obey the rotation laws. They are rotated relative to the Z axis of the view or screen. They are rotated by the amount of the active angle as for 2D.

Text and Dimensioning

Text and dimensioning may be placed in 3D files without problems. Generally, it is better that the model file is free of text. Model files sometimes become large and complicated. The last thing that we want is for updates, tentative points and the like, to be slowed by the text contained in the file. An option, with display updating, is to set the text display to fast text. A better option is to have the text placed in another file with the model referenced to it. If you do need to place text in the model file, it should be kept on a separate level to that of any elements forming the model. This applies especially to projects where other users may want to reference the model (but not the text).

Drawings can be produced from the 3D design using the EDGES hidden line removal utility. Any text placed in the model file, and that is required for drawings, should be located so that parts of the model don't obscure it when we use the hidden line removal utility.

Text

MicroStation has 2 types of text, the standard text and VIEW INDEPENDENT text. Standard text is two-dimensional and we place it on a plane parallel to the screen in the current view. When viewed edge on, normal text appears as dashes.

VIEW INDEPENDENT text, may be placed in any view and is still able to be read in all the other views. View independent text is particularly useful for noting 'monuments' such as survey or trig points, which may be required to be used or referenced in any view.

Dimensioning

Dimensioning in 3D files is similar to 2D. The main difference is that, we have to consider on which plane we want the dimensioning to be located. With version 3.3 of MicroStation, dimensioning is placed on a plane parallel to the screen. It is at the active depth of the data point for the origin of dimensioning.

3D models can be used to produce drawings from different aspects of the model. It is a good idea to keep the dimensioning data on a completely separate level to any elements. This can be done with the 'LD=' key-in. Entering 'LD=63' ensures that all dimensioning data goes to level 63 though the active level is something else. In this way we can place elements on the active level, but any dimensioning would go to level 63 without us changing level.

Associated Dimensioning (Version 4)

With version 4 of MicroStation we have access to another form of dimensioning. This dimensioning is attached to or associated with the element it dimensions. It operates in 3D files like it does with 2D. Any changes to the size of the associated element causes a change in the dimension to reflect the change.

This element is not compatible with other versions of MicroStation or IGDS. We can specify that we want the earlier style dimensioning with the 'SET COMPATIBILITY ON' key-in. When we want the ASSOCIATED DIMENSIONING we use the 'SET COMPATIBILITY OFF' key-in. We can set the association lock on/off by keying in 'LOCK AS ON/OFF'.

We have three axis choices for the orientation of dimensioning. They are VIEW, DRAWING and TRUE, as for 2D. Figure 3.12 shows examples of each. The dimensioning is in the ISO view to show, more clearly, the difference in the 3 choices.

View

Aligns dimensioning data with the X and Y axes of the view or screen. That is, all dimensioning data will be either horizontal or vertical, relative to the screen. Because we are working in 3D, the dimensions will be on different planes. The first data point identifying the dimension origin determines the plane for each group of dimensions.

True

As the name suggests, gives the actual dimension, not the 'projected' dimension. The dimensioning elements, here, are placed in the alignment of the points defining the dimension. They are not necessarily aligned with any of the axes, either view or model.

Drawing

Aligns dimensioning data with the axes of the drawing (i.e., the design cube). With the axis set at DRAWING we can place the dimensioning using any view because it always aligns with a plane of the model.

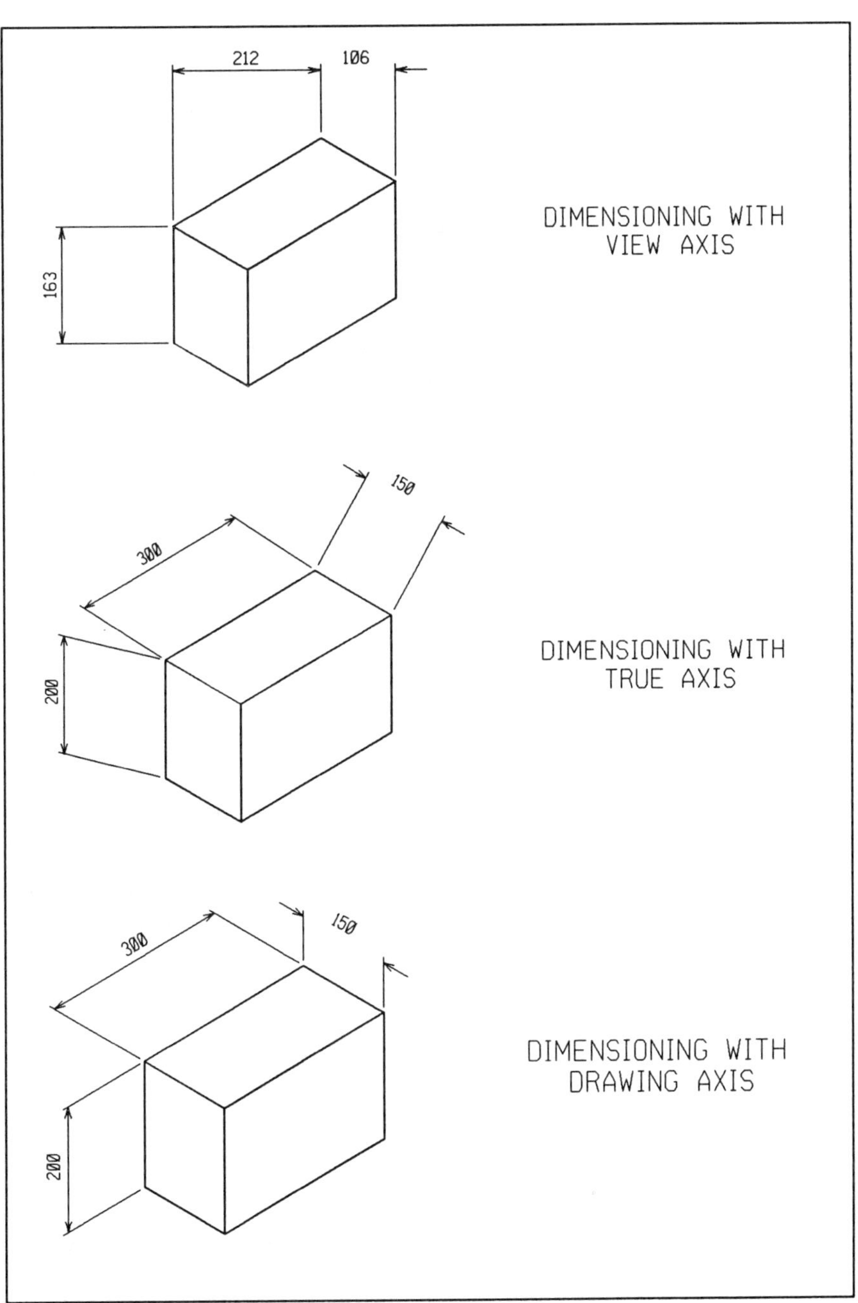

Figure 5.23 Dimensioning in 3D

Digitizing

Often, we have information available in hard-copy format. One option is to digitize the information into a design file. With a 3D file this information can be digitized with any orientation we require. For example, if we have cross-sections looking north, we can digitize them into a FRONT view.

To start, we set up view 1 with the orientation we require. We then use the DIGITIZER SETUP procedure as we do for 2D digitizing. With three-dimensional files, the difference is that there is a Z value associated with the monument points. When we digitize, we vary the X,Y screen coordinates as we trace the information. In 3D files, the same applies.

Where different elevations are required for various sections of the information we can employ a user command that is delivered with MicroStation. This user command, 'DIGZ.UCM' allows us to vary the Z value without re-defining the XY coordinates of the area being digitized.

To digitize contour maps for example, we can digitize the contours at their correct elevation (Z value) in the TOP view. To begin we would set up the TOP view to digitize the first contour. If we continued on digitizing the other contours, they would all have the same Z-value (elevation) as the first contour. Here, rather than use DIGITIZER SETUP to laboriously vary the Z-value for each separate contour, the user command DIGZ.UCM can be used.

6 : Creating 3D Models

By now you should be able to work reasonably happily in 3D design files. In this chapter we will look at methods for creating models. This entire chapter is in the form of an exercise in creating a 3D model. You will not always be told which view to use, or when to zoom in/out or fit a view. Remember to use the different views to help you. If something does not go as expected, try to work out what happened and why. Then, if it happens again when you are in a panic to get a job done, it won't be as traumatic. A mistake that is made quite often is to project an element without checking that the active scale and active angle are set correctly. The results sometimes look interesting but aren't very useful. With version 4, using the *Construct Surface/Solid of Projection* tool in the 3D palette, this is less likely to happen. When the tool is selected, we are shown the current status of both.

In general, the methods used in 2D are valid for 3D as well. This includes using level control to separate the data and reduce the amount of information being displayed on the screen. Using colors to differentiate between various items, or pieces of equipment, is another good practice. In 3D, parts of the design may be located in front of or behind, as well as adjacent to, our area of interest. Colors can help us to mentally separate different parts of the model, even when the view is zoomed out.

In many instances our base information is in the form of two-dimensional drawings or survey type data. We can digitize these existing 'hard copy' plans into MicroStation 2D or 3D files. Where the existing information is in another

CAD format, we can often rely on DXF transfers to translate the drawings into MicroStation format. Most CAD packages support the DXF protocol for translations. If we already have MicroStation 2D files of the relevant project then we can convert them into 3D design files with the 2D to 3D translator that comes with MicroStation. With version 4 this is not so important as we can reference a 2D file to our 3D model file. Both the DXF translator and the 2D to 3D conversion programs are easily accessed with version 3.3, via the MCE menu. These are discussed in chapter 13 (Utilities).

View Set Up

Before we start in a design file, we should set up the views as we want them. You may have your own preferred set-up. If not, set them up as we did for the tutorial in chapter 1. That is, views 1-4 set to be ISO, TOP ,FRONT and RIGHT respectively.

Version 3.3

Use view 1 as the working view. We can then WINDOW AREA from the other views into view 1 as necessary.

With a dual screen configuration, the second screen can be set with one view to use as the working view.

Version 4

Use view 1 as the working view. We can WINDOW AREA into this view as required. View 1 can then be re-sized to fill the screen, when needed. When the other views are required, we can 'Tile' or 'Cascade' the screen. Also we can reduce view 1 to a small window, while using the other views.

With a dual screen configuration, the second screen can be set with one view to use as the working view.

Getting Started

We will first look at methods of using two-dimensional information to help us create our three-dimensional model. The first example is that of an office layout. We will draw the layouts as 'flat' plans in the three-dimensional file. From these plans we will project the various elements to form the model.

We will create the 3 floor, office layout in two distinct parts:

1. The building layout, which is the same for each floor
2. A partitioning layout that varies for each floor.

In preparation, create a new 3D design file and call it LAYOUT1.DGN. Use the metric mapping seed file, as for the training file. Set the views up as you prefer, or as we have just discussed in the previous section. Create a new 3D cell library also, or use the cell library you created for the beginning exercises.

Figure 6.1 shows the dimensions and layout of the building. The columns are 1 unit square. We will use single lines for the walls, and the partitions. Elevations for floors 1,2 and 3 are 10.5, 16.0 and 21.5 respectively. Other dimensions which we will need later are the depth of the floor slab, which is 0.5 and the height of the columns, which is 5.0.

We will place the building layouts for floor 1 on level 1. Floors 2 and 3 will be placed on levels 2 and 3 respectively. This will allow us to deal with each floor separately. We could use display depth also, to separate them.

Our first task is to place the floor layout in the file.

- Make the active level 1.
- Use a TOP view, and draw the floor layout. Use blocks for the floor slab and the columns. Use linestrings in other cases where two or more lines are joined together. This makes it easier to project them later.
- Elevation of the first floor is 10.5 so place all the elements at a Z value of 10.5 in the TOP view.

Figure 6.2 shows the floor layout, as it should look, in ISO view.

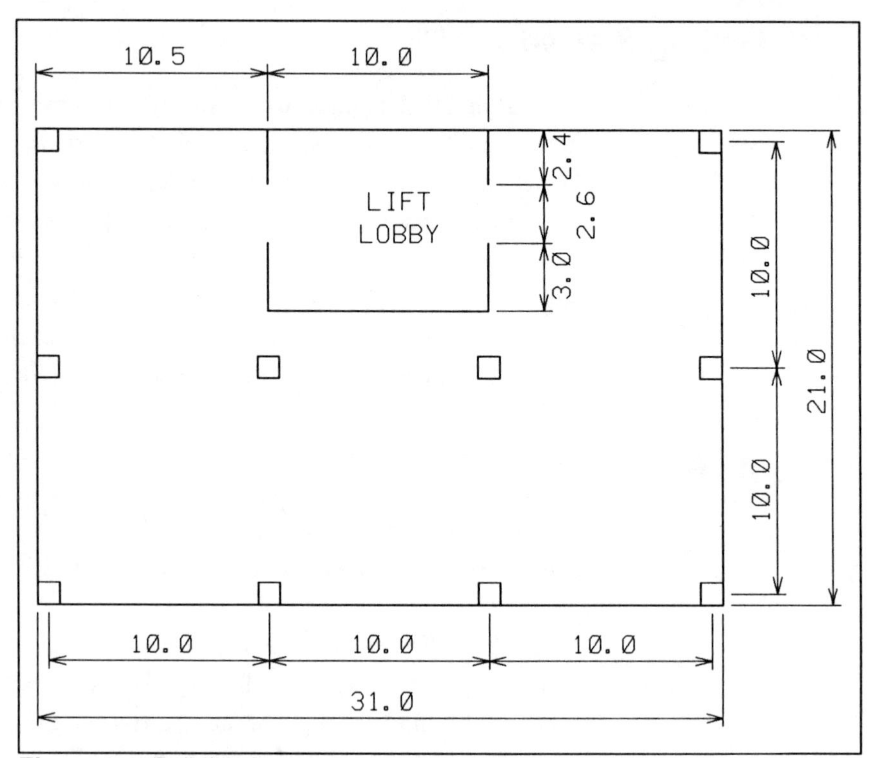

Figure 6.1 Building Layout

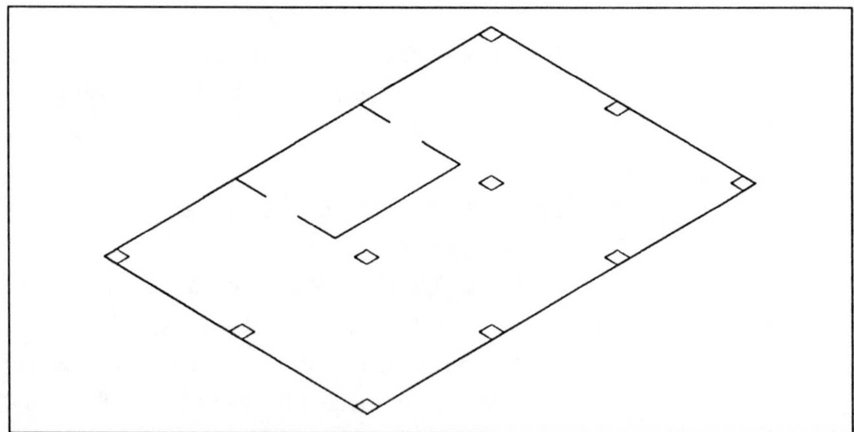

Figure 6.2 2D Floor Layout in ISO view

Figure 6.3 shows the layout of the partitioning for the first floor. The partitioning for floor 1 can be placed on level 11. Use level 12 for floor 2 and level 13 for floor 3. This will keep the partitioning separate from the building layouts which we be on levels 1, 2 and 3. The partitioning layout for floor 2 is shown in figure 6.4 and that for the third floor in figure 6.5. Floor 3 partitioning is a mirror image of the first floor.

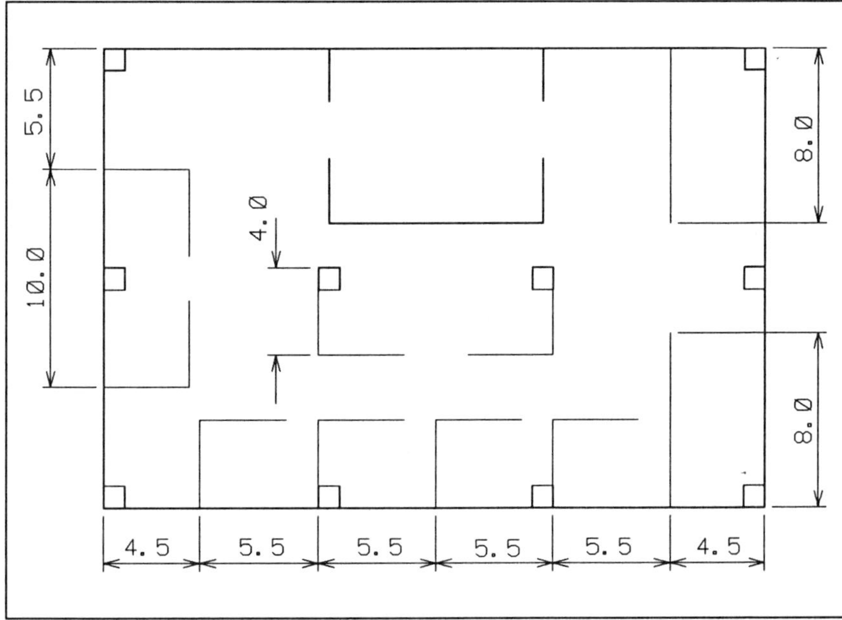

Figure 6.3 Floor 1 Partitioning

As the partitioning is generally made up of 4x4 modules, we can draw one in each orientation and copy it to create the others. We could also draw one, go ahead and project it, then copy the three-dimensional partition to the other locations. You can choose the method that suits you best. Once one item is in the correct location, we can use either DX or DL to copy it for the others. Most of these operations are similar to those you would use in a 2D design file.

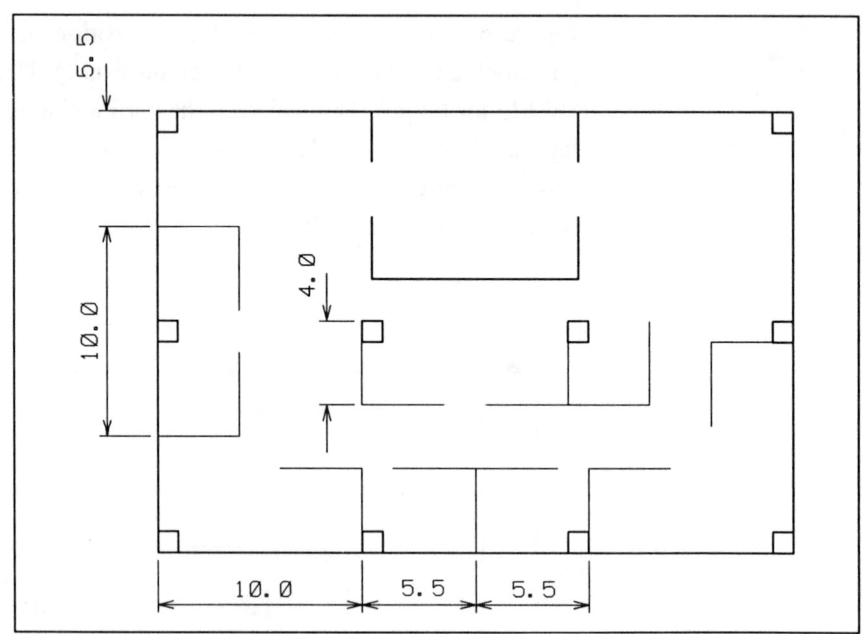

Figure 6.4 Floor 2 Partitioning

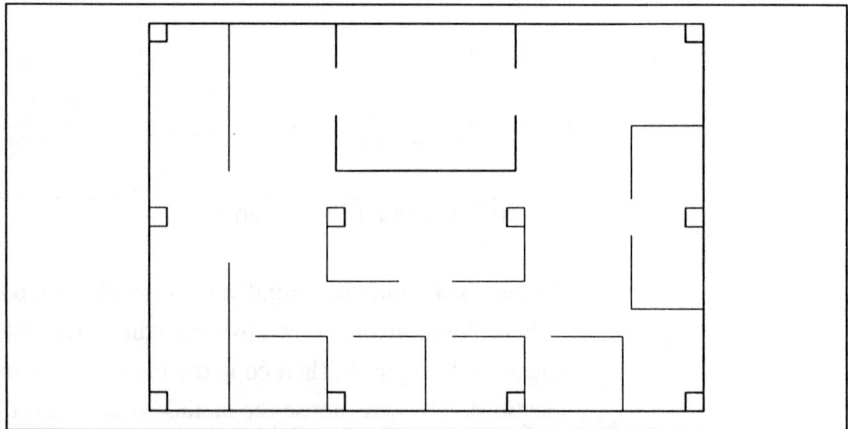

Figure 6.5 Floor 3 Partitioning (Fl.1 mirrored)

When the layouts have been set up in our design file, we can begin to project the various shapes and linestrings to produce three-dimensional objects. We will start with the floor, columns and lift lobby. For the floor slab, we can use the ISO view.

To project the floor slab downwards:

- ° Make sure that ACTIVE ANGLE is 0 and ACTIVE SCALE is 1 (ver.3.3).
- ° Set ACTIVE CAPMODE to ON (ver.3.3).
- ° Select *Construct Surface/Solid of Projection* tool. Version 4 users can check the ACTIVE ANGLE, ACTIVE SCALE and set the Type to Solid.
- ° Identify the block depicting the floor slab, then key in DL=,,-0.5 for the distance. The minus sign indicates downwards in real world co-ordinates.

We will now project the walls of the lift lobby, then the columns. Version 4 users can complete this task with a fence project, but first we will look at the version 3.3 method.

Version 3.3

To project the walls of the Lift Lobby upwards:

- ° Check that ACTIVE ANGLE is 0 and ACTIVE SCALE is 1
- ° Select *Construct Surface/solid of Projection* tool from the menu.
- ° Use the ISO view, and in turn, project the 3 elements that make up the lobby, 5.0 vertically. Use DL=,,5 to specify the distance. The 5 in this case is positive, indicating upwards in real world co-ordinates.

To complete the first floor, project the columns upwards:

- ° Identify one of the shapes depicting the columns and project it 5.0 vertically, again using DL=,,5 to specify the distance.
- ° Copy the projected column to each of the other locations to form the other columns.

While version 4 users can use the method described above, the following is a more efficient way of achieving the same result.

Version 4

We can use the 'FENCE SURFACE PROJECTION' key-in which allows us to project the contents of a fence. With this method, we can complete the construction with one operation.

Note!

While the complete key-ins are shown here, remember that these key-ins can be abbreviated. Here 'FENCE SURFACE PROJECTION' can be abbreviated to 'FEN SUR', or even 'FE SU'.

- In a TOP view, place a fence just inside the outline of the floor slab. That is, so that some part of each column is included in the fence (refer to figure 6.6).
- Set the fence lock to OVERLAP.
- Key-in 'FENCE SURFACE PROJECTION'
- Place a data point in the TOP view, then key-in DL=,,5 to complete the construction.

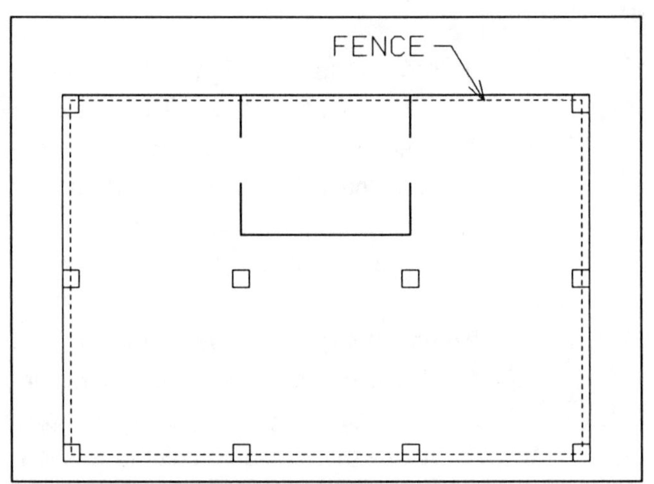

Figure 6.6 Projecting with a Fence

Figure 6.7 shows an ISO view of how your model should look now, with the elements projected.

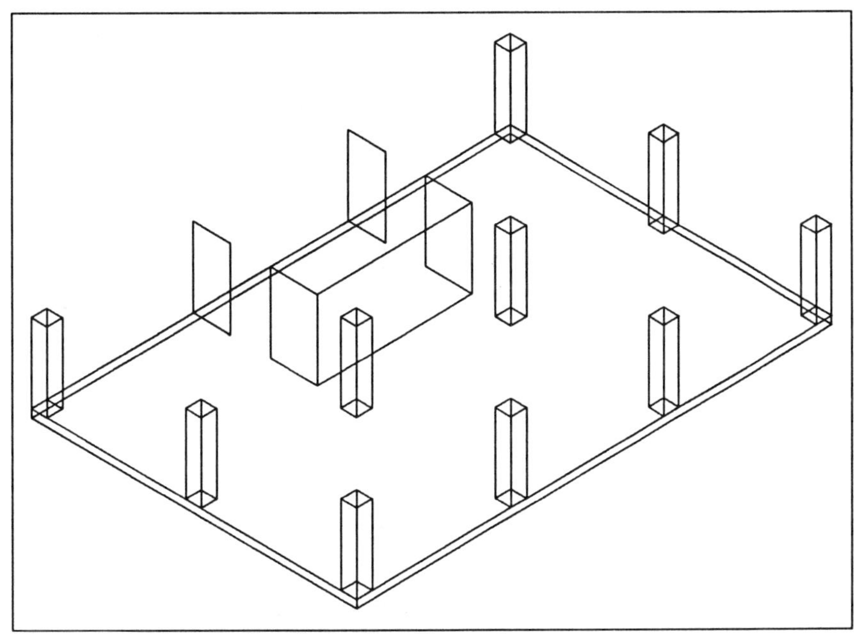

Figure 6.7 Floor 1 after Projecting Elements

Building layouts for the second and third floors of our office model are identical, so we can copy floor 1 to levels 2 and 3. We can then move each to its correct elevation, along with its relative partitioning layout. We must be sure that we only move the elements that we want. We can do this by only having the relative levels on when we perform the operations.

To start, make the active level 1 and turn off all other levels.

- Fence copy floor 1 to level 2.
- Make the active level 1 and turn off level 2. This is to make sure that we don't get duplicate copies on level 3.
- Fence copy floor 1 to level 3.

We have the respective building layouts for the three floors on separate levels, but at the same vertical elevation. We now have to move them to their correct elevation. When we do this we should move the partitioning layouts for each floor at the same time. Again, this task is simple, because we have separated them onto different levels.

Before moving the floors to their correct elevations make sure that the relevant levels are displaying. Floor 2 uses levels 2 and 12, floor 3 uses levels 3 and 13. The elements can be moved within a fence. Make sure that all the elements are displayed and are contained by the fence. Fitting the view before placing the fence will ensure that all elements are displayed.

- Make the active level 2.
- Set the levels in the working view so that only levels 2 and 12 are displayed. The working view is the view in which the fence will be placed.
- Fit the working view.
- Place the fence around the elements.
- Select the *Move Fence Contents* tool.
- Move the contents 5.5 vertically (data point followed by DL=,,5.5).

Follow a similar procedure for the third floor. Move the elements on levels 3 and 13 by 11.0 vertically.

Once the floor layouts have been located correctly, we can project the partitioning for the floors. Each partition is 1.6 high. Apart from the difference in the height, the procedure is identical to that used for the general floor layout. Again, version 4 users can use the 'FENCE SURFACE PROJECTION' key-in and project the elements as a group in a fence.

Version 3.3

Make the active level the same as the level that the originating elements are on. This ensures that the projected surfaces can be separated. Projected surfaces are produced on the active level, irrespective of the level of the original element.

- Make the active level 11 and project each partition by 1.6 vertically (DL=,,1.6).
- Where there are identical partitions, with the same orientation, then we can project one and copy it to the other locations.
- Repeat the above for the other floors. Make the active level 12 for floor 2 and 13 for floor 3.

Version 4

We can project the partitioning for each floor either of two ways.

- We can work on each floor separately, using a fence with the 'FENCE SURFACE PROJECTION' key-in.
- Alternatively, we can have the partitioning for all three floors displayed (levels 11, 12 and 13).

If we use the latter method, then all the projected partitions will be placed on the active level. These may be moved to other levels later, if required.

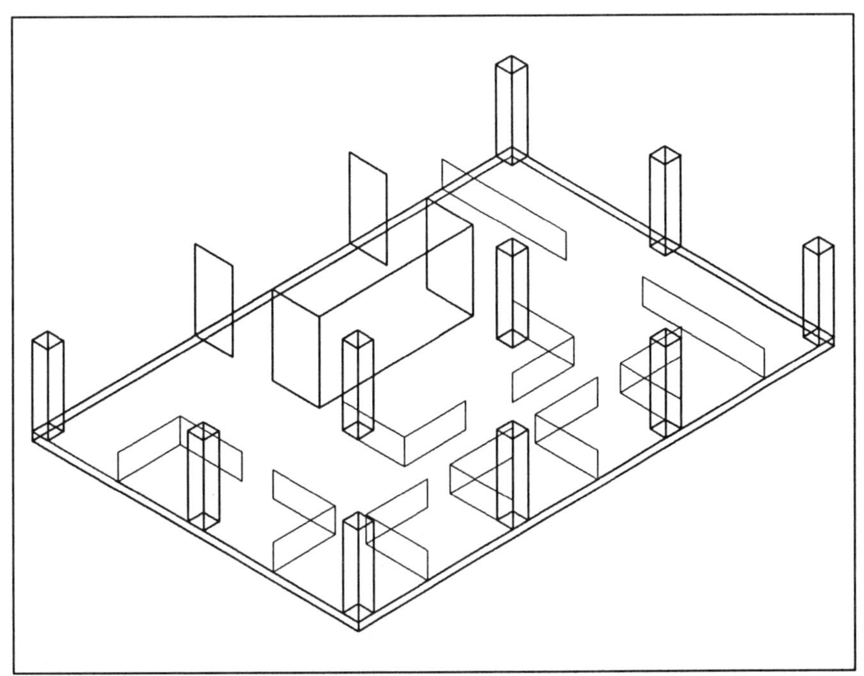

Figure 6.8 Floor 1 Completed

Figure 6.8 shows floor 1, in wireframe form, after the partitioning has been projected. In the following diagram (figure 6.9), we can see the three floors, both individually and 'assembled', after hidden line removal.

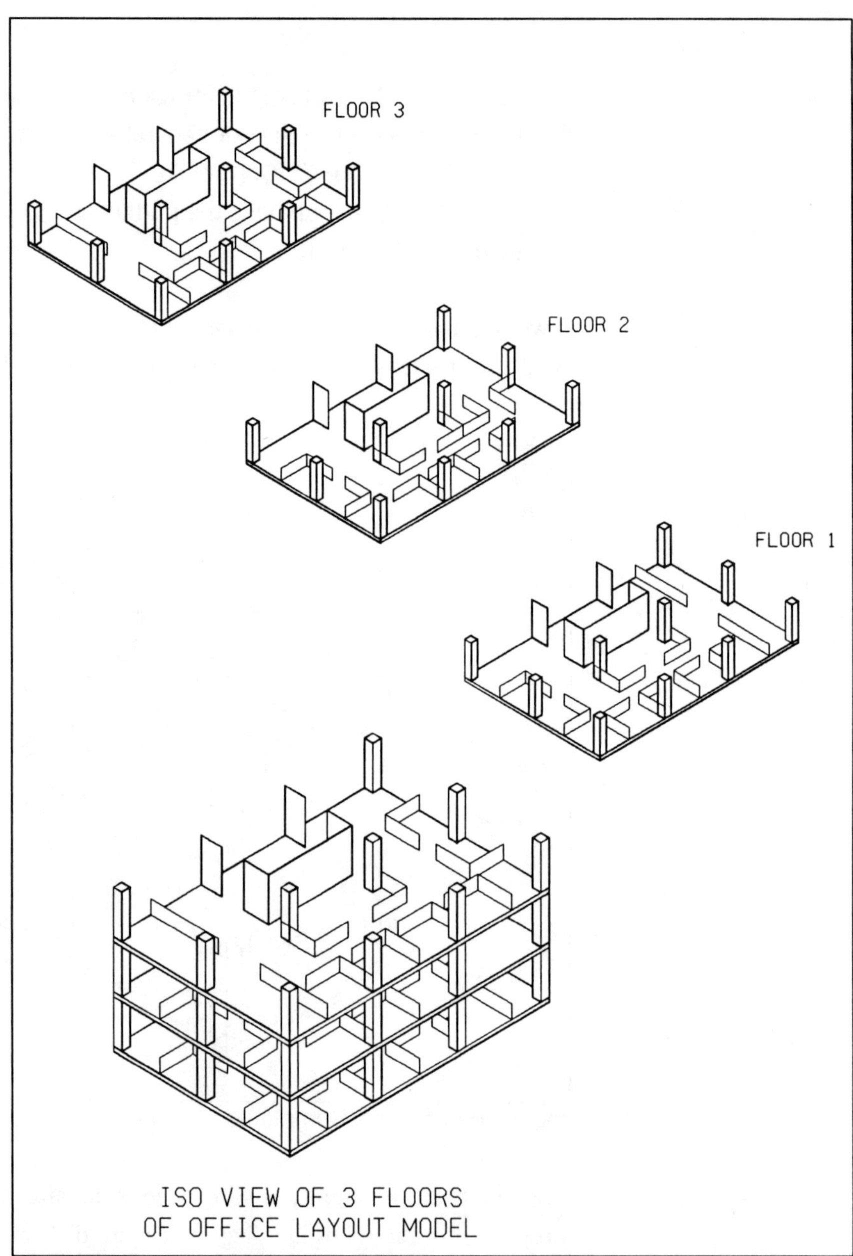

Figure 6.9 The Office Layout Model

The model has been created with the floors on separate levels. This makes it simple to perform any modifications to individual floors. Where it is not as clear cut as this example, we can use DISPLAY DEPTH to select the part of the model that we want to work with.

It is well worth the effort to separate the model with levels, or reference files, where possible. Documentation also should also be kept, indicating where various parts of the model are located in the design file (i.e., which levels or reference files). With version 4 we have the option of using NAMED LEVELS. This, method can be used also to document where the different parts of the design have been placed.

To complete the model:

- Copy the floor slab, for the third floor, 5.5 vertically.
- Move the copy to level 4.
- This will form a roof to complete this part of the model.

We will continue by constructing some furniture and fittings for the office. We will use a new file for these models.

- Create a new 3D design file, named FURNISH1.DGN.

This file will be used for creating the various items. As they are finished we will make cells of them to simplify placing them in the office layout model.

We will use the same method as before to construct a desk for our 'office'. We will create a two-dimensional plan from which the three-dimensional objects will be projected.

Figure 6.10 shows the desk we are going to create. The first task is to place the layout of the desk as a two-dimensional plan in our 3D file.

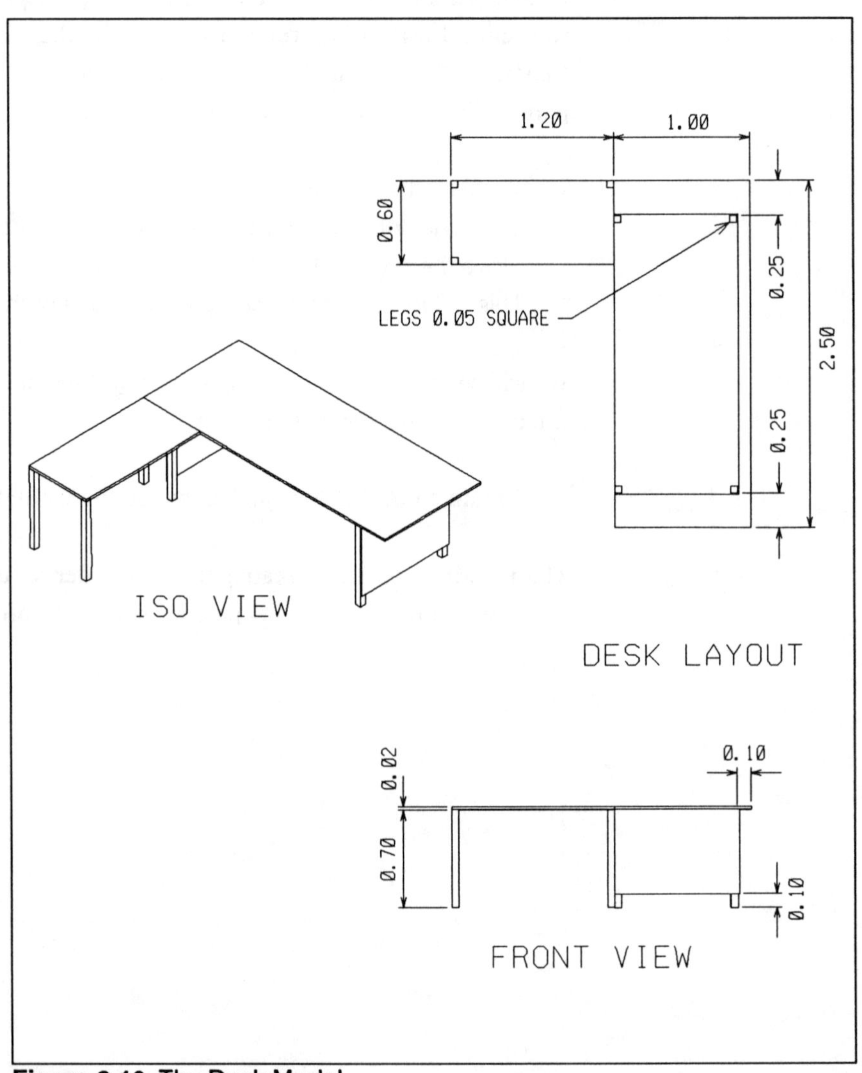

Figure 6.10 The Desk Model

Creating the plan of the desk can be completed as follows:

° Make the Active Level 1.

° Place the shapes and linestrings to form the layout of the desk using a TOP view. Make sure that they are all at the same elevation. That is, they should be co-planar, with the same Z value. The FRONT and RIGHT views will quickly show us if this is true. Once the first element is placed, the others can be placed by using a tentative to the existing elements and then DL key-ins for the other points.

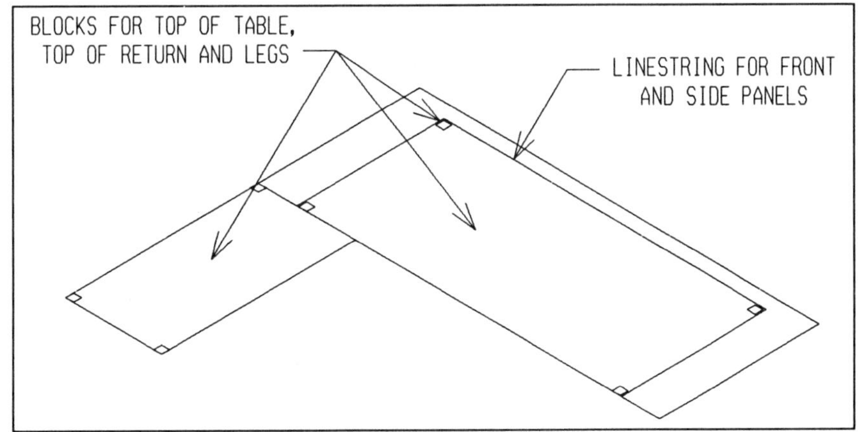

Figure 6.11 Desk Layout Ready to Project

When the layout is ready, we can use the ISO view, as before, to project the elements into three-dimensional components.

Make sure that ACTIVE ANGLE is 0 and ACTIVE SCALE is 1. With version 4, this can be checked in the dialog box after the *Project Surface/Solid of Projection* tool has been selected in the 3D tool palette:

° Set ACTIVE CAPMODE to ON (set Type to Solid in dialog box with version 4).

° Project the top of the desk and the desk return 0.02 vertically upwards. Identify each block and project using DL=,,.0.02 to specify the distance.

° Project one of the legs 0.7 vertically downwards. Identify a block representing one of the legs and use DL=,,-0.7 to specify the distance.
° Copy the projected leg to the other locations for legs.
° Project the linestring for the front and side panels 0.6 vertically downwards. Identify the linestring and key-in DL=,,-0.6 for the distance to project.

Version 4 users again can use a fence with the 'FENCE SURFACE PROJECTION' key-in to complete the construction of the legs in one operation.

Your desk should look like figure 6.12 in wireframe form.

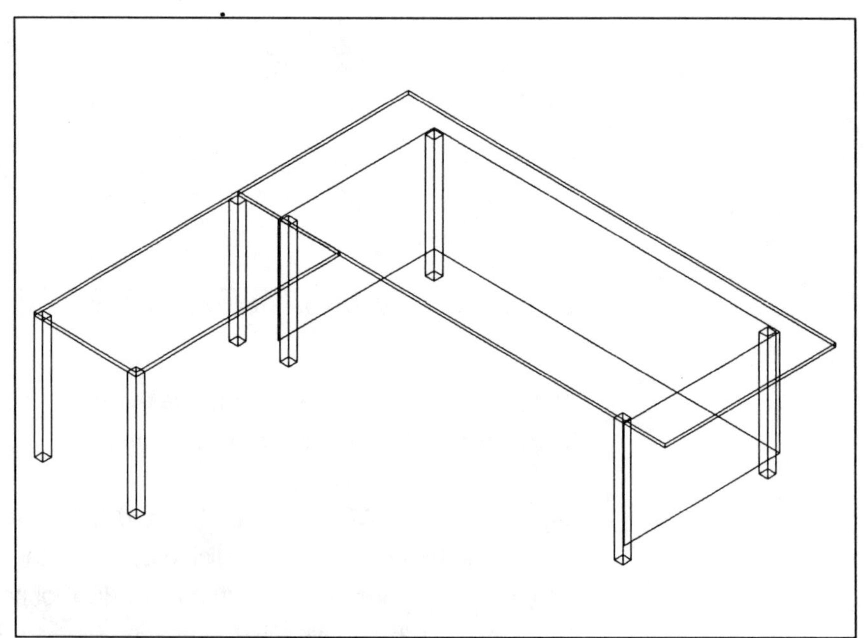

Figure 6.12 Finished Desk, ISO view - wireframe

The method we used to create both the office layout and the desk quickly converts the two-dimensional layouts into three-dimensional models. Our models, so far, are simple. In a working environment the models will most likely be more complex. Even so, the initial work on the model often can be started by using two-dimensional layouts. Once there is some part of the project in the three-dimensional file it is much easier to place new items. They can be placed relative to existing elements using DX= and DL= precision inputs.

Though we are working in three-dimensional files, we can still use two-dimensional information. In some instances, only parts of a project may need to be fully three-dimensional while other parts may remain in two-dimensional form.

As we will want to place the desk in the office a number of times we will make a cell of it. It is probably easiest to place the desks in the TOP view so we will create the cell in the TOP view. If we make the origin at the bottom of one of the legs it will be simple to place the desks at the correct levels for each floor.

Set up your views so that:

- The whole desk is displayed in a TOP view.
- One of the legs, at least, is displayed in another view where it is easy to tentative to the bottom of the leg. This view will be used for defining the origin.
- Attach your 3D training cell library to your design file

Create the cell as follows:

- Place a fence around the desk in the TOP view.
- Define the origin at the bottom of the leg displayed in another view.
- Key in CC=cellname just as in 2D to create the cell and add it to the cell library.

For extra practice at using views, and the other techniques discussed so far, try to create the chair to go with the desk. This also can be added to the cell library. The chair is shown in front and side elevation in figure 6.13, and the components are shown in figure 6.14.

As you can see, this model is a little more complex to create than the desk and the office layout. We could still make a two-dimensional layout of each component in one view and then project them. We would then have to move and rotate the parts to assemble the chair. It is easier to use the views. This model can be created using different views to get the correct orientation of the components for assembly.

In situations such as this, you have to look at the model as a number of separate parts. The first step is to break the model into these separate components.

Figure 6.14 shows an exploded view of the chair with descriptions of the elements forming the parts. Using standard views makes the task of creating the chair much more simple.

You should be able to construct this model with the information supplied in figures 6.13 and 6.14.

After placing a center line, as a reference, in a FRONT or RIGHT view the other components can be created as follows:

- The seat in a TOP view.
- The backrest in a FRONT view.
- The back support in a LEFT view
- A leg in either a FRONT, BACK, RIGHT or LEFT view. This is then copied and rotated 90 degrees to form the other legs.
- The support shaft and roller castors are cylinders and can be constructed in any view. We can use cylinders with radius. All that is needed is a center line to snap to. Only one castor needs to be created this way. The others are rotated copies of the original. Remember when rotating elements or copies that the axis is always the Z axis of the view or screen. This is identical to how elements rotate in 2D files.

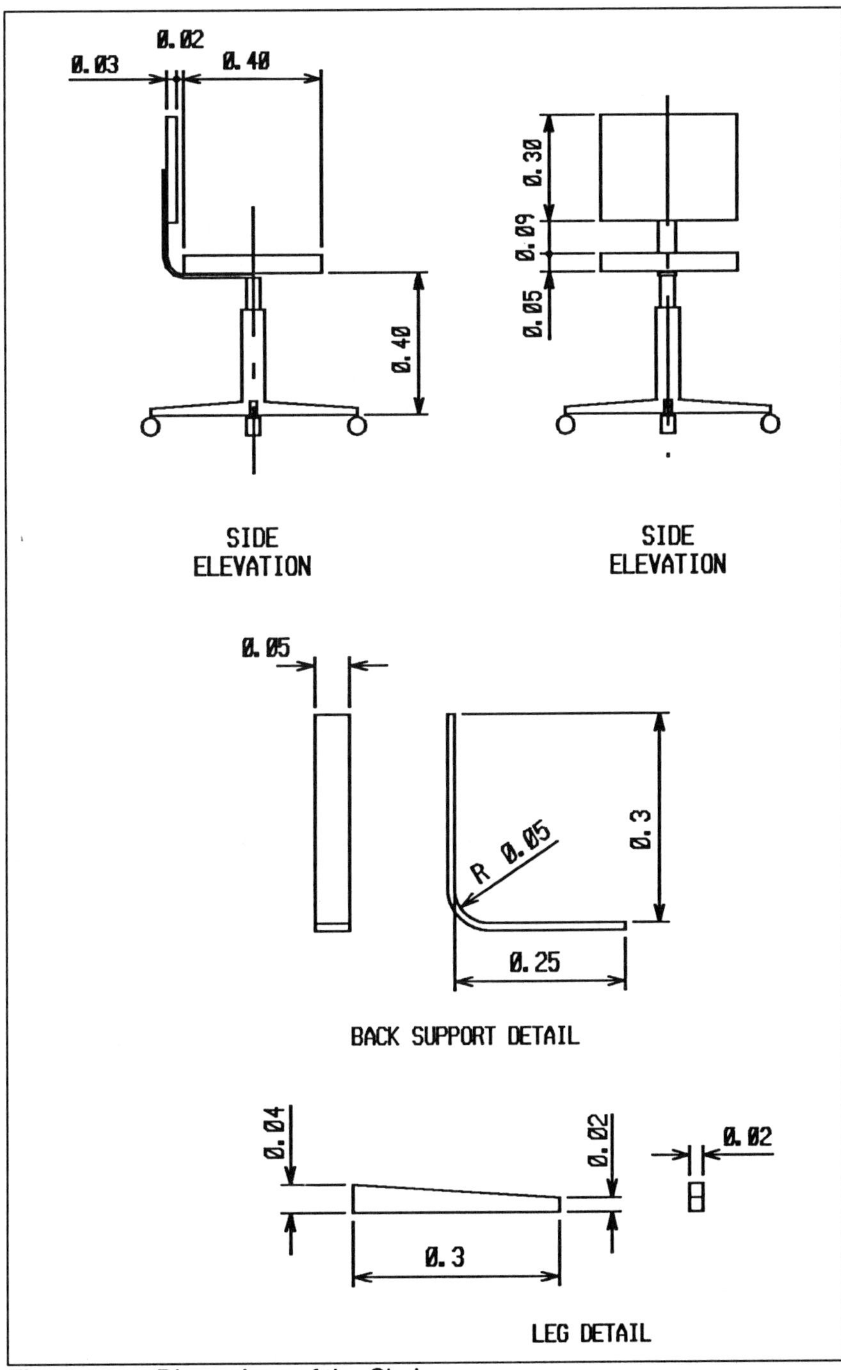

Figure 6.13 Dimensions of the Chair

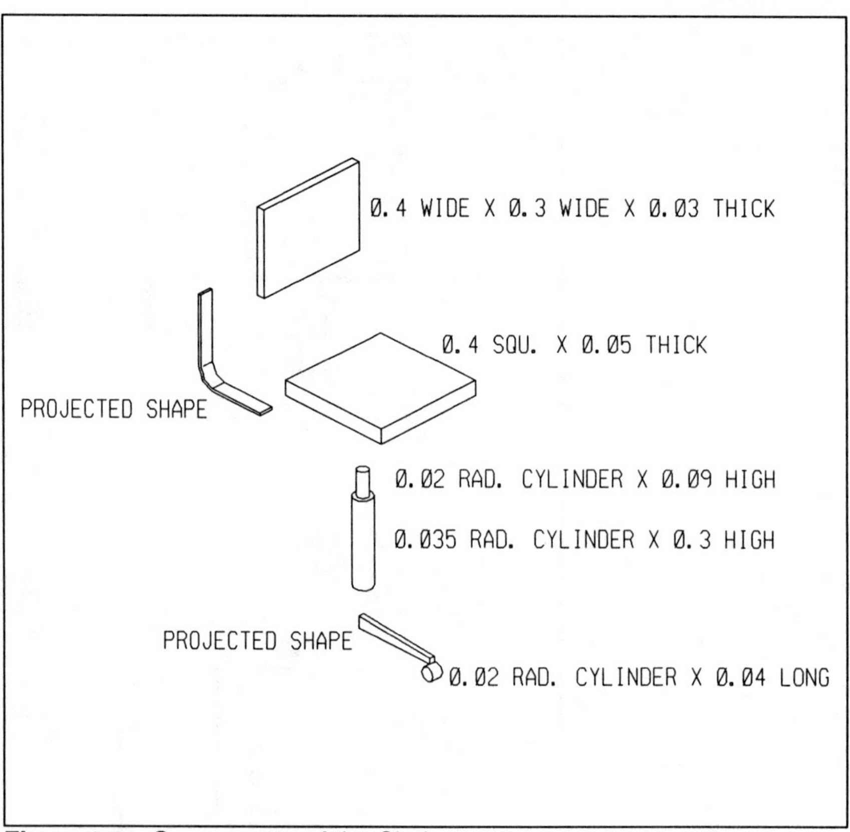

Figure 6.14 Components of the Chair

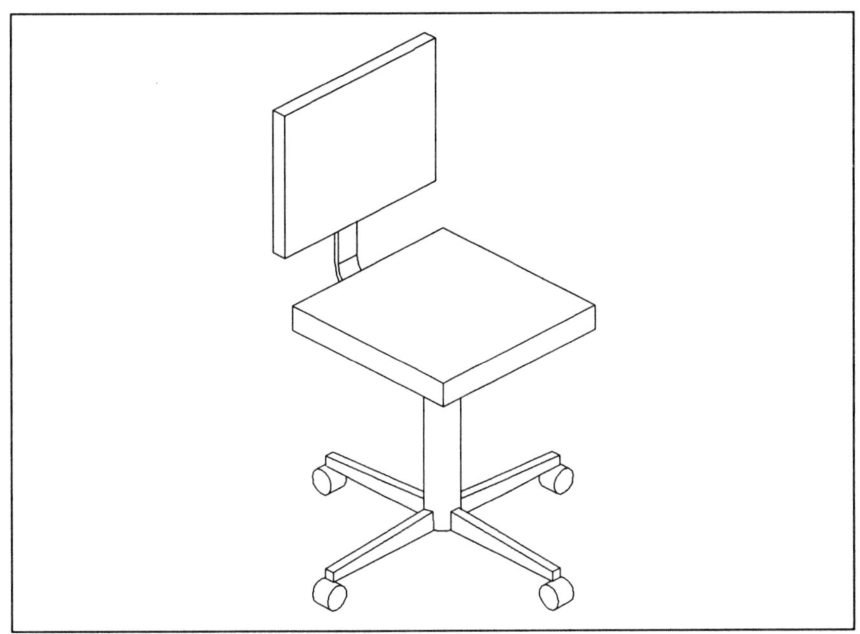

Figure 6.15 The Finished Chair

When you have completed the chair, make a cell of it and add it to your library.

- When you create the cell, think about where it would be best to put the origin.
- Also, set the display depth to just include the elements for the cell.

Figure 6.15 is a view (after hidden line removal) of the finished chair.

Now that you have cells of the desk and chair, go back to the office design file and place furniture in the offices on the 3 floors.

- Place the cells on a different level to the partitioning and floor layout.
- Where the offices are facing the same direction, fence copy can be used to place the furniture in identical positions.
- You can separate the floors with display depth, or with level control.

That completes this exercise (and chapter). You should be getting quite confident with 3D work now. You probably will have realised that most objects can be simplified into a number of basic components. If you are proficient at constructing the basic components, it is only a small step to being able to solve the more complex problems that arise from time to time.

In the next chapter we will be looking at advanced techniques. If you have progressed through the book from the start, then you will already know most of the tools that we will be using.

7 : Advanced Techniques

In chapter 6, we looked at some methods for creating simple 3D models. Here, we will look at techniques for creating complex models. That is, models that require more than the simple 3D operations.

We will look at, in more detail, techniques for projecting with active angle and active scale. These techniques are important for MicroStation version 3.3 users. With version 4, we have B-spline constructions, which can be used in most cases, instead of projecting with active angle or scale.

Creating Complex Models

To help illustrate advanced 3D modelling techniques, we will first create the model shown in figure 7.1. While the model looks simple, it requires some complex 3D operations to complete. We will construct it using methods and tools available with version 3.3. Then we will look at alternatives that could be used with version 4. There is, of course, more than one way to create this model. The methods shown here are not necessarily the 'right' way, nor are they the only way. If you think of other methods, try them. The idea of this book is to open the doors to 3D modelling with MicroStation. The right way to do things is the way that you feel most comfortable with, that gets the desired results.

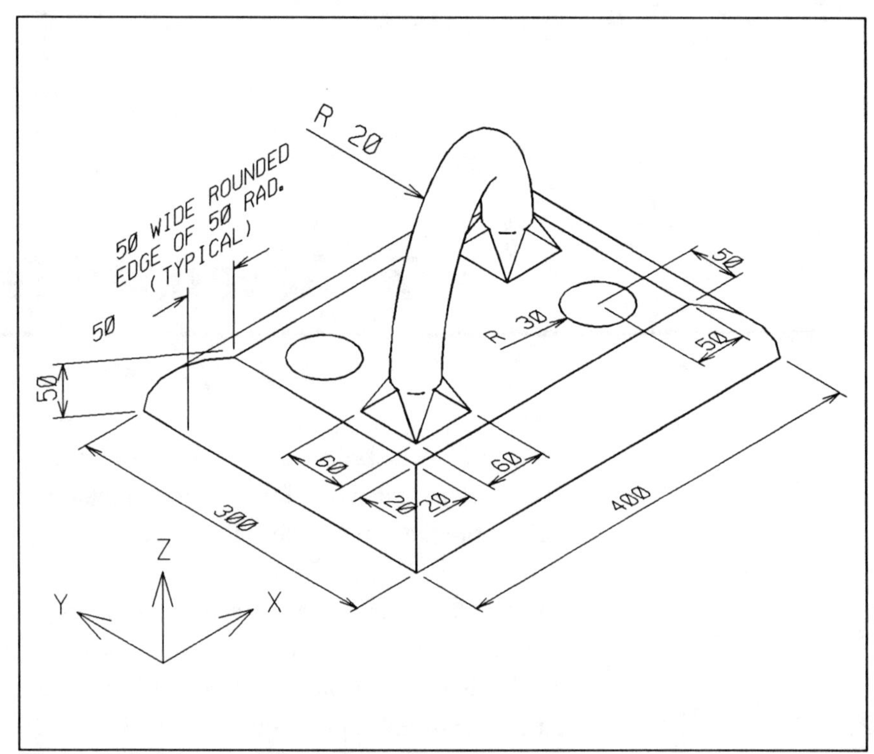

Figure 7.1 Advanced Techniques Exercise

For this model we will be working within the limits of the 1000x1000x1000 cube defined by the 2 active points in your training file. This will allow you to use your training file if you wish. Version 3.3 users can leave level 1 on, with the text to indicate the views. You will need to ZOOM IN or use WINDOW AREA at times to see various parts of the model more clearly as you work. If you decide to go ahead without the aid of the active points, just remember that a FIT may be required now and then. Use FIT when elements disappear partly, or do not appear at all, in a particular view.

As we did with creating previous models, the first thing to do is set up the views as you want them. This is not mandatory but it is a good habit to adopt. If you always set up your views before starting, there is less chance that you will make the mistake of commencing in a view with the wrong orientation.

Because version 3.3 of MicroStation does not have some of the more sophisticated construction tools of version 4, we have to use work-arounds for creating various shapes. Version 4 users can use these methods also, and they may find them useful for some situations. After this section, using version 3.3 methods, we will construct the same model with version 4 tools.

Version 3.3

To start, we will break the model down into separate, less complex, components as in figure 7.2. We can see that it consists of 5 basic items. One of these, the handle base, will be further simplified later.

We will construct the model in order of MAIN BODY, FRONT EDGE, SIDE EDGE, HANDLE BASE and HANDLE.

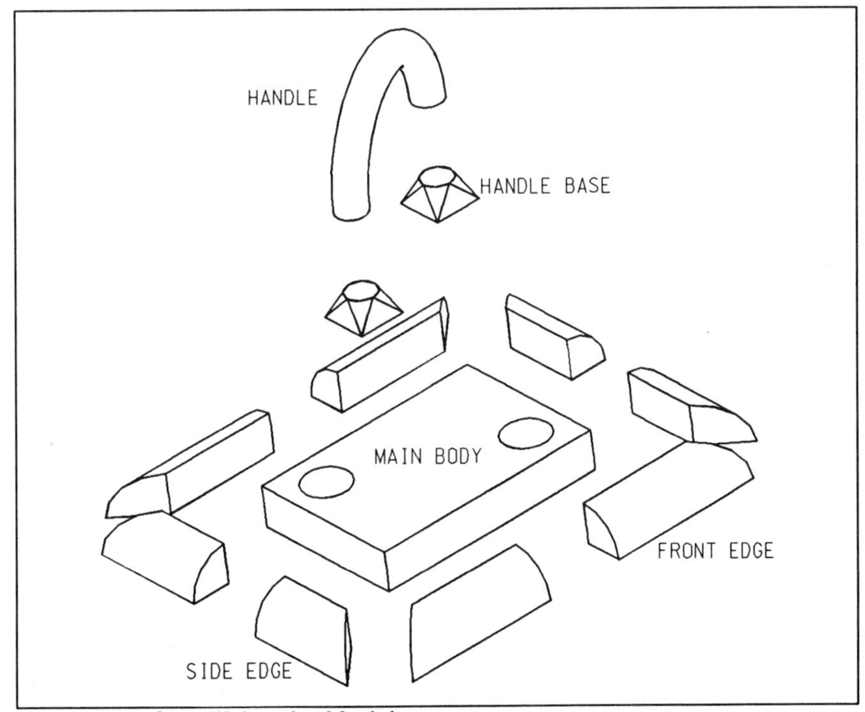

Figure 7.2 Simplifying the Model

The Main Body

Make the active level 20 and turn the display off for all other levels (except level 1 if required) in all views.

Use a TOP view (i.e., make the TOP view current).

- Place a 300x200 block with
  Point 1 - XY=400,400,400
  Point 2 - DX=300,200
- Place 2 circles of radius 30
  Circle 1 - DX=50,-50 from the top left corner of block
  Circle 2 - DX=-50,50 from the bottom right corner of block

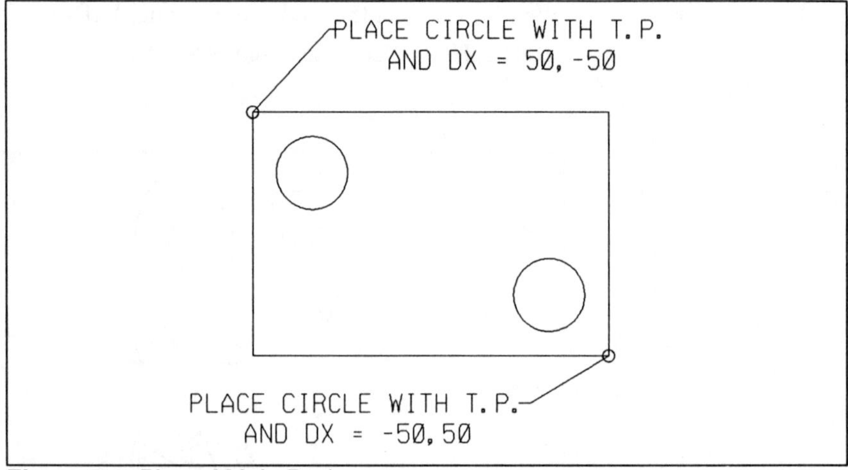

Figure 7.3 Plan of Main Body

Now, use the *Group Holes* tool to make the bottom surface containing the two holes.

Key in GROUP HOLE or select the tool in the Chain palette (ver.4) or from the sidebar menu (Utils - GHole).

- Identify the block as the solid element.
- Identify the 2 circles, when prompted, as the holes.

This creates the orphan cell that we will project to form the main body.

We want to project this cell as a solid (i.e., ACTIVE CAPMODE ON), with ACTIVE ANGLE = 0 and ACTIVE SCALE = 1.

With version 4 these parameters can be checked, in the dialog box, after selecting the tool. With version 3.3, the parameters are set prior to selecting the tool.

Select the *Construct Surface/Solid of Projection* tool from the menu.

° Identify the cell in any view (with a data point only, if you find that a tentative followed by a data point doesn't work).

° Key in DL = ,,50 for the projected distance and direction (the direction being in the positive Z direction of the model).

The result should be the main body, as in part of figure 7.4 (ISO view).

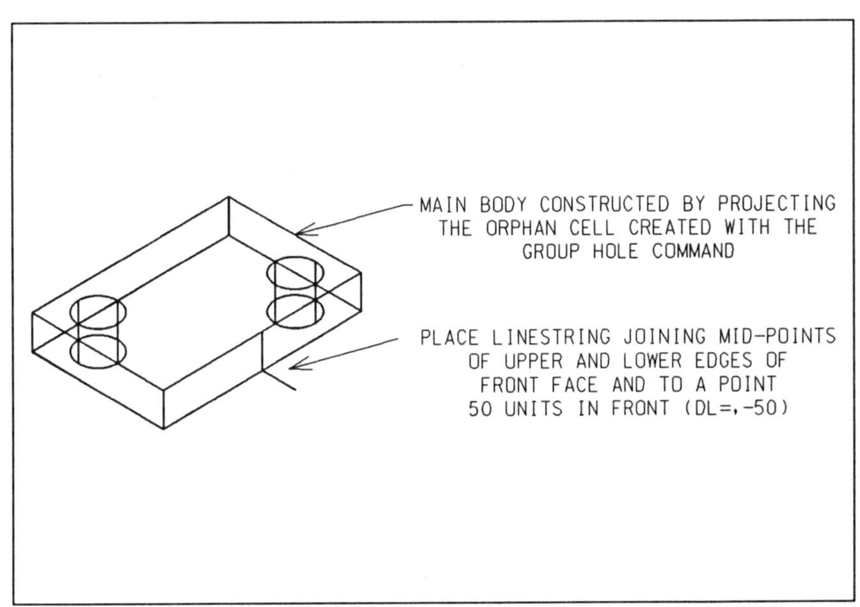

Figure 7.4 First Part of Model

The Front Edge

We form the front edge section from a projection of a complex shape. We can project the shape with an ACTIVE ANGLE and an ACTIVE SCALE to make the mitred corners.

- Set the active level to 21 (leave level 20 on as well).
- Set the SNAP DIVISOR 2, (KY=2), to allow for snapping to the middle of elements.
- Place a linestring from the mid-point of the upper front edge to the mid-point of the lower front edge. From here, to a point 50 units in front of the front edge (DL=,-50). Use figure 7.4 as a reference.

We can now place an arc by center with 3 points. This may be done in any view because an element like this requires a minimum of 3 points to place it. The 3 points define both the arc and its orientation.

Select the *Place Arc by Center* tool.

- Identify the outer end of the linestring for point 1.
- Identify the mid-point of the linestring for point 2 (the center).
- Identify the third point of the linestring for the other end of the arc.

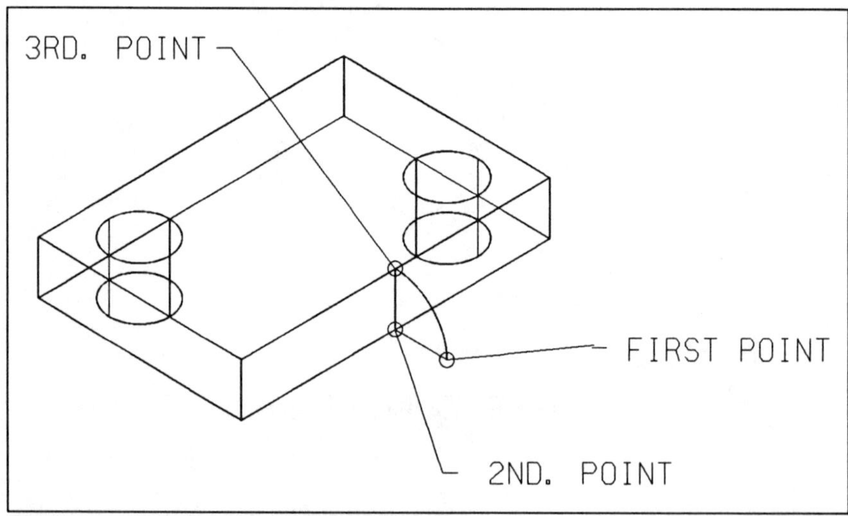

Figure 7.5 Placing an Arc in the ISO view

To construct half the front edge, we chain the arc and the linestring together, to form a complex shape. We can then project this shape to form the solid.

Select the *Create Complex Chain* tool.

° Chain the arc and the line together.

We will project this chained element from the center of the front (where it is now), to the left end of the same edge of the main body, as seen in a TOP view.

The problem is that we want to form a mitred corner. To do this we need to:

– Specify an ANGLE through which the element rotates as we project it.
– Set an ACTIVE SCALE so that the element increases in the X and Y directions to form half the corner. We do not want any increase in the Z direction (height).

Setting the Scale Factor

We want the element to rotate through half the total angle of the corner. Half the total angle, here, is 45 degrees. It is easy to calculate, mathematically, that the scale factor is 1.4142. Sometimes it is not so obvious. In these situations, we can let the system set the scale by distance.

For our example, we could have done this as follows (use figure 7.6 as a guide):

° We know that the offset of the sides is 50 units in both directions, so place a 50 x 50 block.
° Rotate the view by 3 points, so that a diagonal of the block is parallel to the X-axis of the screen.
° Place a 50 long horizontal line from one corner of the horizontal diagonal. This will represent the existing length for setting the scale.
° Key in ACTIVE SCALE DISTANCE or select from a menu.
° Set the scale using the two diagonal points and the end of the line.
° This will set the scale to 1.4142 on all axes. Set the Z scale back to 1 with the key-in ZS = 1. We don't want the element to be scaled in the Z direction as it is projected, we want it to retain its current height.

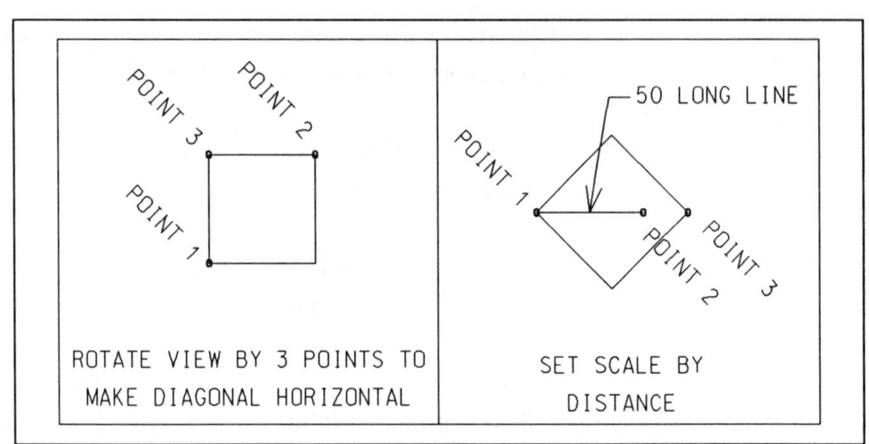

Figure 7.6 Set Scale by Distance

Projecting with Active Scale and Active Angle

We are now ready to create the edge. We have to project the element in a TOP or BOTTOM view. This is because the axis about which we want the rotation is parallel to the Z-axis of the model. While we can identify the element in any view, we must use the correct view for selecting the projection distance. This is because we are rotating and scaling the element as we project it. Remember, *all scaling is about the X, Y and Z axes of the screen* and *all rotations are about the Z axis of the view or screen.*

Do the following, using figure 7.7 as a guide:

- WINDOW AREA close to the model in the ISO and TOP views to help with identifying elements.
- Set the ACTIVE ANGLE to -45 degrees (AA=-45). This can be set after selecting the tool in the 3D palette, with version 4. We will create the left side first and we know that rotations about the Z axis are negative in the clockwise direction (right hand rule, chapter 2).
- Select the *Construct Surface/Solid of Projection* tool.
- Version 4 users, set Active Angle to -45 degrees.
- Identify the chained element, in the ISO view, with a tentative to the end of the line touching the lower part of the main body.

° In the TOP view, tentative to the bottom left corner of the main body. Check in the other views to ensure that the bottom of the main body highlights. When the correct element highlights, a data point will complete the construction.

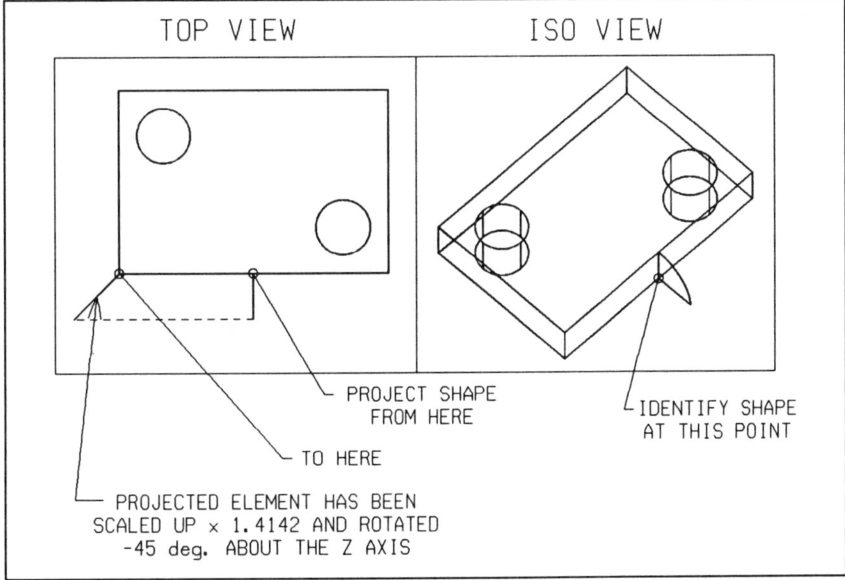

Figure 7.7 Projecting with Active Angle & Scale

This forms half of the front edge of the model. We can MIRROR COPY this element to form the other half of the front edge and the back edge.

Using the TOP view to perform the MIRROR COPY operations:

° Create the other half of the front edge using MIRROR COPY VERTICAL about the center of the main body.

° For the BACK EDGE, use MIRROR COPY HORIZONTAL, either individually, or in a fence, about the center of the main body again.

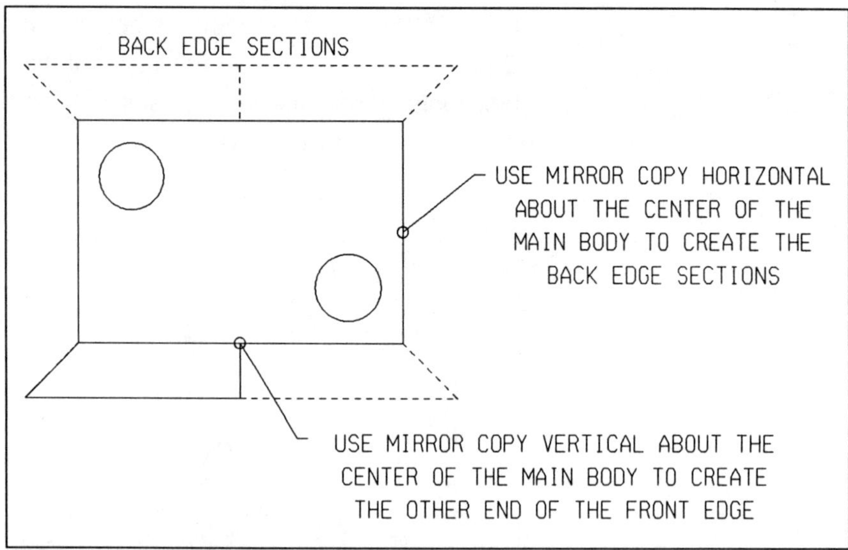

Figure 7.8 Using Mirror Copy

The Side Edge

The SIDE EDGE is similar in construction to the FRONT EDGE and should be no trouble. Rather than create a new shape for the section, use the same shape that was used for the front edge.

Some hints:

- Rotate the shape used for the front edge through 90 degrees.
- Move it to the center of the right edge of the main body.
- Again, use the TOP view for the actual projection.
- The SCALE FACTOR will remain the same.
- The only other point to watch is which way to project the shape. You have to ensure that it rotates the correct way as it is projected. If you make a mistake, UNDO is always there.
- Once you have created half the right edge, use MIRROR COPY to create the other end of the right edge and then the whole left edge.

Once you have finished, it is a good idea to reset the ACTIVE ANGLE to 0 and the ACTIVE SCALE to 1.

The Handle Base

This part of the model can be further broken down, to simplify construction, as shown in figure 7.9. The base has a block, 4 triangular shapes and 4 arcs that we project with a scale factor. We place the circle, initially, as an aid for creating the arcs. Later we will use it to create the handle.

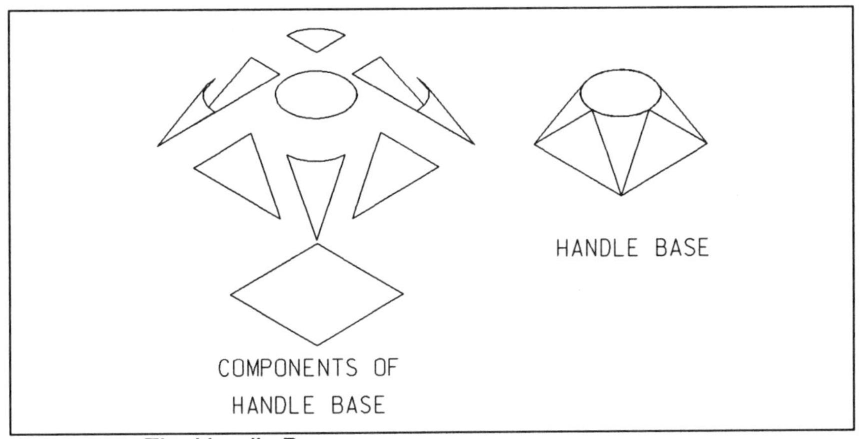

Figure 7.9 The Handle Base

To start:

- Make the active level 22 and turn off level 21. We don't need to see the edges of the model. It doesn't make much difference to us in a model this small and simple. When we are working in more complex designs it reduces confusion to have any unnecessary levels turned off. Screen update times, also, improve by keeping the least amounts of information displaying, while creating the design.

In a TOP view, place a 60 x 60 block on the top surface of the main body 20 units clear from the left and front faces.

Select the *Place Block* tool from the menu:

- Tentative to the left front corner of the upper face of the main body. Check in other views to be sure that the tentative is to the upper face (or check that the Z value of the read-out is 450).

- Key in DX or DL=20,20 for the first point (TOP view has the same axes as the model, so DX and DL are identical).
- Key in DX or DL=60,60 for the second point.

Now place the 20 radius circle, 30 units above the center of the 60 square block. Still use the TOP view.

Select the *Place Circle by Keyed-in Radius* tool:

- Key in 20 for the radius.
- Tentative to the front left corner of the block in the TOP view.
- Key in DX or DL=30,30,30 to locate the center of the circle.

If necessary WINDOW AREA in closer to the handle base, in both the TOP and ISO views, to make it easier to see what you are doing. Place the triangular shape forming the front face of the base.

Select the Place Shape tool:

- Identify the left front corner of base block for point one.
- Tentative to center of circle followed by DL=,-20 for second point.
- Identify right front corner of base for third point, then back to the left front corner to close the shape.

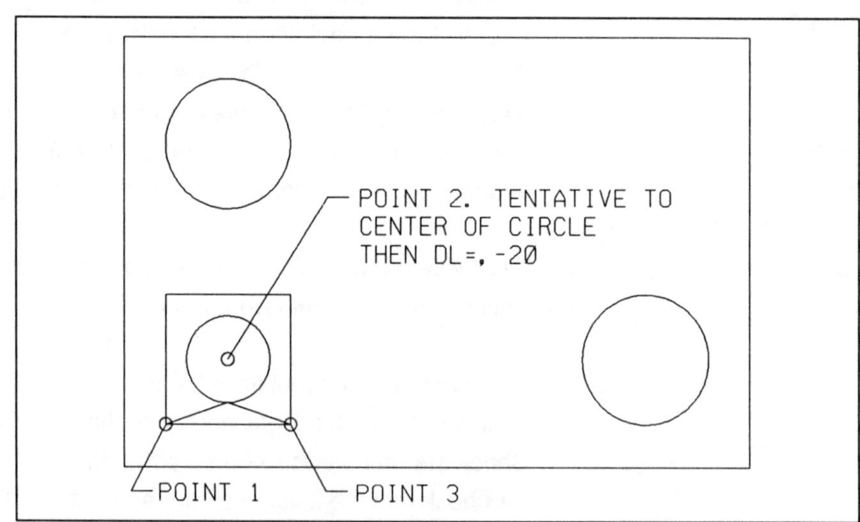

Figure 7.10 Placing the Triangular Shape

Note:

We could have used a snap to the key-point of the circle for the second point. This is O.K. where the key-points of the circle align exactly with our orientation, but it may not always be the case and inaccuracies could result. We may have, for example, placed the circle in a TOP view that had been rotated a few degrees about the Z axis, for some reason. The key-points of the circle, then, would be slightly out of square for our construction.

To form the other 3 sides, use the TOP view.

- Set ACTIVE ANGLE to 90 degrees, and check that ACTIVE SCALE is 1.
- Select *Rotate Element by Active Angle (Copy)* tool.
- Identify triangle (in any view).
- Tentative to the center of the circle in TOP view for the AXIS OF ROTATION. When the circle highlights, instead of using a data point to accept, key in DL=0, three times. This instructs the system to use the tentative point as the axis to rotate and copy the element 3 times. Alternatively, DOS users can use DL=0|3, where '|' is the DOS pipe symbol (usually found on the same key as the backslash on PC's).

To place the arc we can use the ISO view and the *Place Arc by Center* tool. Arcs require 3 points to construct which also define the plane. The arc, therefore, can be constructed in any view.

As some of the elements appear collinear, rotate the ISO view 15 degrees about the X axis (enter RV=15 followed by a data point in the ISO view). Also, make the active color something other than that of the circle. Then, when the arc is placed on top of the circle, you will be able to see it.

Select the *Place Arc by Center* tool:

- Identify the top point of the left triangle for point one.
- Identify the center of the circle for point two.
- Identify the top point of the front triangle for the 3rd. point to create the arc.

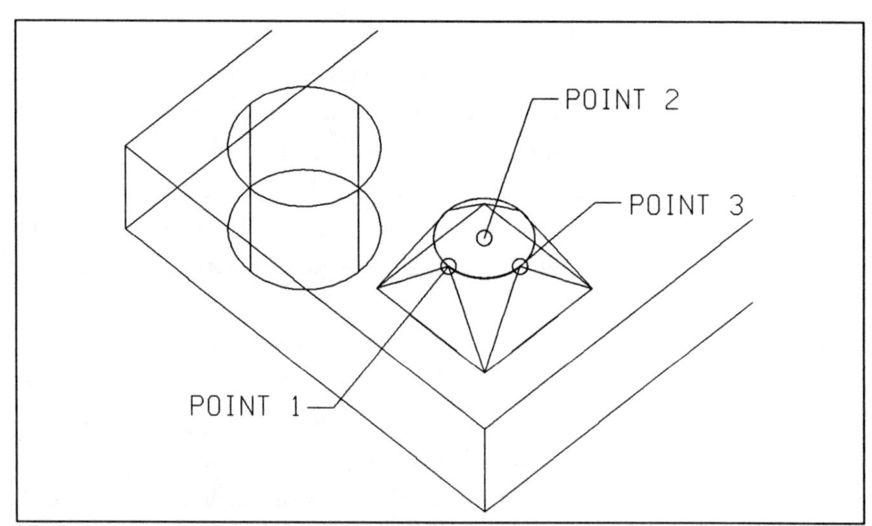

Figure 7.11 Placing the Arc in rotated ISO view

We are ready to construct the curved corner section that fills in the gaps between the triangles. A simple way to do this is to make the ACTIVE SCALE very small. When we project the arc it will be extremely tiny at the other end of the projection. We can use any view to perform the projection. There will be no active angle set and the scaling is in all directions (X, Y, and Z).

Make sure that ACTIVE ANGLE is 0. Set ACTIVE SCALE 0.0001.
NOTE: If your data readout is set to 3 decimal places, then the scale factor will show up as 0.000 on the screen.

Select the *Construct Surface/Solid of Projection* tool:

- Identify the arc at one of its extremities.
- Specify the distance for the projection by identifying the point where the two triangles meet at the bottom in the front left corner of the block, as figure 7.12 shows.

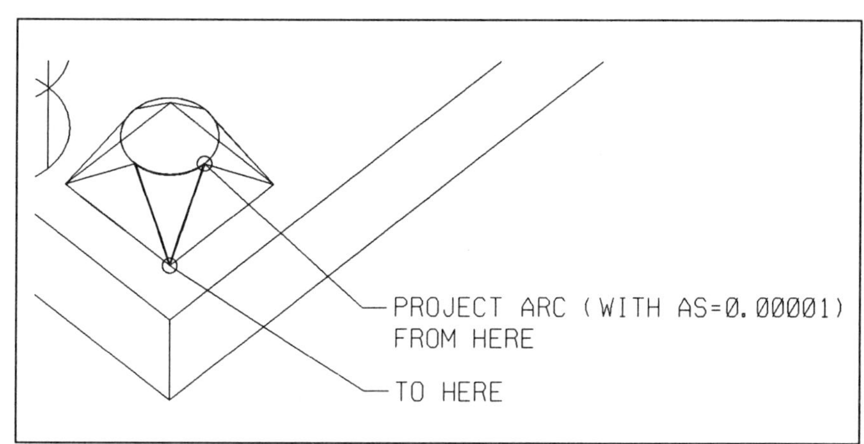

Figure 7.12 Projecting with Active Scale

Now, using a TOP view and an ACTIVE ANGLE of 90 degrees:

° Select the *Rotate Element by Active Angle (Copy)* tool.

° Copy the newly created surface into the other 3 corners, just as we did for the triangular surfaces.

Once you are sure that the base is correct, fence copy it to form the other base. The TOP view is the simplest to use for placing the fence so that it doesn't contain any other unwanted elements. Any view can be used to specify the distance and direction to move the contents. One way would be to ZOOM OUT in the ISO view so that you can see the main body. Then use this view to specify where the copy is to go.

To complete this operation:

° Place a fence, in the TOP view, around the handle base.

° Select the *Move Fence Contents* tool.

° Tentative to the back right corner of the block at the base of the handle support to specify the 'from point'.

° For the 'to point', tentative to the back right corner of the upper surface of the main body, then key in DL=-20,-20 for the actual location. We require DL here, because we are specifying model coordinates in a view other than TOP.

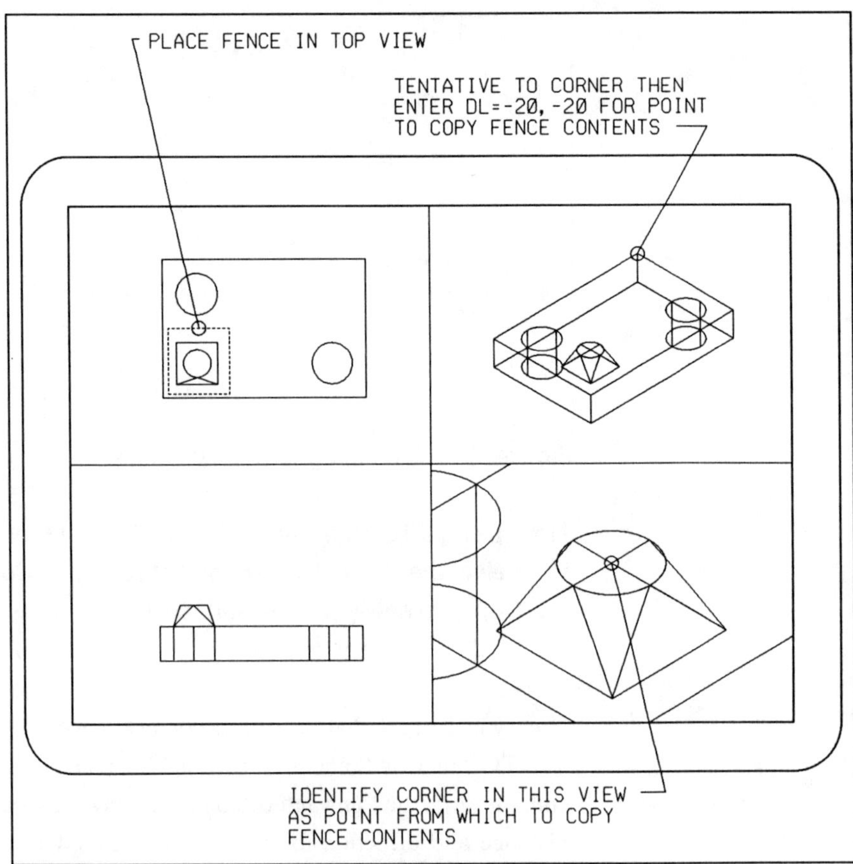

Figure 7.13 Using the Views to Fence Copy

Figure 7.13 shows a typical screen set up for performing this operation. By using the views we can complete the fence copy quickly and efficiently. As has been discussed before, view manipulation is very important in 3D work.

The Handle

We will construct the handle with a SURFACE OF REVOLUTION from the circle that we placed on the handle base. First, we need to place a construction line that we can use to locate the AXIS OF REVOLUTION.

Whether you use a construction class element or a primary class (normal) element is up to you. Many operators use primary elements always. The reason is that they don't have to remember to change from primary to construction and vice-versa. They normally have specific levels where they place their construction elements. These can be turned off or deleted when they are no longer required.

For the construction element, here, place a line joining the centers of the circles on the handle bases:

° Make your Active Level 60 for placing the line and turn off level 21.

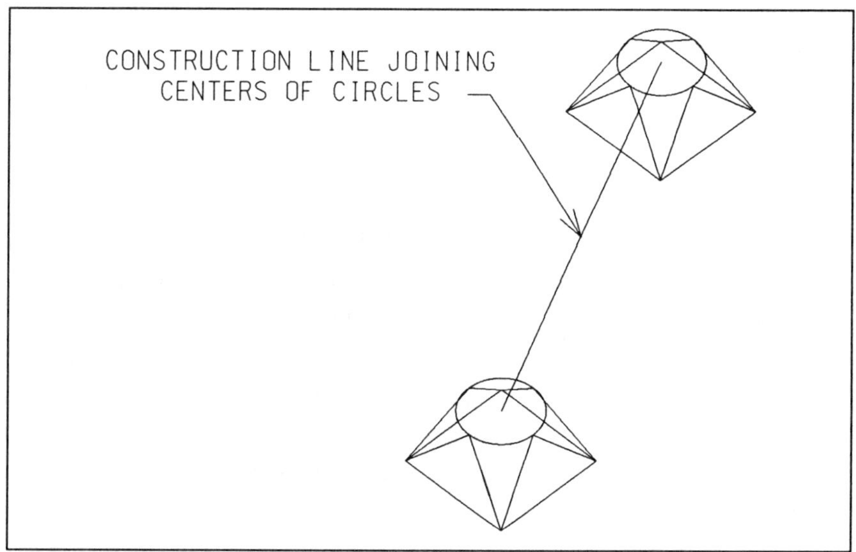

Figure 7.14 Placing the Construction Line

Surface of Revolution in a Non-Standard View

To create the handle with a surface of revolution we first have to rotate the view. This is to orientate it so that the AXIS OF REVOLUTION will be parallel to the Z-axis of the view or screen. On this occasion, we cannot use a standard view, because the required axis does not align with any of the X, Y, or Z axes of the design cube.

Key in ROTATE 3PTS or ROTATE VIEW POINTS (version 4):

- Identify one end of the construction line for the first point and the other end for the second point that determines the direction of the X-axis.
- Key in 'DL=,,figure' for the direction of Y. The 'figure' is arbitrary. We can use any figure, to give a direction for Y (e.g., DL=,,10).
- Now select the view/quadrant of the screen that you will be using to perform the surface of revolution. Select any view except the ISO, as this is a good view to use for identifying the circle later.

It does not matter which end of the line you select first and second. The difference between the two options is that the model, in the rotated view, would be viewed from opposite sides. This makes no difference to the surface of revolution.

Create the handle:

- Check that SNAP DIVISOR = 2 (i.e., key in 'KY=2').
- Select the *Construct Surface/Solid of Revolution* tool.
- Key in 180 for the angle.
- Identify the circle on the base on the right hand side of the rotated view. Because we keyed in 180 (i.e., plus 180) for the angle, the surface of revolution will be created in an anti-clockwise direction.
- Select the AXIS OF REVOLUTION by snapping to the center point of the construction line we placed between the two circles.

That completes the construction of this model. Running Hidden Line Removal should show up any mistakes you may have made with creating surfaces.

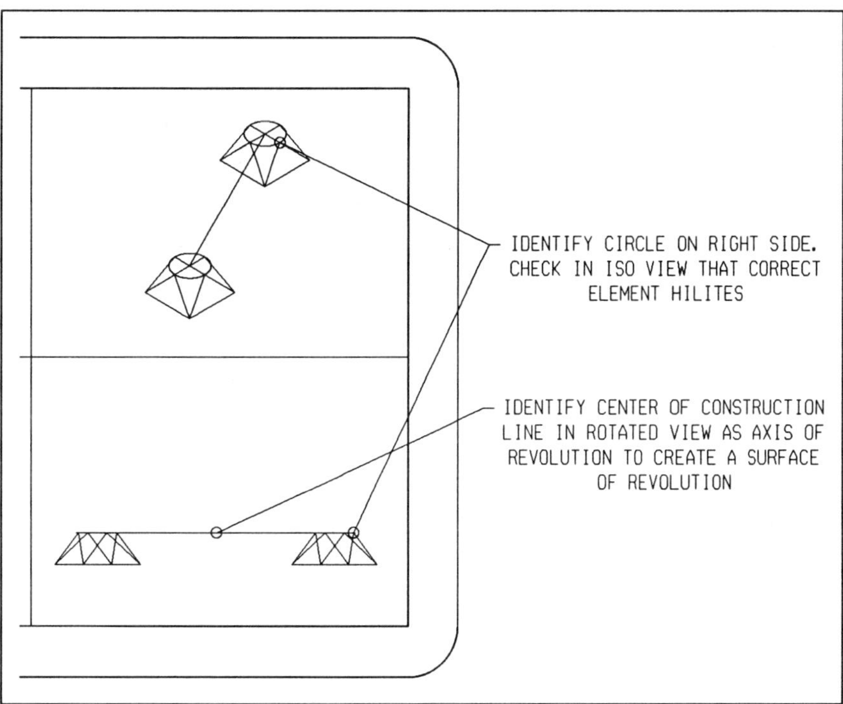

Figure 7.16 Creating the Handle

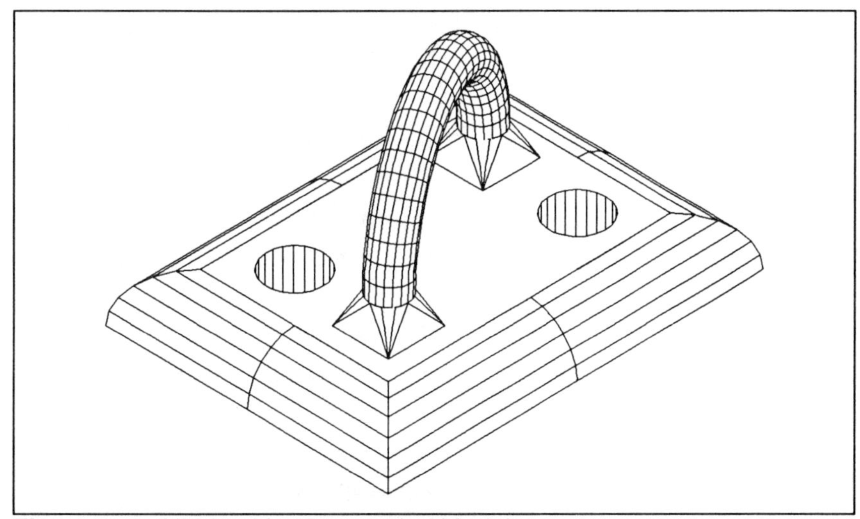

Figure 7.15 Hidden Line Removal of Model

Version 4

To construct the same model with version 4, we can use different techniques. Again, we simplify the model by reducing it to a number of basic items as in figure 7.17. You will notice that we don't have as many separate parts as we did for version 3.3. The edge sections can be constructed as single pieces, for example, as can the handle bases. For both of these we can use B-spline surface constructions.

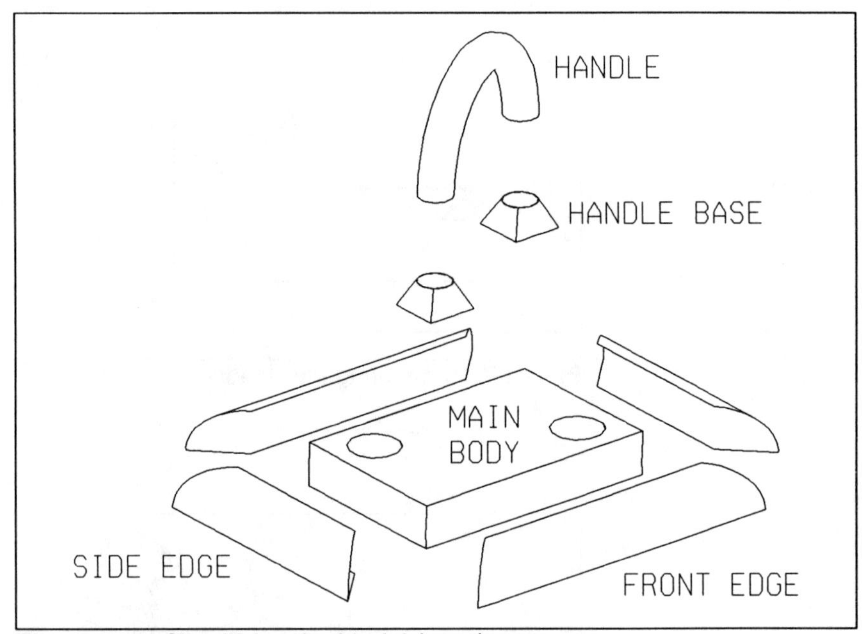

Figure 7.17 Simplifying the Model (ver.4)

Version 4

Set up your screen as before with views 1 to 4 as ISO, TOP, FRONT and RIGHT. Make the active level 30. Turn all other levels off (except level 1 if required).

Make the TOP view current.

- Place a 400x300 block with
  Point 1 - XY = 350,350,400
  Point 2 - DX = 400,300

For the main body use the same method as we did for the version 3.3 method shown on page 7-4 and 7-5.

Still using the TOP view, place a 60x60 block for the handle base, on the top face of the main body (as we did with the version 3.3 method).

- Tentative to the left front corner of the upper face of the main body. Check the other views to ensure the upper face highlights.
- Key in DX=20,20 for the first point.
- Key in DX=60,60 for the second point.

Place a 20 radius circle, for the top of the handle base, 30 above the center of the 60x60 block.Place the circle in the TOP view as follows:

- Tentative to the front left corner of the block.
- Key in DX=30,30,30 to locate the center of the circle.

We now have the basic elements from which we will construct the rest of the model. Figure 7.18 shows the model as it should look now, ready to commence construction of the sides, handle bases and handle.

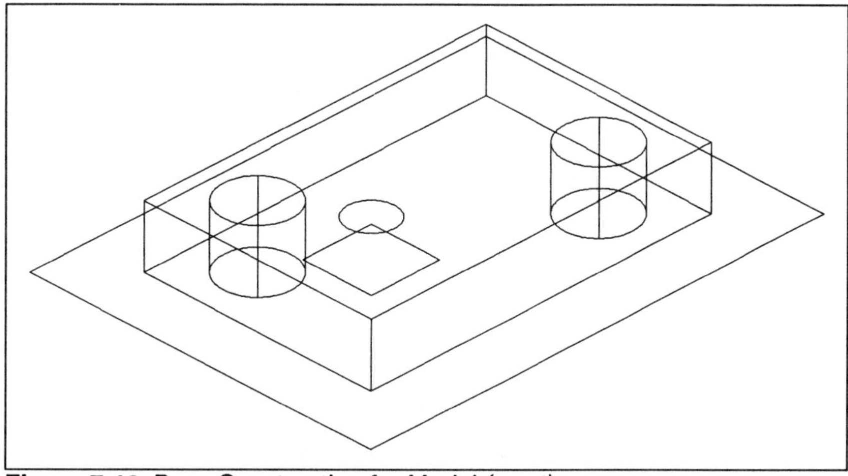

Figure 7.18 Base Construction for Model (ver.4)

To construct the front edge we will use one of the B-spline constructions that are available with version 4. We will construct the corner sections as chained elements formed from a line and a quarter ellipse.

Set the active level to 31, and using an ISO view:

- Place a line joining the lower front right corner of the main body to the corresponding corner of the first block that we placed (figure 7.19 (A)).

Using the same ISO view, place a quarter ellipse:

- Select the *Place Quarter Ellipse* tool
- Place the ellipse as indicated in figure 7.19 (B).

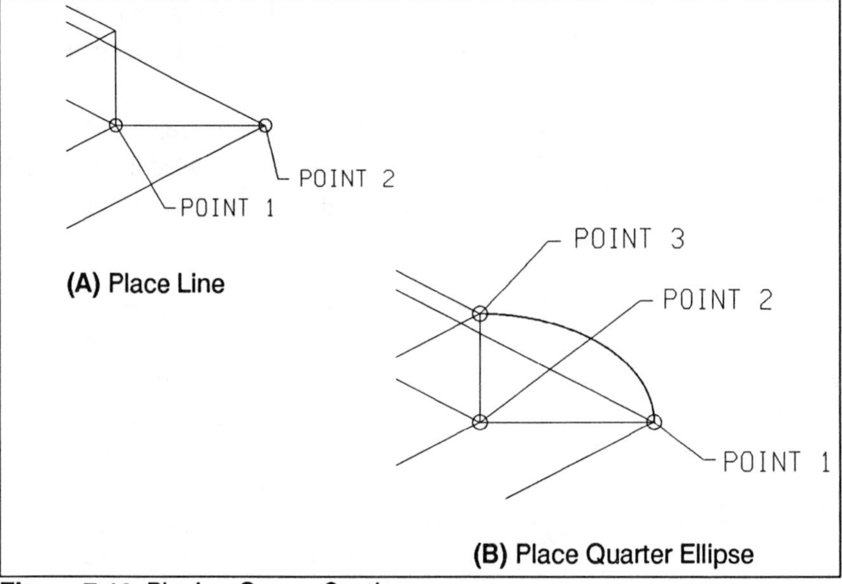

Figure 7.19 Placing Corner Sections

To make the corner section, chain the line and the quarter ellipse together.

- Set keypoint snap to 2 (KY=2).
- Using *Mirror Element About Vertical (Copy)* and *Mirror Element About Horizontal (Copy)* tools, in a TOP view, copy the chained element to the front left and back left corners of the model. Snap to the center of the main body to copy.

Front & Side Edges (version 4)

We now have three corner sections in place. To create a surface between them we can use the *Construct B-spline Surface by Edges* tool. This tool is found in the 'Derived Surfaces' palette which is chosen in the '3D B-splines' sub-menu as shown in figure 7.20.

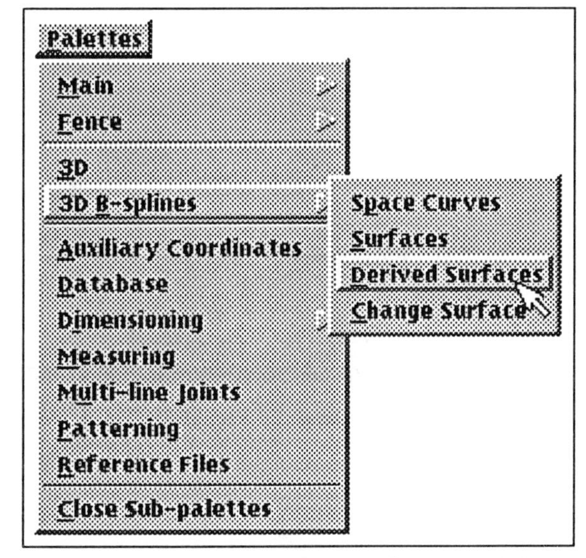

Figure 7.20 Selecting Derived Surfaces Palette

The *Construct B-spline Surface by Edges* tool is on the far right of the 'Derived Surfaces' palette. With this tool, one option is to create a ruled surface between two selected elements. We will create a surface between two of the corner sections.

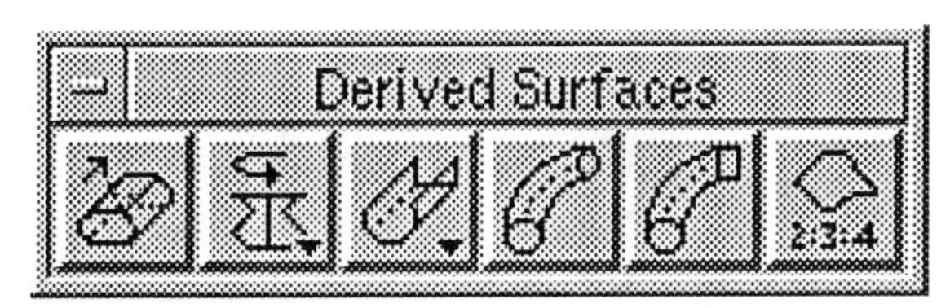

Figure 7.21 Derived Surfaces Palette

With the active level 31, turn off all other levels in the ISO view, to make the element selection easier.

° Using the *Element Selection* tool, select the two chained elements forming the corner sections of the front edge.

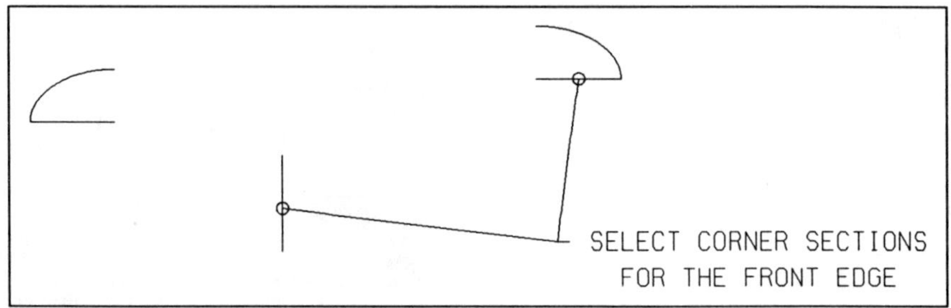

Figure 7.22 Selecting Sections

° Select the *Construct B-spline Surface by Edges* tool from the palette. Notice the surface that is constructed between the selected elements.

° Enter a data point to complete the construction.

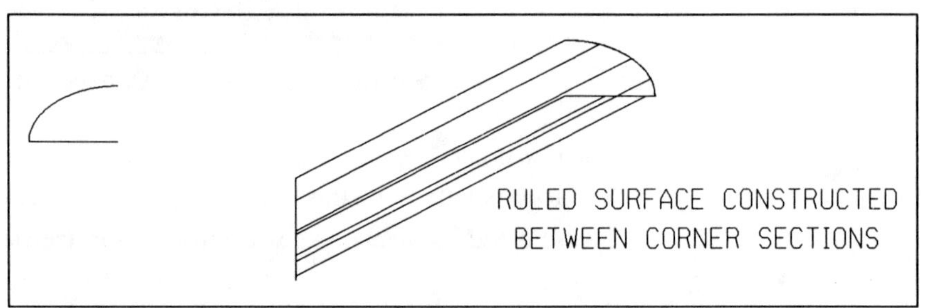

Figure 7.23 Constructing Surface

Use a similar technique to create the left edge.

° Use the *Element Selection* tool to choose the two corner sections for the side edge.

° Select the *Construct B-spline Surface by Edges* tool from the palette.

° Enter a data point to complete the construction.

Use *Mirror Element About Horizontal (Copy)* and *Mirror Element About Vertical (Copy)* tools, in a TOP view, to create the back and the right edges.

The Handle Base (version 4)

Again, we can use the *Construct B-spline by Edges* tool to create the handle base. We can construct a B-spline surface between the block, at the bottom of the base, and the circle at the top. Before we do, there are two points that we need to know.

- The ruled lines of the surface are constructed between points on each element.
- These points are in a particular order, starting from a key-point.

Because of the above, it is possible that our surface will have a twist in it, unless the starting points are directly opposite one another. Figure 7.24 shows the handle base the way it should look when completed.

Because of the way we placed the circle, our construction will not look like this initially.

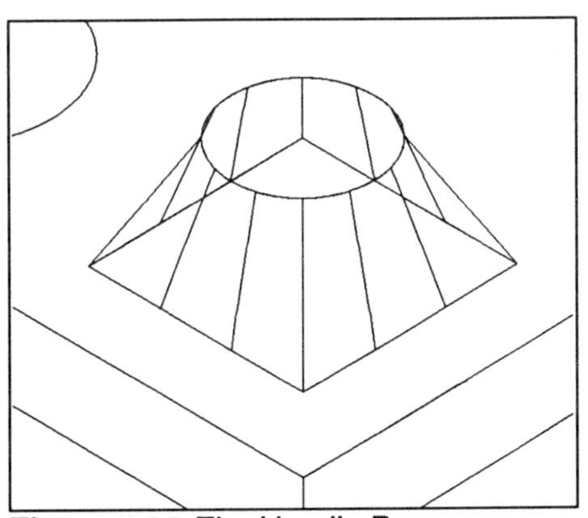

Figure 7.24 The Handle Base

Try constructing the surface for the base as follows:

Window area in closer to the handle base in the ISO view:

- Using the *Element Selection* tool, select the block and the circle forming the top and bottom of the handle base.
- Select the *Construct B-spline Surface by Edges* tool.
- Notice how the ruled surface is drawn between the elements. Enter a data point to complete the construction.

Your ruled surface for the handle base may look like that shown in figure 7.25. Here we can see two views of the construction, a TOP view on the left and an ISO view on the right. We can see the ruled lines, indicating the surface, joining the two elements. Also shown are the starting points for the two elements which were used to create the ruled surface.

As can be seen, the starting point for the circle is almost on the other side of the circle. It is rotated 135 degrees from the starting point of the block. To solve the problem we will rotate the circle minus 135 degrees.

Delete the ruled surface.

- Set Active Angle to -135.
- Select the *Rotate Element by Active Angle (Original)* tool.
- Identify the circle in a TOP view, at the circle center.

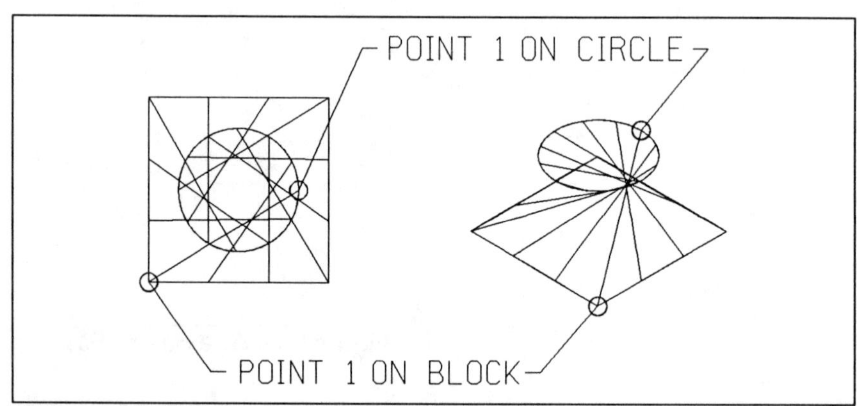

Figure 7.25 Ruled Surface with a Twist

This time, when you construct the ruled surface between the circle and the block, the two starting points should align.

° Construct the ruled surface again as we did previously.

Like the previous diagram, figure 7.26 (part A) shows two views of the ruled surfaces. This time the first points are aligned with one another.

In part B of the figure we see another alternative. While the starting points are aligned, the ruled surface is constructed between points taken in opposite directions about each element. Where this occurs, we can press RESET and the other alternative will be shown. A data point will then complete the construction.

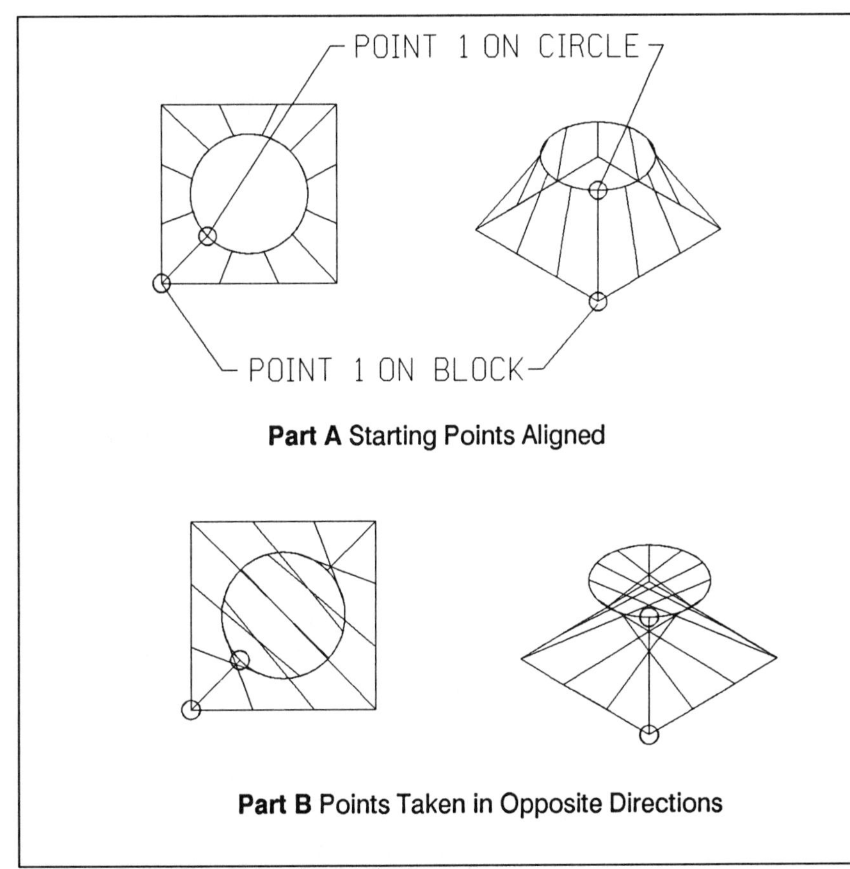

Part A Starting Points Aligned

Part B Points Taken in Opposite Directions

Figure 7.26 Alternative Ruled Surfaces

We will now look at, in more detail, projecting with active angle and active scale. You should know the basics from the previous section. B-spline surface constructions, available with MicroStation version 4, take away much of the need for using this 'work-around'. The techniques may be useful, however, for other problems that you may have to solve.

Projecting with Active Angle

When we project with active angle, the projected element is rotated by the specified amount (the active angle). Surfaces are created between the original and projected elements. For the exercise, assume that we have to create a chute, 800 long, with a 40 degree twist.

- Place a 600x400 block, in the TOP view, with its origin at XY=350,300,200.
- Set active scale to 1 (AS=1), active angle to 40 (AA=40) and Active Capmode to OFF (i.e., set Type to Surface with version 4).
- Project the block from the front left corner, in the TOP view, by a distance of 800 vertically (DL=,,800).

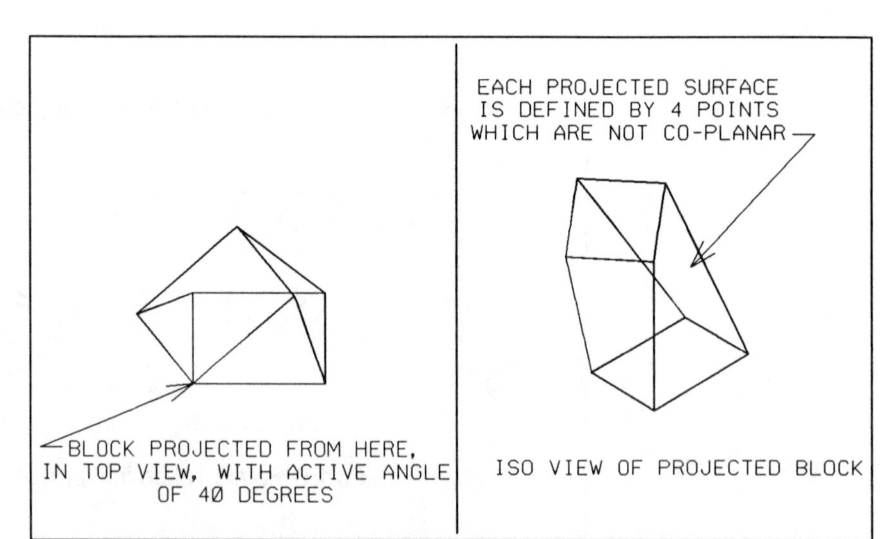

Figure 7.27 Projecting with Active Angle

Our original shape has been copied vertically 800 and, at the same time, rotated by the amount of the active angle, 40 degrees (see figure 7.27). The resulting surfaces between the original and projected elements, however, are not planar elements. The four key-points defining each of the projected surfaces are not co-planar. In other words, the surfaces are warped. They all have a twist in them, but it is not indicated. We could triangulate each surface as shown in figure 7.28. This may not be accurate enough, in which case we have to use a different method.

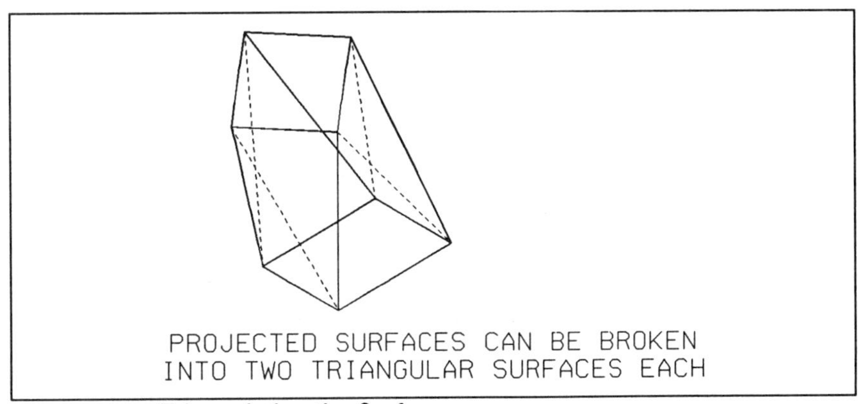

Figure 7.28 Triangulating the Surfaces

With a little extra effort we can show the surfaces more accurately, with a twist, by using the following technique.

First, copy the original block to another level and turn off the level with the original block and its projection.

- Make sure the active scale is 1 and active capmode is OFF.
- Make the active angle 4 degrees. We are going to project the element in 10 increments to create the full 40 degree twist.
- Project the block, from the front left corner, using the TOP view, by keying in DL=,,80. This is repeated 9 times. We can use the UP ARROW cursor key to recall the DL=,,80 key-in for the 9 repetitions. Another way is to use the DOS pipe symbol with the initial key in to specify the total repetitions. That is, for the project distance key in DL=,,80|10.

Our resulting projection is a truer representation of the model with twisted, or warped, surfaces. All we have done is broken the overall angle into smaller segments. This form of construction does take more disk space, but the result is a more realistic model.

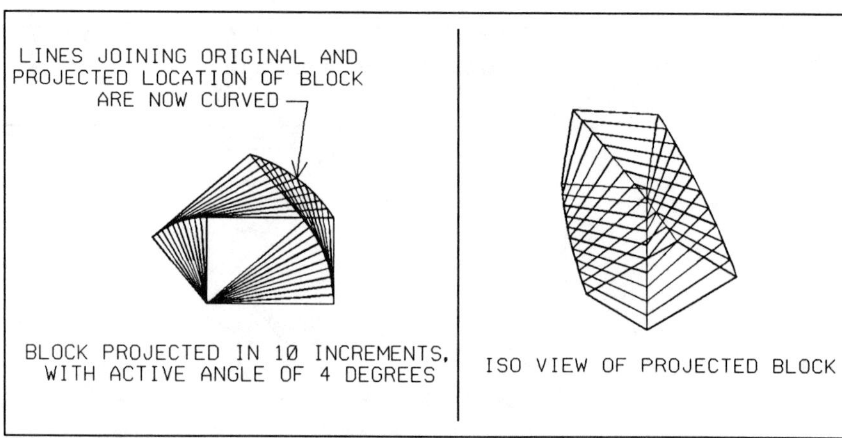

Figure 7.29 Increasing the Accuracy of the Model

When the twist is about a key-point of the shape, the above method is fine. If the twist is about a point that doesn't fall on the element we have to use a different technique again.

One way is to create a cell of the element with the origin located at the point about which we want the rotation to take place. When we project the cell we make sure that we identify the cell at its origin. To demonstrate we will project our block with a rotation about a point 250 in from the left and 150 back from the front edge.

Note:
This technique is necessary for version 3.3 users only. With version 4 we can use the FENCE SURFACE PROJECTION key-in to project the contents of a fence. We can specify the origin of the fence with a tentative to the element, followed by 'DX=250,150'.

Version 3.3

Copy the block to a vacant level. Make sure that you have a 3D CELL LIBRARY attached to the file.

° Set the active scale to 1 and the active capmode to off.
° Set the active angle, initially, to 0 degrees.
° Create a cell of the block with the origin at the point about which we want the rotation. That is, specify the origin with a tentative to the front left corner of the block followed by the key-in, DX=250,150.

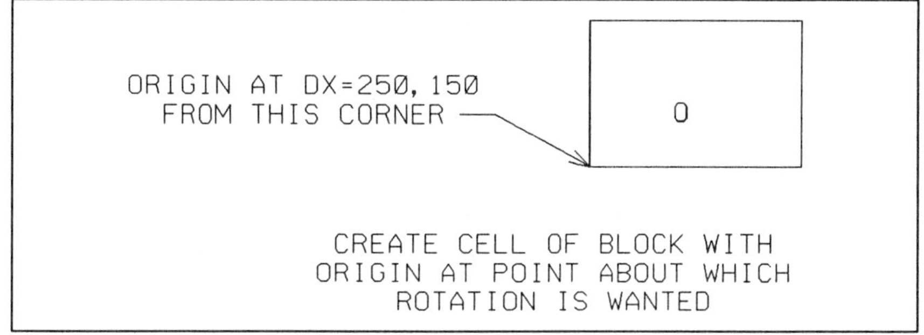

Figure 7.30 Create Cell for Projection

° Place the new cell, just created. The active angle was set to 0 degrees for placing the cell.
° Now, set the active angle to 4 degrees.

We are going to project the cell, from its origin using 'DX=„80', 10 times, as we did before. First, we have to make sure that we identify the cell at its origin. Here, we know that the origin is at DX=250,150 from the front left corner.

To project the cell from its origin:

° Select the *Construct Surface/Solid of Projection* tool.
° Tentative to the front left corner and then key in DX=250,150.
° Key in DX=„80 for the projection. Repeat this key-in another 9 times (or use the DOS pipe symbol key-in, DX=„80|10, initially).

The key-in DX=250,150 after the tentative to the corner of the block made the identification point at the origin of the cell. This method works when the origin of the cell is within the limits of the graphical elements of the cell. If the origin had been outside the block then we would not be able to use this method.

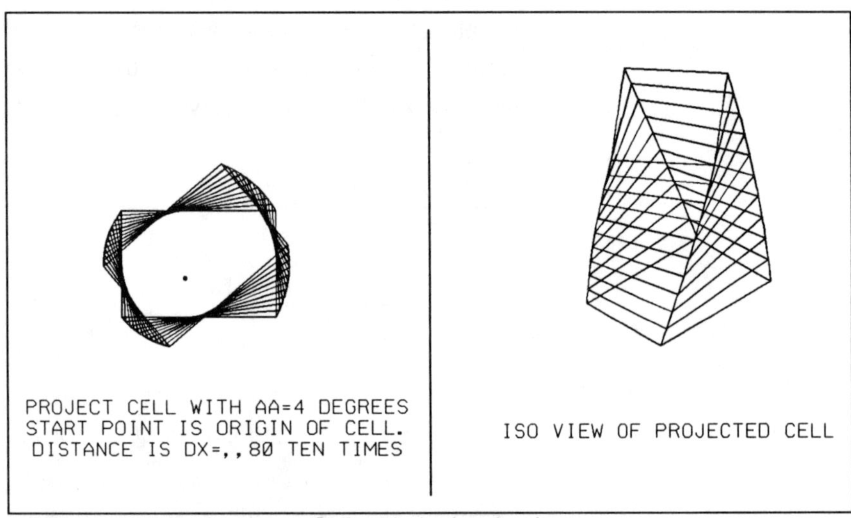

Figure 7.31 Projecting a Cell with Active Angle

Again, with version 4, we would use 'FENCE SURFACE PROJECTION'. With this key-in we are not restricted with the placement of the origin of the fence. We can specify the origin of the fence to be any point we require.

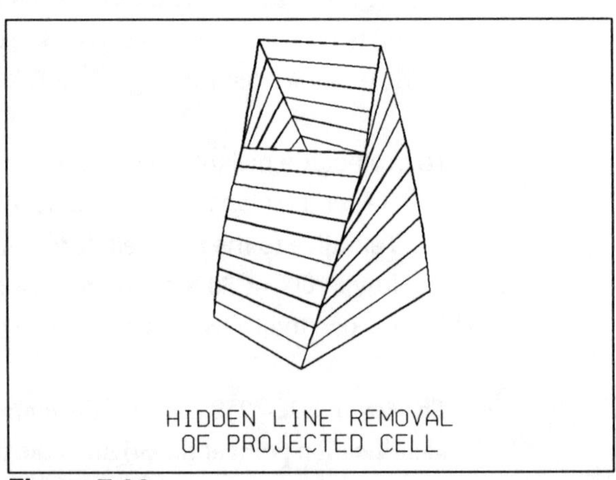

Figure 7.32

When using this form of multiple projection there may be times when the error message ‘element too large’ appears. To quickly overcome this, break the model into smaller sections. Another way is to break the projections into smaller parts. For example, a 360 degree twist could be made of two segments with a 180 degree twist.

Also, when we project cells they take the color of the first element placed in the cell. If colors are important then the cell should be separated into its separate colored sections before creating the projections.

Active points should not be placed in cells to indicate the origin as they would be projected as well. If you need to, an active point can be placed in the active file and the cell placed at this point. To identify the cell, tentative to and accept the active point. From there, reset to identify the cell, which will highlight.

Figure 7.33 is an illustration of how greater accuracy can be gained by using smaller steps when projecting with active angle. This cell consisted of three adjacent circles to form 3-core twisted wire. With version 4, the same results could have been achieved by projecting the three circles as the contents of a fence using the FENCE SURFACE PROJECT key-in.

The first variation, projecting in one step, is similar to having 3 tubular balloons and twisting the ends 180 degrees. They squeeze down to next to nothing in the middle.

In the other examples we can see how the model can be made to look more realistic by projecting in smaller steps.

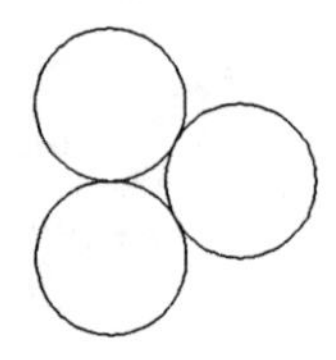

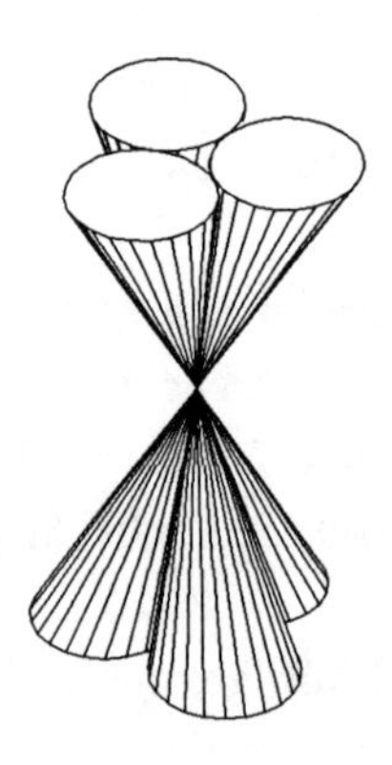

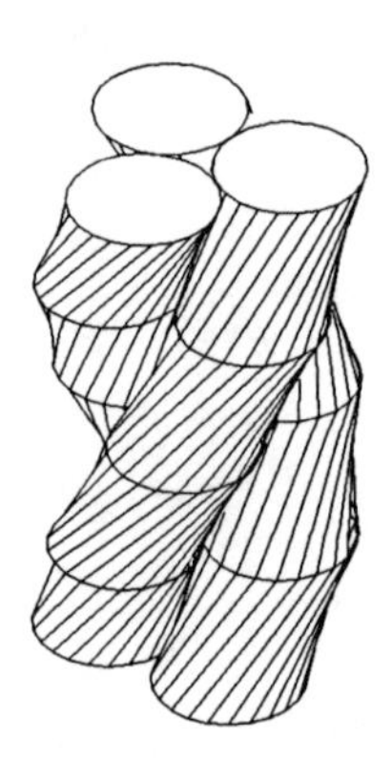

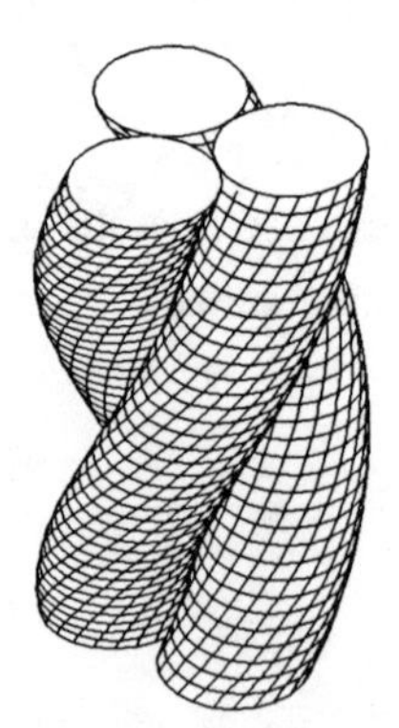

Figure 7.33 Projecting Cells with Active Angle

As we discussed, projections of cells, work fine with one restriction. The projection origin, the origin of the cell, must be within the limits of the graphical elements of the cell. When the origin is outside these limits, a tentative with a DX= key-in does not find the cell. So, we have to use a different work-around. In fact, we don't create cells. We project the elements themselves - individually if there is more than one.

With version 4, projecting the contents of a fence can be used no matter where the projection origin is.

For the example we will use a typical problem that comes up in underground mining. Refer to figure 7.34 which shows a TOP view of the geometric layout of part of an underground drive (tunnel). The vertical section of the drive is 4x4.

At first glance this seems straightforward. The drive is descending at about seven percent. Where we have a problem is the arc. Because the drive is descending constantly we have to create a helical arc.

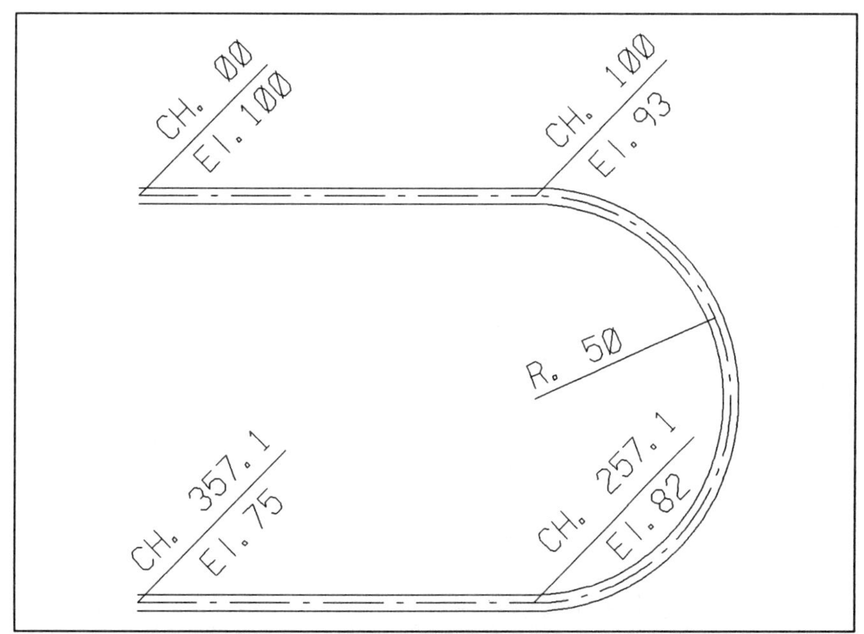

Figure 7.34 Geometry of Underground Drive

That is, we have to project a 4x4 block, representing the section of the drive, vertically downwards, as well as in a circular direction horizontally. We can create this relatively simply with version 4. With version 3.3 the construction is more involved. We will deal with each version separately, after the initial construction, which is the same for both.

The first step is to draw the center line. We only need to draw the straight sections.

- Using a TOP view place two 100 long lines at an elevation of 100 (Z-value). The lines are parallel and 100 apart. We are starting with a two-dimensional center-line layout which we will modify in the Z direction.
- Using the figures on the geometric layout as a guide modify the ends of the straights to their correct elevations (i.e., modify them in the Z direction).

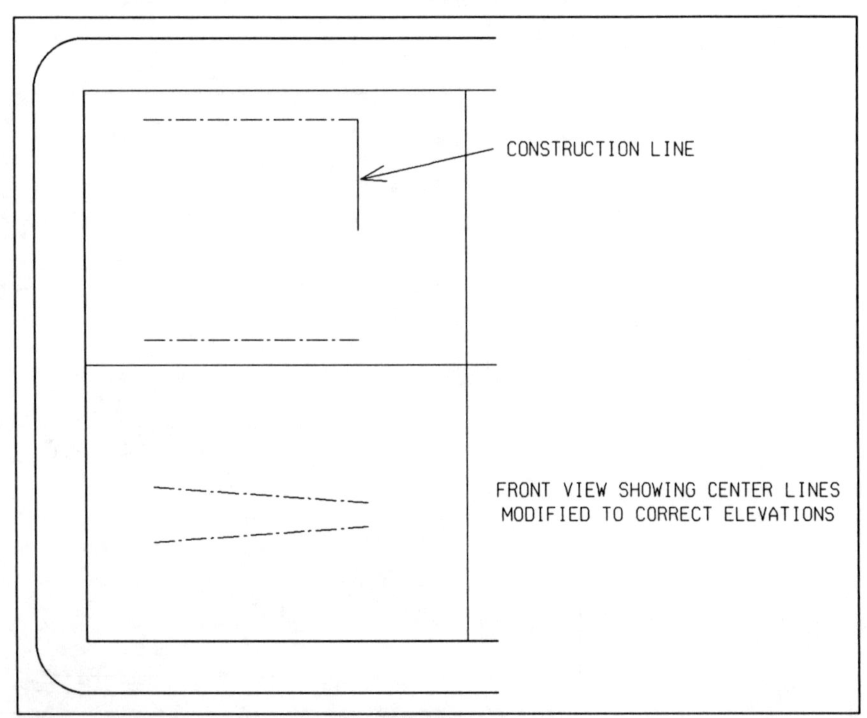

Figure 7.35 Centerline after Modification

Still in a TOP view, place a line from the point at CH.100 to the center of the arc.

° Identify the line at the CH.100 point then key in DX=,-50 for the other point. This will be a construction line for creating the helical drive.

We know that the vertical section of the drive or tunnel is 4x4. We now have to place a block which we will project to form the drive.

° Place the section in a RIGHT or LEFT view as in figure 7.36, which is a RIGHT view showing the block being placed.

° Copy the block across to the other end of the arc. We will need it to produce the straight section of the drive on that side.

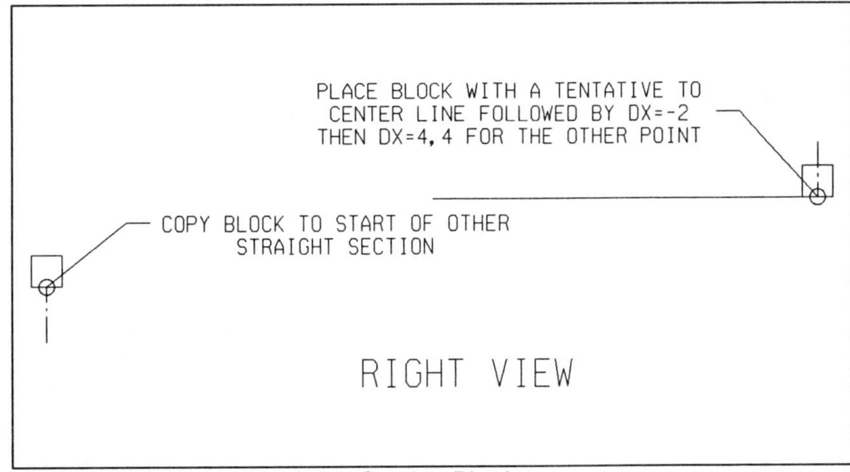

Figure 7.36 Placing the Drive Section Block

We will now go through the steps required for the version 3.3 construction. Then we will look at the version 4 alternative. For version 4 users, your section starts on page 7-40.

Version 3.3

We will be projecting a block to create the drive or tunnel. Initially we have to create points around the arc to project the block. We know that the difference in elevation from one end of the arc to the other is 11, from the geometric layout. For this example we will project the line in 20 sections of 9 degrees. Each section, therefore, will be 0.55 lower than the previous to make a total of 11 for the 180 degrees of the arc.

Version 3.3

In a TOP view, project the line as follows:

- Check that the ACTIVE SCALE is set at 1
- Set the ACTIVE ANGLE to -9 degrees.
- Select the *Construct Surface/Solid of Projection* tool.
- Identify the line at the end that is the center of the arc.
- For the distance to project key in DL=,,-0.55. This key-in has to be repeated 19 times for a total of 20. Alternatively, if you are using a PC, the DOS pipe symbol can be used. That is, DL=,,-0.55|20.

Looking at the various views, you will see that the outer edge of the line does form a helical arc, in the form of short chords.

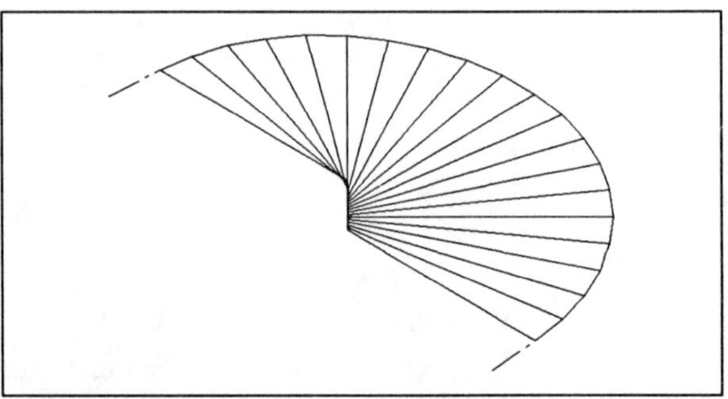

Figure 7.37 Iso View of Projected Line

We will now project the block that we placed first, with an active angle of minus nine degrees. We will use a TOP view and project it by identifying, in turn, the ends of each segment of the projected construction line. As we project each section the block section will be rotated by 9 degrees clockwise, due to the active angle. Each part of the projection, also, will descend in the Z direction as we require.

Set the TOP and ISO views as shown in figure 7.38. We will use the TOP view to specify the distance to project. The ISO view will be used to identify the block in the first place.

To produce the projected surface, around the arc, do the following:

- Make sure that the ACTIVE SCALE is 1.
- Set the ACTIVE ANGLE to -9 degrees.
- Set SNAP DIVISOR to 2 (KY=2).
- Select *Construct Surface/Solid of Projection* tool.
- Identify the center of the base of the block in the iso view.
- When prompted for the distance to project select the end of the line dividing the first and second segments of the projected construction line.
- For the next and subsequent projections select the ends of the other dividing lines in turn until the full 180 degrees has been completed.

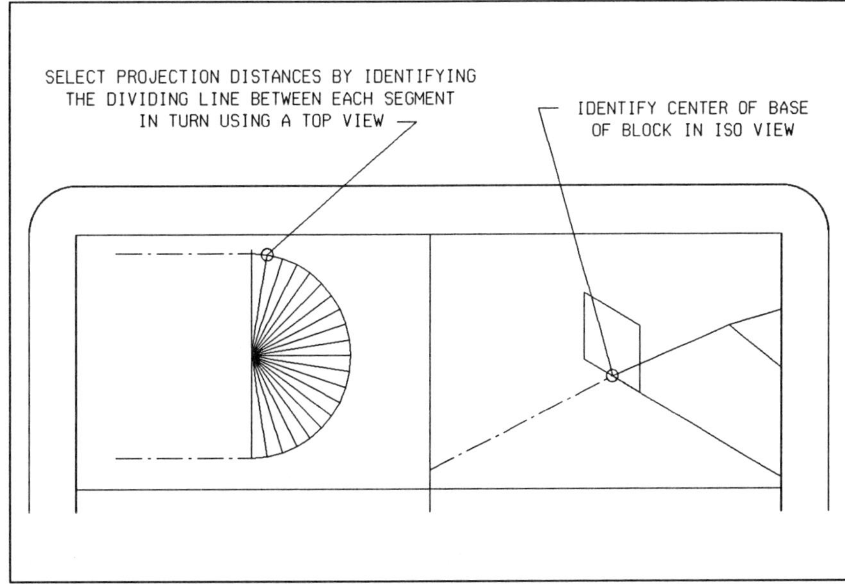

Figure 7.38 Projecting Block Around the Helix

Version 4

With version 4 we can produce the helical construction with a projection of the contents of a fence. We will create the 180 degree projection in 20 steps of 9 degrees. As the total drop from one end of the arc to the other is 11, this means that each 9 degree section will have to drop 0.55 in the vertical or Z direction.

Version 4

In preparation:

- Set the active angle to minus nine degrees (AA=-9).
- Make sure that active scale is set to 1 (AS=1).

Using a TOP view, and figure 7.39 as a guide, do the following:

- Set fence lock to Inside.
- Place a fence around the block located at the top of the view.
- Key-in FENCE SURFACE PROJECT.
- For the origin of the fence manipulation, identify the construction line at the end located in the center of the arc (in the TOP view).
- For the distance to project, key in DL=,,-0.55. This has to be repeated another 19 times. Alternatively, DOS users can key in DL=,,-0.55|20 to complete the task in one hit.

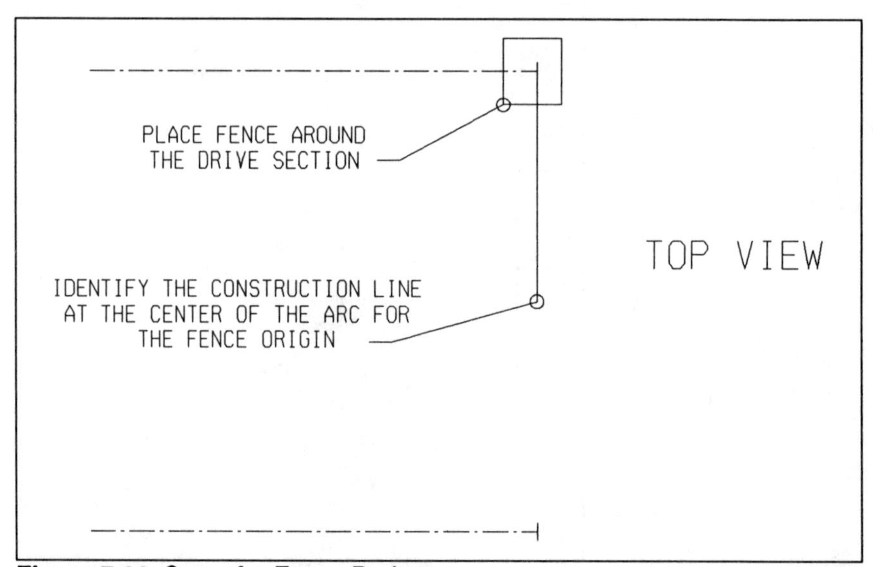

Figure 7.39 Setup for Fence Project

Both Versions

When you have finished that part of the model, all that remains is to project the straight sections. These can be projected in any view. Before doing so make sure that the **ACTIVE ANGLE** is set back to 0.

Figure 7.39 shows two views, after hidden line removal, of the finished drive or tunnel.

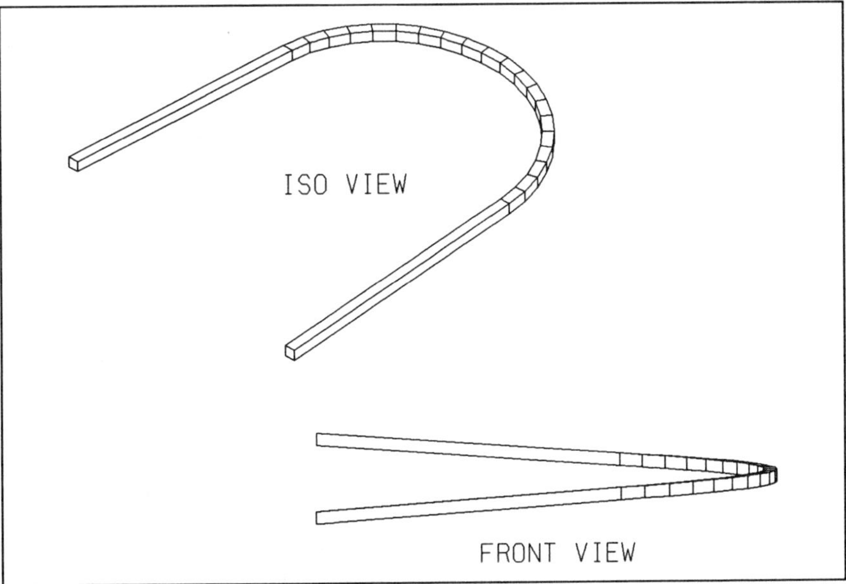

Figure 7.40 Views of the Finished Drive/Tunnel

To get more practice in three-dimensional modelling, try creating models of simple items around your desk. It doesn't matter how simple they are - any practice is better than none. It is through practice that the various tools become familiar. If you have followed all the examples and exercises through to here, you should be getting a good grasp on working in 3D.

Pictured below are two examples of common items to give you some ideas. MicroStation version 3.3 was used for both.

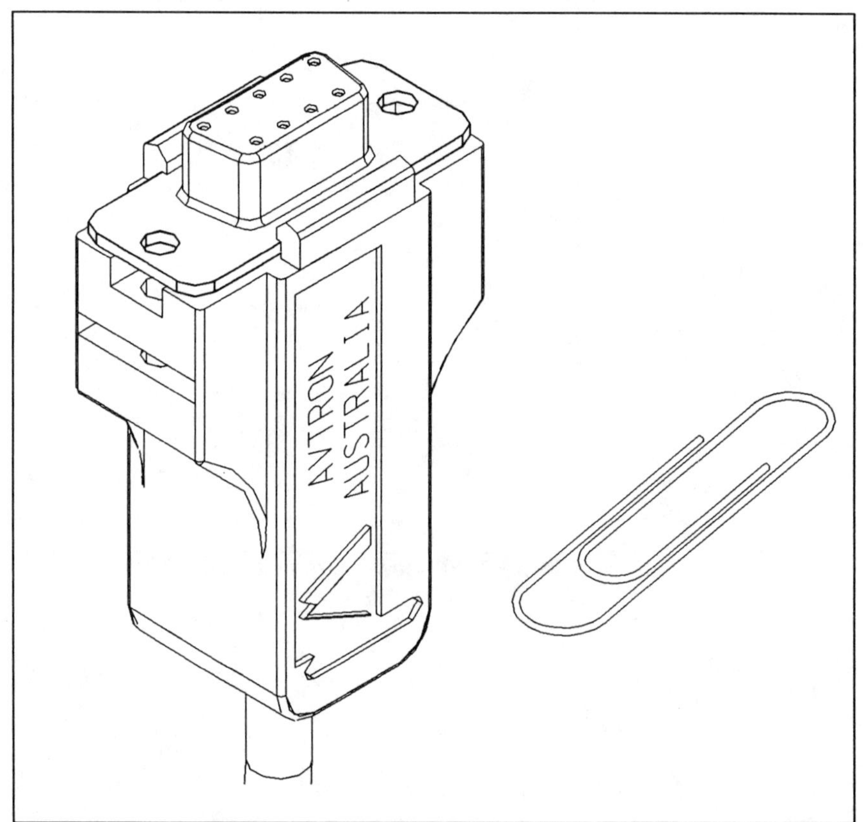

Figure 7.41 Models of Common Items

8 : B-spline Surfaces version 4

Often, in three-dimensional work, we have to create an object that is difficult to break down into simple 3D elements or shapes. We can overcome many of these problems by using standard projections and surfaces of revolution. In other cases we can project with active angle, active scale or both.

Version 4 of MicroStation provides us with another group of powerful 3D tools. These tools are used for creating B-spline surfaces. In the previous chapter (Advanced Techniques) we looked at one form of B-spline surface construction. There, we created ruled surfaces between elements. In this chapter we will look at some other examples.

You may be familiar with the tools for creating B-spline curves. These are contained in the 'B-splines' palette (figure 8.1). This palette is opened from the 'Main' sub-menu of the 'Palettes' pull-down menu.

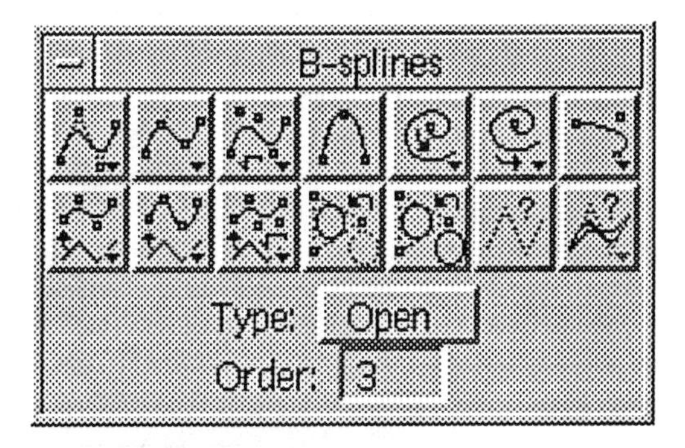

Figure 8.1 B-splines Palette

B-splines Palette

Tools in the 'B-splines' palette are used for:

- Placing B-spline curves, spirals and parabolas.
- Constructing B-spline curves from existing elements.
- Converting existing elements to B-spline curves.
- Enabling/disabling the B-spline control polygon/net.
- Changing the order of B-spline curves.

These tools also may be used equally in two-dimensional drawing and three-dimensional modelling. In addition to these, we have a suite of strictly three-dimensional B-spline tools.

3D B-splines Palette

Tool palettes for 3D B-spline constructions are opened from the '3D B-Splines' option of the 'Palettes' pull-down menu. We have a choice of four palettes, as figure 7.20 in the previous chapter showed. The selections are 'Space Curves', 'Surfaces', 'Derived Surfaces' and 'Change Surface'.

Space Curves Palette

Included in the 'Space Curves' palette (figure 8.2) we have the tool for placing helixes (second from right). Chapter 5 (Placing & Manipulating Elements) has a section on using this tool. For our purposes here, we are more interested in the B-spline surface tools with which we can create complex surfaces.

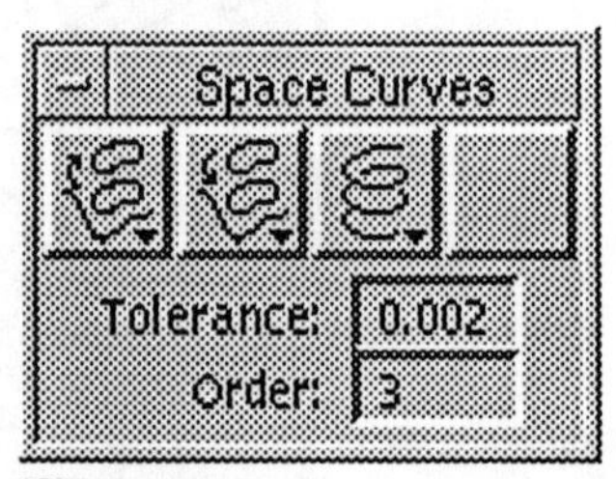

Figure 8.2

B-spline Surface Tools

B-spline surface construction tools allow us to construct surfaces from data provided by the user, or from existing elements. We also have tools for creating B-spline surfaces of projection and surfaces of revolution. All are grouped into two palettes. Here, we will discuss these tools and look at some examples.

Surfaces Palette

Tools contained in the 'Surfaces' palette (figure 8.3), from left to right, are:

- *Place B-spline Surface by Poles*
- *Place B-spline Surface by Points*
- *Place B-spline Surface by Least Squares*
- *Construct B-spline Surface by Poles*
- *Construct B-spline Surface by Points*
- *Construct B-spline Surface by Least Squares*

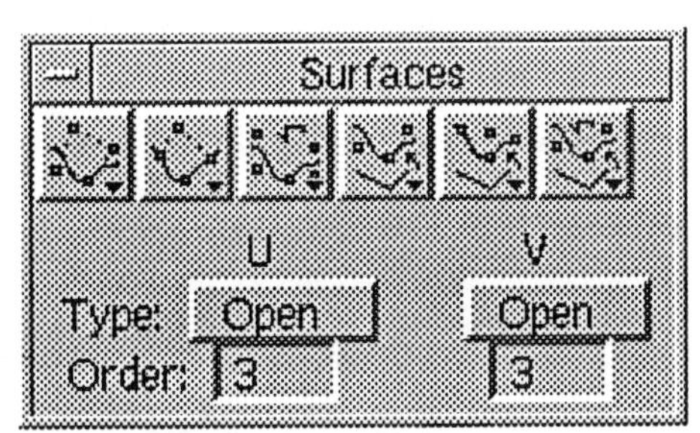

Figure 8.3 Surfaces Palette

We have both placement and construction tools for three variations of B-spline surface. While the placement tools rely on user inputs to define the surfaces, construction tools create the B-spline surfaces from selected existing elements.

Construct B-spline Surface by Poles

With this tool we can construct a B-spline surface which uses the vertices of a group of selected elements as the poles of its control net. The control net is the three-dimensional equivalent of the two-dimensional control polygon. We can enable or disable the display of the control net using the appropriate tool in the 'B-splines' palette.

For this tool to function correctly, we must ensure that the number of poles in each direction (U and V) is equal to or greater than the order for that direction. The order of the surface can be set in the pop-down U and V fields, or in the B-splines settings box. For the example we will be discussing here, the number of poles in each direction was determined by:

(a) the number of vertices in the elements and

(b)the number of elements.

Before choosing the *Construct B-spline Surface by Poles* tool, we select the elements that are to be included in the construction. We select them in the order in which we want the surface to be constructed. Figure 8.4 shows TOP and FRONT views of three linestrings. These were selected, in order from top to bottom (in the FRONT view). A B-spline surface was then constructed using the *Construct B-spline Surface by Poles* tool.

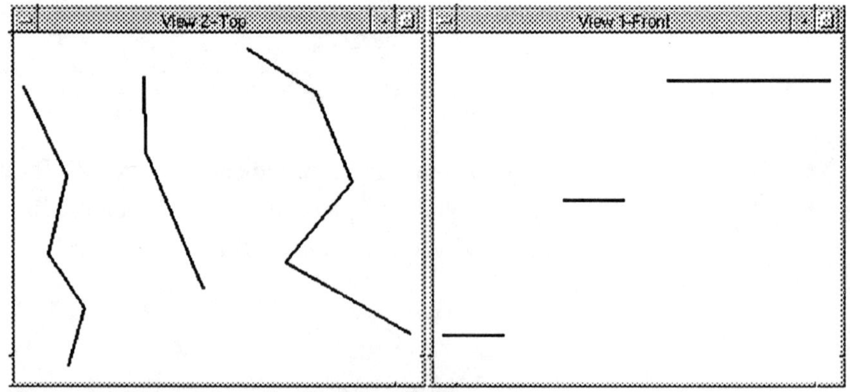

Figure 8.4 Linestrings for B-spline Surface

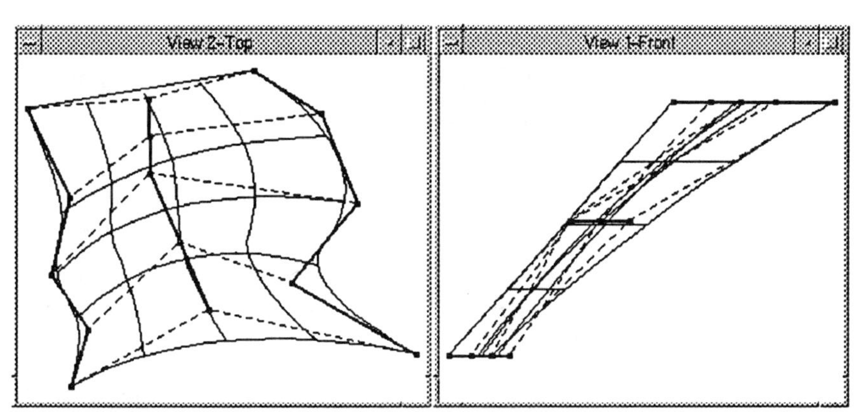

Figure 8.5 B-spline Surface by Poles

Figure 8.5 shows the surface and its control net. It can be seen that the control net has used the selected linestrings to create the B-spline surface. We can see the dashed lines of the control net joining the vertices of the linestrings.

If we had selected the linestrings in a different order, the resulting surface would be different. Say, for example, that we had selected (in the FRONT view) the middle linestring followed by the bottom and top linestrings respectively. The resulting surface would be as shown below (figure 8.6). The control net still joins the vertices of the linestrings, but in a different order. This has, in turn, changed the shape of the B-spline surface.

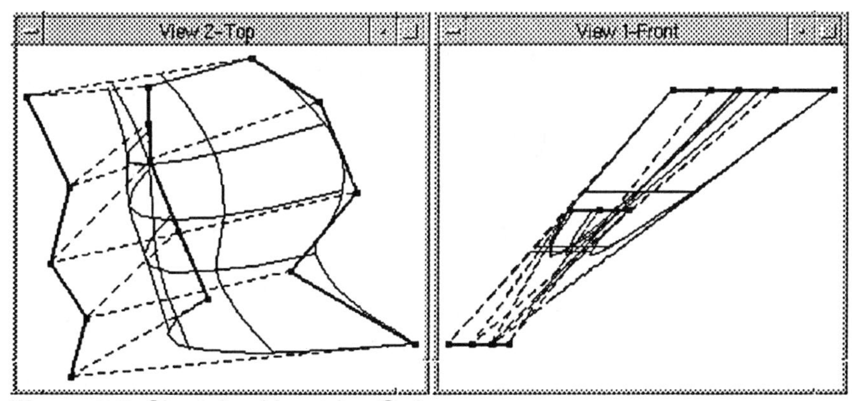

Figure 8.6 Changing Linestring Order

Two other tools in the 'Surfaces' palette, *Construct B-spline Surface by Points* and *Construct B-spline Surface by Least Squares*, work in a similar fashion. Where they differ is the method used for calculating the B-spline surface.

Figure 8.7 shows a B-spline surface constructed from the same three linestrings, but using the *Construct B-spline Surface by Points* tool. Here, the surface itself passes through the vertices of the linestrings. With the previous construction it was the control net that passed through these points.

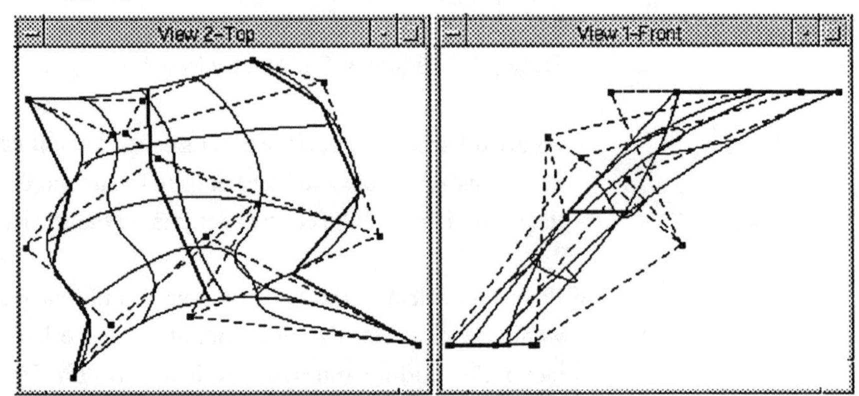

Figure 8.7 B-spline Surface by Points

Figure 8.8 shows the other alternative, constructed with the *Construct B-spline Surface by Least Squares* tool. Again, the same linestrings were used.

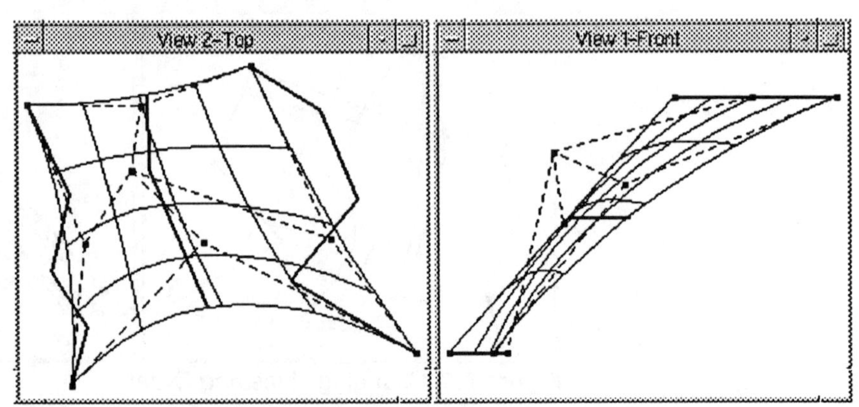

Figure 8.8 B-spline Surface by Least Squares

We have three other tools in the 'Surfaces' palette. These are for placing B-spline surfaces, as against constructing surfaces from other elements. With these tools, we define the parameters with data points.

Place B-spline Surface by Poles

To construct a B-spline surface with this tool we define the poles with data points. Again, the number of poles in each direction (U and V) must be equal to or greater than the order for that direction. When we have defined the points for the first row, we enter a reset. We can then start on the next row. When we have defined the same number of points as that for the first row, a new row is started automatically. When we have defined the final point in the last row, we again enter a reset to initiate the generation of the surface.

Figure 8.9 shows the beginning of a B-spline construction using the *Place B-spline Surface by Poles* tool. Points were defined by identifying vertices of the linestrings. In the figure, the first row has been completed and the second row is being entered. The control net can be seen (attached to the cursor), indicating the points that have been defined.

We have two other tools that perform in a similar manner. They are *Place B-spline Surface by Points* and *Place B-spline Surface by Least Squares*. Refer to the previous section to see the difference between the three types of generated B-spline surfaces.

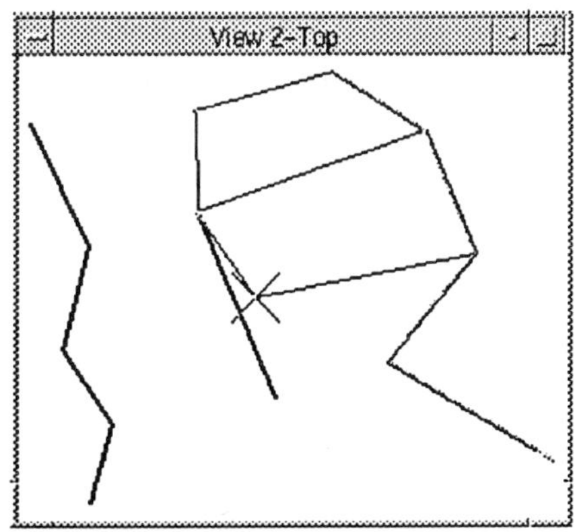

Figure 8.9 Using Data Points

Derived Surfaces Palette

With the 'Derived Surfaces' palette, we again have a selection of six tools. The palette is shown in figure 8.10, with the tool being (from left to right):

- *Construct B-spline Surface of Projection*
- *Construct B-spline Surface of Revolution*
- *Construct B-spline Surface by Cross-Section*
- *Construct B-spline Surface by Tube*
- *Construct B-spline Surface by Skin*
- *Construct B-spline Surface by Edges*

Figure 8.10

B-spline Surface of Projection

We use the *Construct B-spline Surface of Projection* tool to create a B-spline surface by projecting a planar element. While the operation is similar to the *Construct Surface/Solid of Projection* tool, the resulting surface is a B-spline surface. For most situations we would use the latter tool, which creates a ruled surface/solid of projection.

An advantages with the B-spline surface, over the ruled surface, is that it can be modified.

B-spline Surface of Revolution

Immediately to the right of *Construct B-spline Surface of Projection* is the *Construct B-spline Surface of Revolution* tool. There is a significant difference between this and the *Construct Surface/Solid of Revolution* tool. With the B-spline tool we can specify the axis of revolution in any view. We specify this axis with two points, which also define its orientation.

When we construct surfaces/solids of revolution, the axis of revolution is always perpendicular to the view in which it is selected. Figure 8.11 shows an ISO view containing a shape from which a surface of revolution is to be produced about the line to its right. That is, the line represents the axis of revolution.

When we use the *Construct B-spline Surface of Revolution* tool, we identify the element first, followed by two points to define the axis. These may be defined in any view. Here, the two points defining the axis of revolution are at each end of the line.

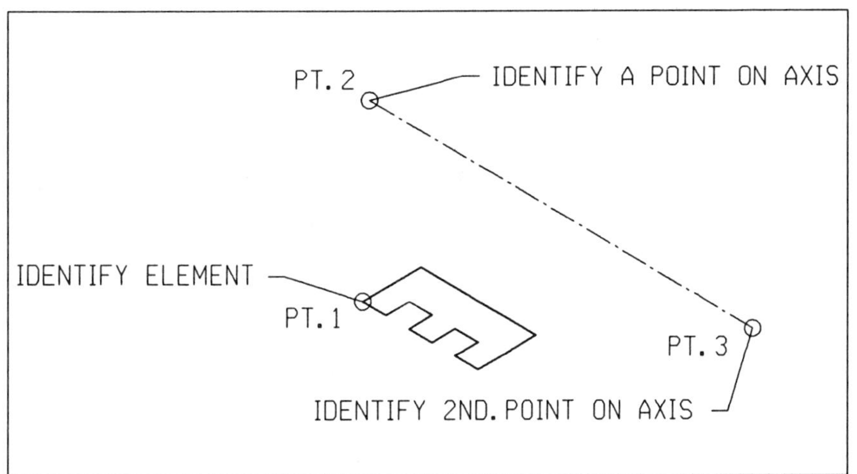

Figure 8.11 B-spline Surface of Revolution

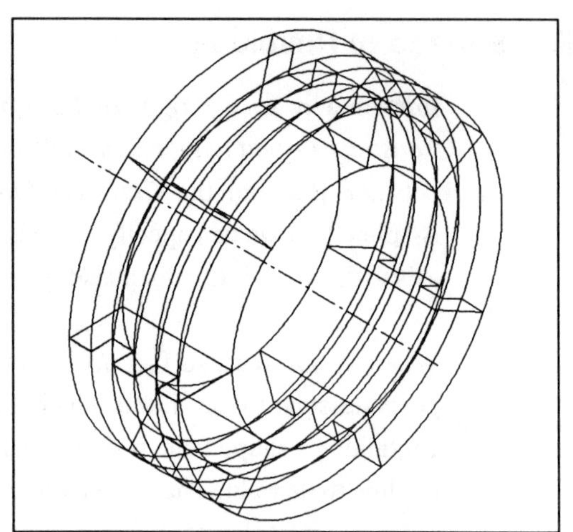

Figure 8.12 Completed Surface

We can see, in figure 8.12, that the B-spline surface of revolution has been created correctly.

B-spline Surface by Cross-Section

This tool, which is located third from the left in the 'Derived Surfaces' tool palette, allows us to construct a surface that is transformed between sections.

For our discussion on using the *Construct B-spline Surface by Cross-Section* tool, we will look at how the dinghy illustrated in figure 8.13 could be modelled. Similar methods, to those we discuss for this model, could be adapted for other applications that you may have.

Looking at the side elevation and the sections we can see that the hull is a complex curved surface. One method that can be used, is to create one half of the hull and mirror copy this to produce the other half.

Our first task would be to set up a three-dimensional 'skeleton' of the hull (figure 8.14), with the section lines in place. These could be placed in the file as curves, or digitized, in their correct location. These form the basis for the construction of the B-spline surface 'skin'.

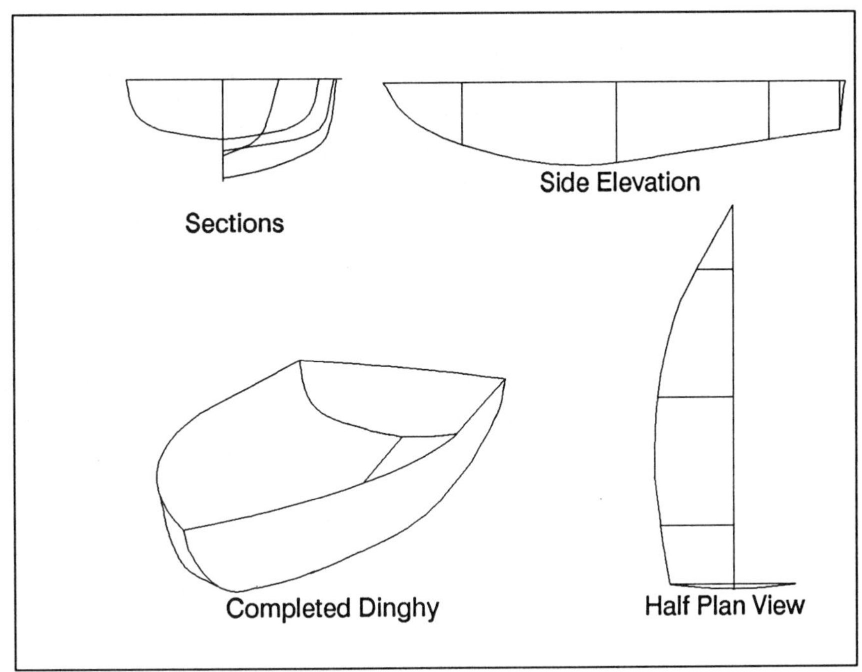

Figure 8.13 Dinghy Model Layout

We would want to make the B-spline surface of the hull finish at the nose of the dinghy. To do this we would place a short curve at the nose. This would be included with the lateral sections when we select the elements to be used.

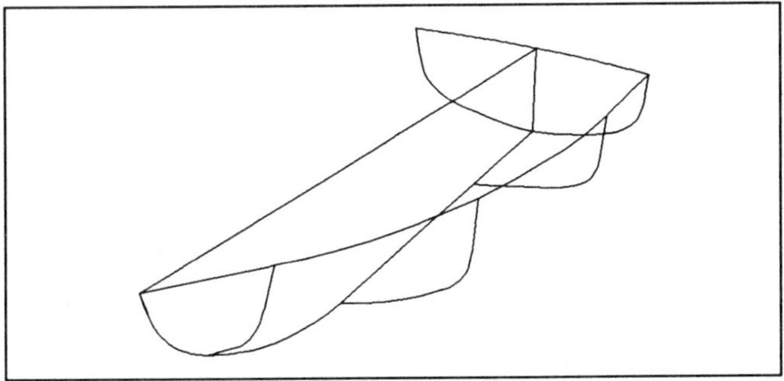

Figure 8.14 ISO view of Initial Hull Skeleton

To construct the B-spline surface, we first select the sections and nose piece with the *Element Selection* tool. Here, the number of sections must be equal to or greater than the order in the V direction. This can be set in a pop-down field from the 'Derived Surfaces' sub-palette. Having checked this, and selected the sections, we then choose the *Construct B-spline Surface by Cross-Section* tool. A surface is created between the sections (figure 8.15). At this point we can accept the surface, or reset to abort.

Using a TOP view, a mirror copy of this half of the hull forms the remaining side. We could use the *Construct B-spline Surface by Edges* tool to place the fill in section at the rear of the dinghy (transom). The edges used would be the two half-sections at the rear and the arc joining the outer ends, across the top.

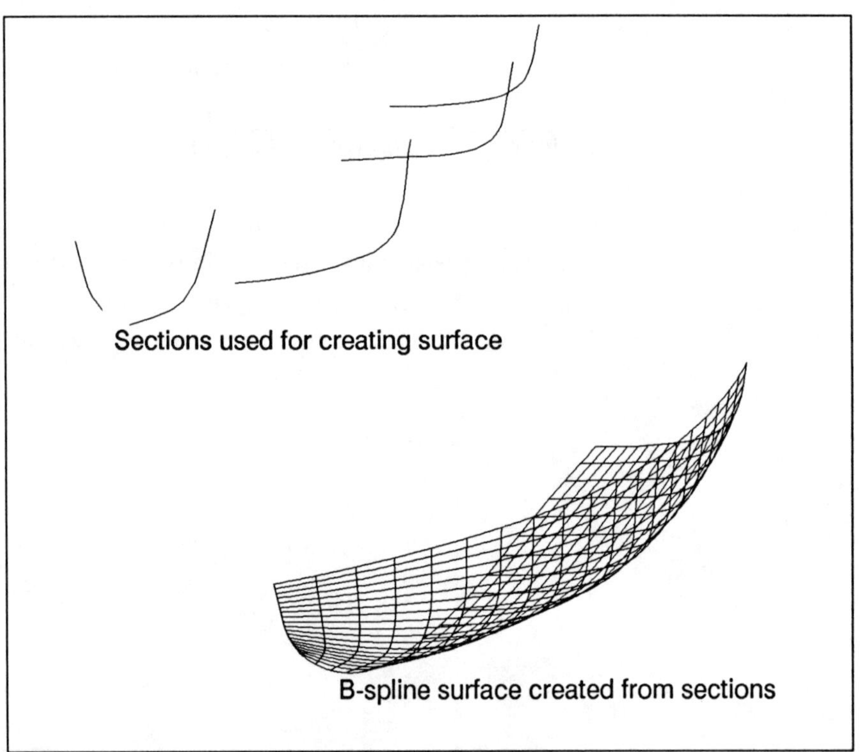

Figure 8.15 B-spline Surface by Cross-Section

There are many applications for both the *Construct B-spline Surface by Cross-Section* and *by Edges* tools. Among these could be body panel design in the automotive industry, ore-body simulation (from geological sections) in mining, and other situations where a complex curved surface is involved.

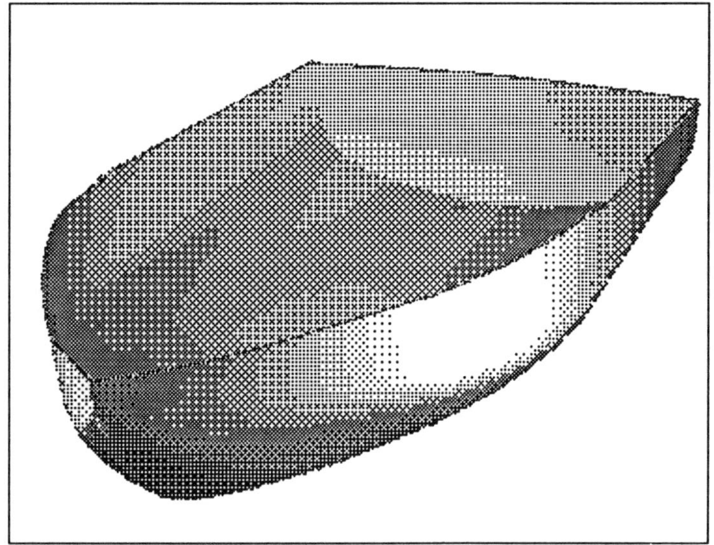

Figure 8.16 Shaded Image of Dinghy Model

B-spline Surface by Tube

We use the *Construct B-spline Surface by Tube* tool to 'project' one element (the section) along another element (the trace). Here, the section is projected along the entire trace, its orientation varying continually to follow the orientation of the trace.

Figure 8.17 shows a TOP view of a block (section) that we will project around the centerline (trace). We will attach the block at the center of its bottom edge. In this example, the centerline is a chained element. Other valid elements for both sections and traces are lines, linestrings, complex shapes, arcs, ellipses and B-spline curves.

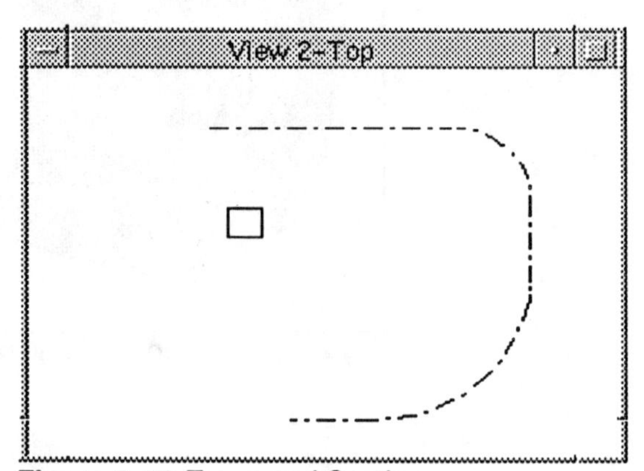

Figure 8.17 Trace and Section

When we select this tool, we are prompted in turn to:

- Identify the trace.
- Identify the section.
- Identify the point, on the section, or the plane of the section, that we want attached to the trace. That is, the section does not have to be attached physically to the trace, it can be offset.
- Enter a data point to define the rotation of the section.
- Accept (or reject) the B-spline surface.

After selecting the *Construct B-spline Surface by Tube* tool, we can identify the trace in any view, as we can the section. We require the correct view however, when we come to identify the point, on the plane of the section, to attach to the trace. This is true also, for defining the rotation of the section.

Figure 8.18 shows TOP and RIGHT views as the rotation of the section is being defined. We can see the exact rotation of the section in the RIGHT view. As we move the cursor, the rotation changes. We can see also, in the TOP view, that the section is attached to the very end of the trace. When we have the correct rotation of the section, a data point initiates the generation of the B-spline surface. We then enter a data point to accept, or a reset to reject the surface. The surface is shown in the ISO view in the figure.

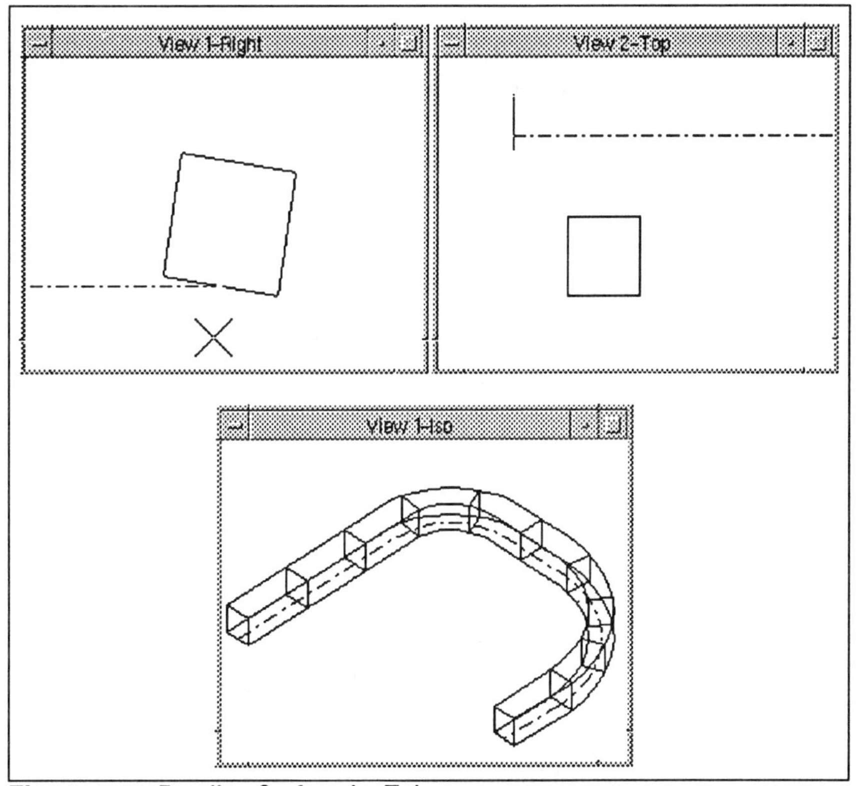

Figure 8.18 B-spline Surface by Tube

B-spline Surface by Skin

This tool, *Construct B-spline Surface by Skin*, functions similarly to the previous tool. Where it varies, is that we can have different sections along the trace. These are transformed from one section to the next as the elements are 'projected' along the trace.

A similar procedure is used to that of the *Construct B-spline Surface by Tube* tool. Because we have more than one section element, however, we have to define the location (on the trace) and rotation for each.

Figure 8.19 shows an ISO view (in wireframe) of a surface constructed with this tool. A rectangle has been transformed into a circular section as it was 'projected' through a 90 degree arc.

Elements that were used in the construction, were the initial rectangular section, the final circular section and the arc trace (figure 8.20).

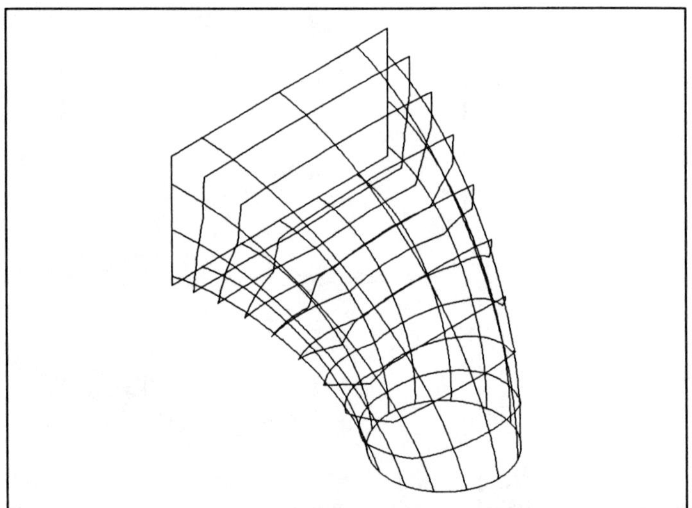

Figure 8.19 B-spline Surface by Skin

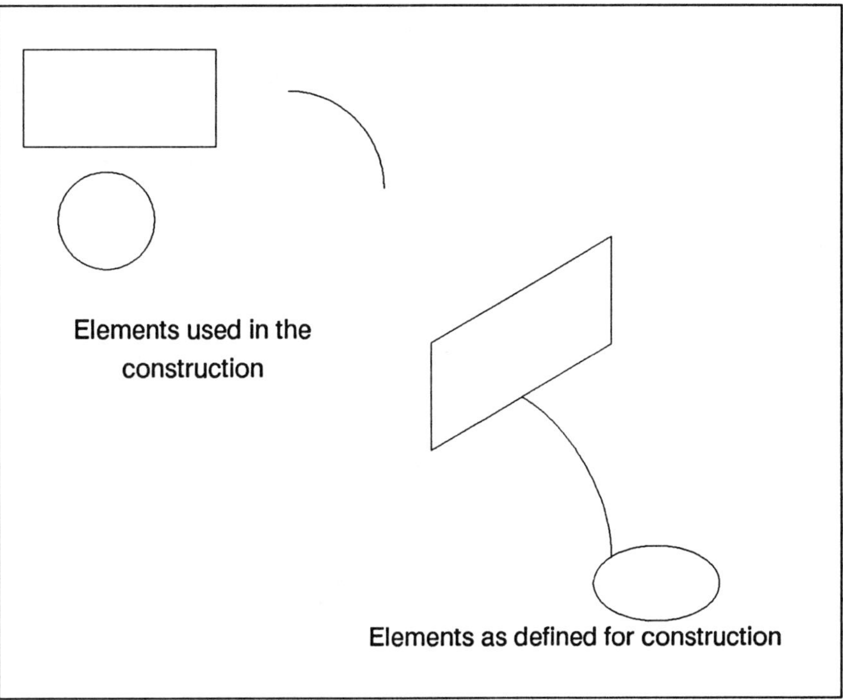

Figure 8.20 Creating B-spline Surface by Skin

Of these, the trace is the only one that needs to be in its exact place in space. This is because it determines the location (and orientation) of the constructed surface.

B-spline Surface by Edges

We will look at now, the *Construct B-spline Surface by Edges* tool. This too can be used to construct surfaces that would normally be very complicated to model. In the previous chapter (Advanced Techniques), for example, we used it to construct a surface between a block and a circle. We used the same tool also, to construct a surface between two identical elements that had different orientations.

Another option we have is to construct a surface between three or four connected elements. In these instances, the existing elements are used to form the outer boundary of the B-spline surface. Allowable edge elements include lines, linestrings, arcs, curves, B-spline curves, and complex chains.

Figure 8.21 shows an object that, with normal techniques/tools, would be very difficult to model. We can see that the front elevation has two semi-circular flutes, which grade to nothing at the rear.

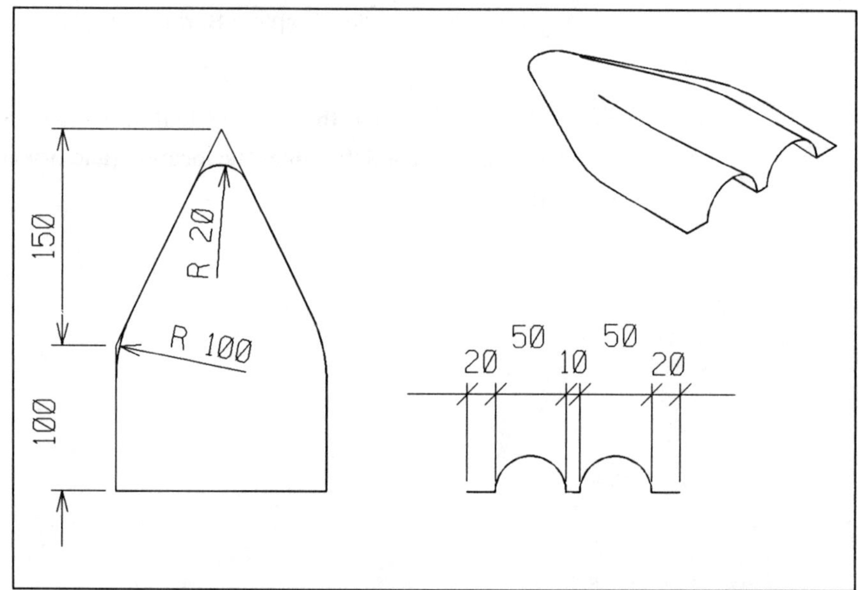

Figure 8.21 Model for Edges Surface

Our first task is to create the edge or boundary elements, which can include chained elements. This can be done with normal 2D techniques, with the sides being placed in a TOP view and the front section placed in a FRONT view. Figure 8.22 shows an ISO view of the finished edge elements. We see them separated, to show the individual components, and connected. Both the sides and front edges are chained elements.

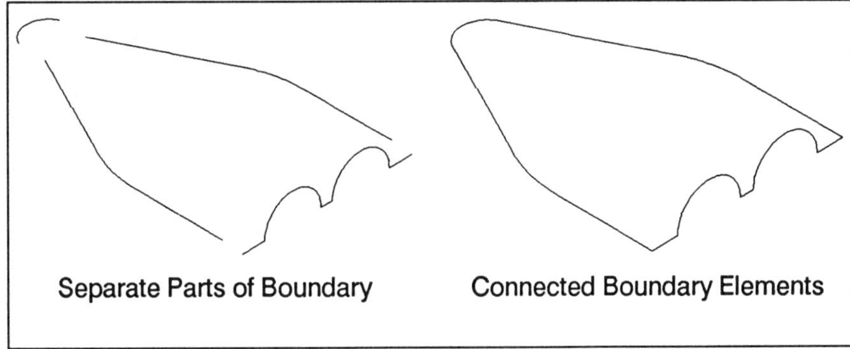

Figure 8.22 Creating the Edges for the B-spline Surface

Having placed the boundary elements, we can now create the surface joining them. Before proceeding, we have to select the elements that are required to form the edge of the surface. We do this with the *Element Selection* tool. When the elements have been selected, choosing the *Construct B-spline Surface by Edges* tool initiates the construction. A data point accepts the construction, as shown in figure 8.23 (ISO view).

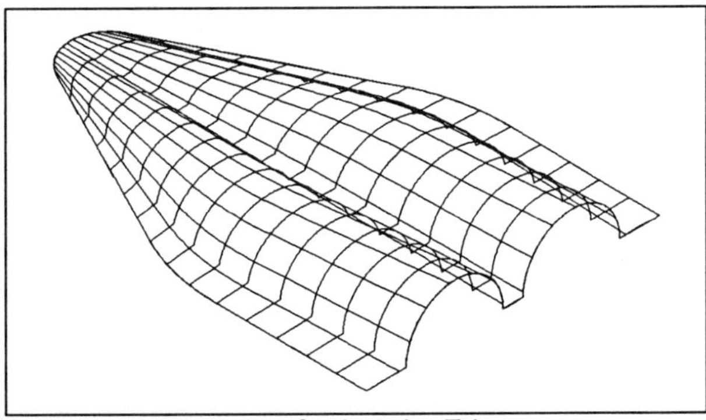

Figure 8.23 B-spline Surface by Edges

Figure 8.24 shows, in ISO view, another example of a B-spline surface by edges construction. Here, the edges each have a 25 high peak in the middle. We can see in the lower part of the figure, that the peak at the center of the B-spline surface is the sum of the two intersecting peaks.

Where we select three or four elements, the generated surface is known as a bilinearly blended Coons patch.

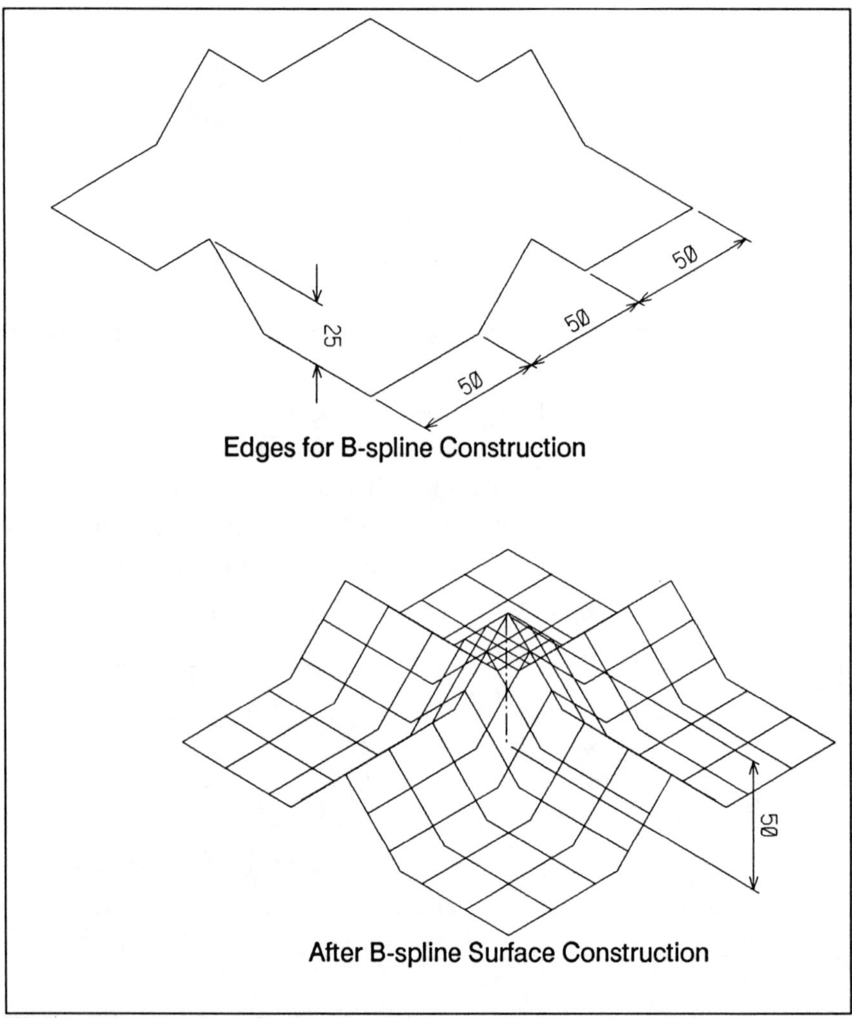

Figure 8.24 B-spline Surface by Edges

Change Surface Palette

Tools contained in this palette are used to change the U and V order and the U and V rules of B-spline surfaces. A pop-down field is provided for each tool to input the setting.

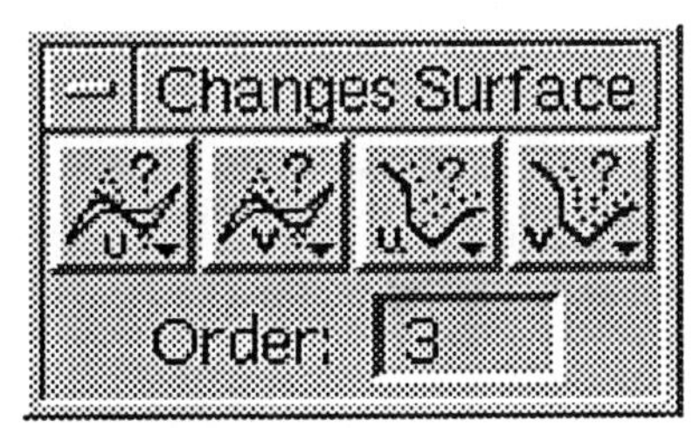

Figure 8.25

In order from left to right in the pallet, the tools are:

- *Change B-spline Surface to Active U-Order*
- *Change B-spline Surface to Active V-Order*
- *Change B-spline Surface to Active U-Rules*
- *Change B-spline Surface to Active V-Rules*

Change B-spline Surface Order

We use the two 'change order' tools to alter the physical attributes (U and V order) of the B-spline surfaces. After selecting the appropriate tool, we set the order in the pop-down field. The next step is to identify the B-spline surface, followed by a data point to accept the change, or a reset to reject it.

Change B-spline Surface Rules

We can determine the way B-spline surfaces display on our screen. We do this with two tools, *Change B-spline Surface to Active U-Rules* and *Change B-spline Surface to Active V-Rules*. U-rules and V-rules are the lines that are used to display B-spline surfaces in wireframe mode. We can use these tools to clarify a wireframe display of a model, where it would be otherwise confusing.

Figure 8.26 shows a view of the dinghy model in wireframe form. This model was built up from three B-spline surfaces. Here, we can see that the U-rules and V-rules vary for each of the surfaces.

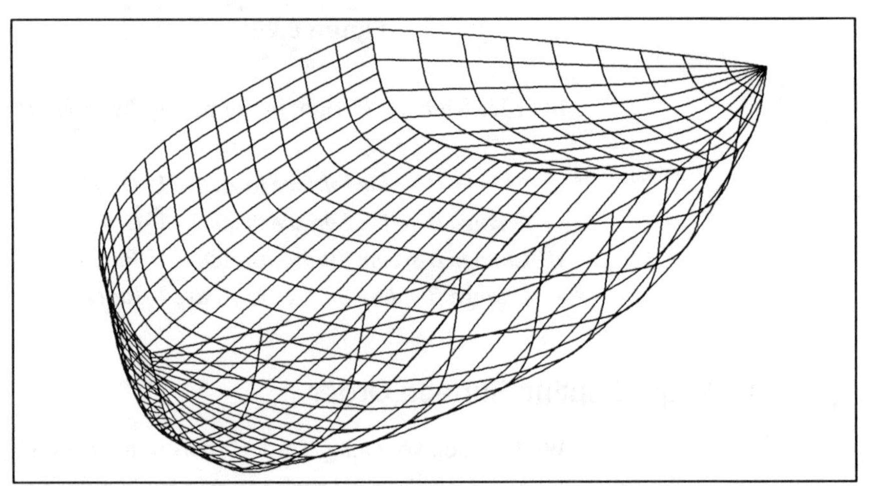

Figure 8.26 B-spline Surface Display

For extra practice, Chapter 14 (Practice Examples) has a simple model, which incorporates a B-spline surface.

9 : Introduction to Rendering

It is not so long ago that three-dimensional CAD with rendering was available on main-frame systems only. With the increase in power and speed of PC's and other micro computers, main-frames are no longer the only option.

Included with the MicroStation package are rendering and hidden line removal utilities. It is with these utilities that we can produce rendered, or shaded, images. We can also produce line drawings of our 3D models with the hidden lines/surfaces removed, using the Edges option.

As well as standard elevations we can generate perspective views of our designs. With version 3.3 the 'SET PERSPECTIVE' key-in is used. Version 4 uses 'SET CAMERA', which allows us to specify a 'camera' position and its target point anywhere within the design file. Version 4 has many improvements and changes to the rendering processes over version 3.3. Because of these differences, each version will be dealt with separately. With both versions, however, we should be familiar with the basic terms.

Rendering

Put simply, rendering is a means by which we can get life-like colored 'pictures' of our three-dimensional wireframe models. The rendering process recognizes elements/surfaces and their properties, colors and position relative to one another and the viewing point. Where objects are hidden, partially or wholly, the hidden portions are not displayed.

Light sources and their effects are also taken into account. The images are produced directly from our three-dimensional design files. They can be saved to disk for later retrieval.

Rendering is a raster process. That is, it is screen-based, rather than design file based. We can't plot our rendered images, like we can a design file. We need special software and color copiers/plotters to get a hard copy of the saved images. An alternative is to photograph the image on the screen.

Hidden Line Removal

Whereas rendering produces shaded, colored images, hidden line removal produces line images with no shading. It, too, is a raster based process. Again, we can't plot the results. The time taken to produce a hidden line removal is less than that for a rendered image. There is no need for the system to read in color or light source information, as these have no effect on the outcome.

Edges Design Files

Edges is a vector based process. It works to the resolution of the units of the design file, rather than the resolution of the screen. With the Edges utility we can produce design files of our model with hidden lines removed. They can be 3D or 2D design files which can be plotted normally. They are a normal design file. The difference is that they are generated by the system from other three-dimensional files. We can use the edges process to produce drawings from our models.

Processing time for edges design files is much longer than that for either hidden line removal or rendering. This is because Edges works to design file units, rather than screen units.

What is Rendering?

Rendering is similar to taking a photograph of our model. In real life, when we take photographs, the lighting is set, the scene is already shaded and the hidden lines/surfaces are not visible. The result is a two-dimensional image of a three-dimensional scene. Depth is given to the photograph by the perspective and the effect of the lighting. More distant objects appear smaller than closer, similar sized objects. Surfaces facing the light source/s are brighter than those facing away. Curved surfaces appear curved because of the smooth change in color from bright (facing the light source) to dim (facing away from the light source). As well, some surfaces may be shiny, and others matte. With our design file the rendering process has to take into account all the above to produce a 'life-like' image of our computer scene.

When we create models, we separate the various parts by using different levels or colors. With level control we can turn off the levels containing those parts that we don't require at the time. This can reduce the confusion of lines that may otherwise be displayed in wireframe form. Where it is necessary to have other parts of our model displayed as we work, then colors are important. For example, we may be working in a part that is colored yellow. In our minds, we filter out the other colors and concentrate on the yellow elements.

A problem arises when we want to create rendered, or shaded, images of the model. More than likely, it will not look natural to have some parts red, others green, or yellow etc. To overcome this we can create a Surface Definition File (version 3.3), or a Material Table (version 4). These instruct the rendering process how to treat the various surfaces.

Version 3.3 Surface Definition Files relate surface definitions to the colors of elements in the design file. For example, all yellow elements could be specified as being shiny metallic.

Version 4 Material Tables allow us to relate surfaces to elements of a particular color, level, or a combination of the two. For example, all yellow elements on level 2 may be chrome. At the same time, all yellow elements on level 10 may be glossy gold.

When we invoke the rendering process the following tasks are performed:

– 1. Three-dimensional data, in the selected view, is converted to a two-dimensional 'picture'. This is similar to a scene being projected onto the film in a camera, via the lens.
– 2. Hidden surfaces are determined and removed. We are only interested in seeing the surfaces that would be visible to us if the model was real.
– 3. The type of graphics card present is determined. With version 3.3, if the graphics card is not capable of displaying 256 colors simultaneously then MicroStation performs a hidden line removal with panel fill. With version 4, no matter which graphics card is present, we get a shaded image. Only the definition of the display varies, for different graphics cards. Saved images, with version 4, are always at the highest resolution.
– 4. Information from SURFACE DEFINITION FILES (version 3.3) or MATERIAL TABLES (version 4) is read. These are used for specifying the color of surfaces and their properties, whether they are smooth, rough, shiny, matte etc.
– 5. Color and brightness of each surface, in the view, is calculated. This takes into account the light sources in the file. With version 3.3 we can have up to 16 default light sources, while there is no limit with version 4. Also, version 4 has a choice of definable light sources.
– 6. Results are displayed on the screen.

Like a camera, rendering depends on light and its reflection off surfaces. In real-life, light rays reflect off objects. The resulting reflections depend on the color and type of surface, and its finish.

The way light is reflected varies from diffuse to mirror-like.

An example of diffuse reflection is the reflection of light from carpet on a floor. Diffuse reflection scatters the reflected light so that the surface appears equally bright from any direction.

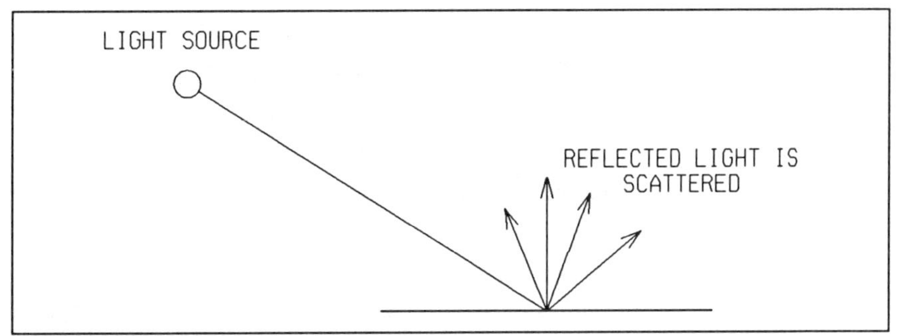

Figure 9.1 Diffuse Reflection

A window reflected in a highly polished floor is an example of mirror-like reflection. The reflection is dependent on the viewing direction. With a mirror reflection the angle of reflection is equal to the angle of incidence (figure 9.2).

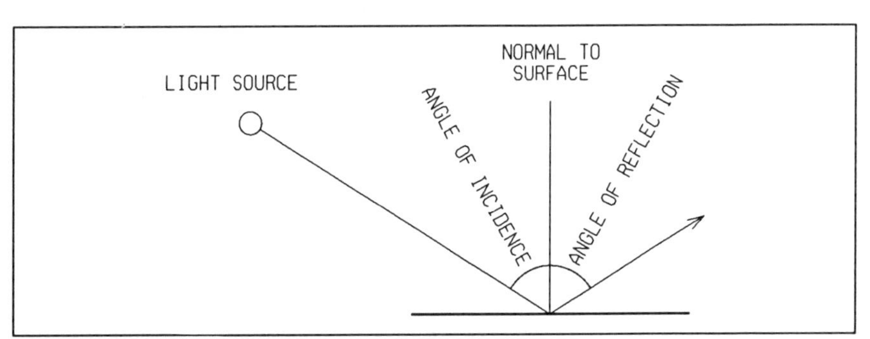

Figure 9.2 Mirror Reflection

Most surfaces reflect light somewhere in between the two extremes of diffuse and mirror-like. With smooth reflective surfaces we get specular highlights. These appear brightest when the viewing point is at, or near, the mirror angle of reflection. They become much dimmer as we move away from that angle or direction. Think of the glint of sunlight reflected off a car. By moving to one side we can reduce the intensity of the reflection dramatically.

With a computer model, we have to input all the information regarding the various surfaces, lighting conditions and viewing directions. We have to tell the rendering process the color of each surface and which ones are smooth, rough, shiny or matte. As was mentioned previously we do this with SURFACE DEFINITION FILES (version 3.3) or MATERIAL TABLES (version 4). Definitions of surfaces in version 3.3 are linked to the colors of elements in the design file. Version 4 allows us to link surface definitions to the colors of elements, the levels they are on, or a combination of the two.

There are various forms of rendering or shading. MicroStation gives us a choice of three with version 4 - CONSTANT, SMOOTH or GOURAUD, and PHONG. With each process, curved surfaces are approximated by a mesh of polygons. The degree to which curved surfaces are broken down is controlled by 'switches' with version 3.3 and 'Stroke Tolerance', in the rendering dialog box, with version 4.

Figure 9.3 shows a section of a smooth curved surface. The arrows are normals to the surface and indicate its instantaneous orientation. The orientation of a curved surface varies smoothly across the surface. Rendering processes use surface normal vectors to determine the color at any given point.

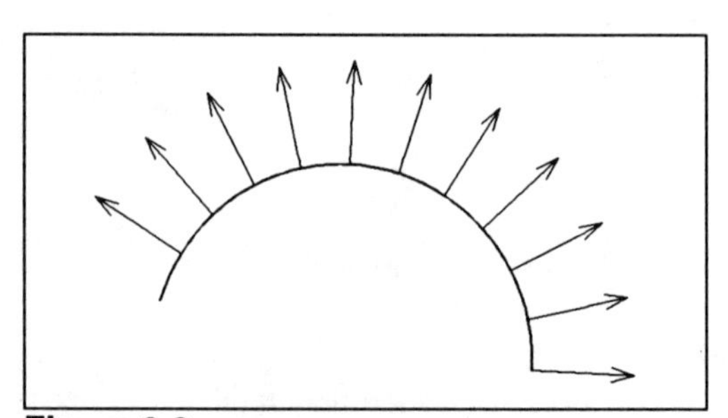

Figure 9.3

While each of the rendering processes uses polygons to calculate color values, it is the treatment of these polygons that differentiates between them.

Constant Rendering

Figure 9.4 shows a section through our smooth curved surface after it has been reduced to a number of polygons by the renderer.

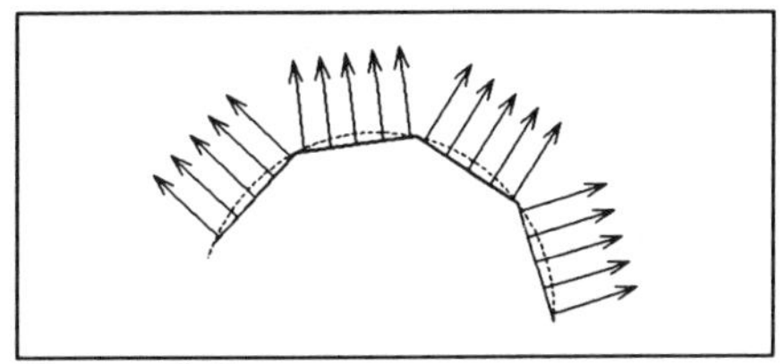

Figure 9.4

With CONSTANT rendering each of these polygons is assigned a single (constant) color. The color is calculated from the surface characteristics and the lighting conditions. Because of the change of color at the edges of the polygons, we get a 'tiled' or faceted effect as in figure 9.5.

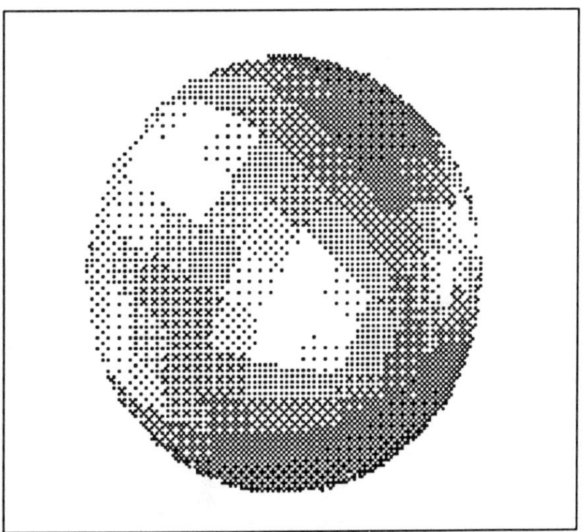

Figure 9.5 Constant Rendering

Specifying finer tolerances reduces the tiled look, at the expense of longer processing time. The sphere shown has highly reflective properties, hence the specular highlights which are reflections of the light sources. With version 3.3, we get CONSTANT rendering, only.

Gouraud (Smooth) Rendering

SMOOTH, or GOURAUD rendering involves linearly interpolating the color values across the polygons. Rather than each polygon being assigned one color, as in CONSTANT rendering, they are assigned a range of colors. First, colors at the polygon boundaries are determined. These colors are then interpolated, or graded, across each polygon. With this smooth grading, we don't get the abrupt color changes at the polygon boundaries of CONSTANT rendering. As figure 9.6 shows, the result is a much smoother effect than that from CONSTANT shading, for the same tolerance setting. We don't get the 'tiled' effect of CONSTANT shading.

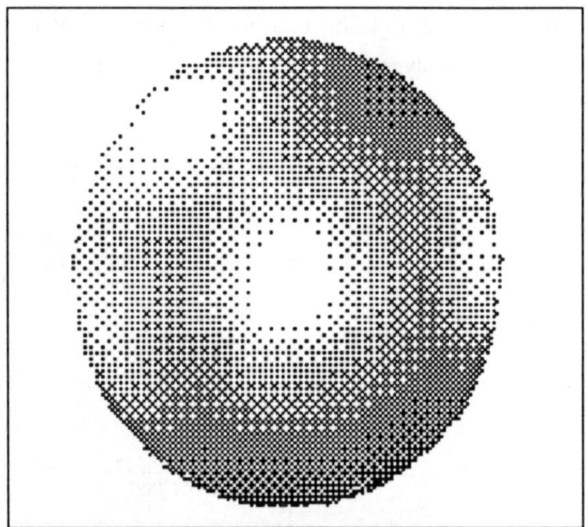

Figure 9.6 Gouraud (Smooth) Rendering

Phong Rendering

PHONG rendering involves the interpolation of surface normals. Here, the surface normals are calculated at the polygon vertices. An interpolation of these surface normals is then performed across the polygon. Color values are then determined from the interpolation of the surface normals. While the resulting image (figure 9.7) is similar to that from SMOOTH rendering there is better definition of the reflection from the light sources.

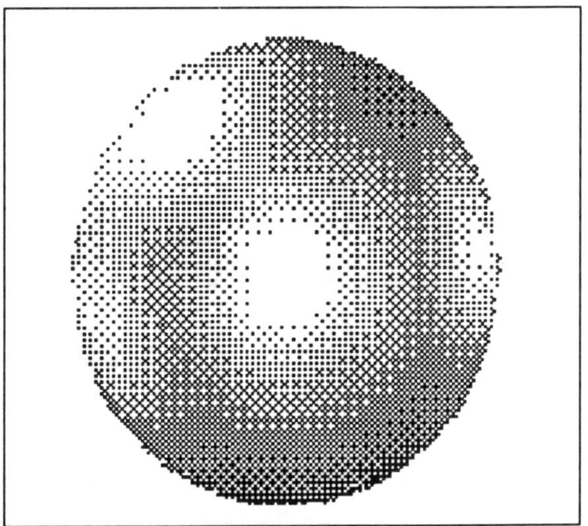

Figure 9.7 Phong Rendering

Where light sources are close to the object and it is important to see the exact location of the beam, then PHONG rendering may be the only way to get the desired results. PHONG rendering produces a more realistic display at the cost of much longer processing time. In most instances, the results are similar to SMOOTH rendering and the extra processing time is not warranted.

In the previous descriptions of the different forms of rendering, the same sphere, with the same tolerances was used. MicroStation version 4 performed the rendering, set to the various options - CONSTANT, SMOOTH and PHONG. Times taken to produce each image were as follows:

CONSTANT - 8 seconds
SMOOTH - 10 seconds
PHONG - 48 seconds

These times are given as an indication only. It is obvious from these figures that Phong rendering takes much longer, relatively, than the other two.

Why Use Rendering?

This question is probably not as relevant as 'why use 3D?'. If three-dimensional modelling is being used, then rendering is a natural follow on, if not a necessity. Wireframe format is fine for the operator to use in the creation of the model, and for initial design work. Rendered images or hidden line removals, however, are more easily understood by everybody.

To get an overall view of a building project or a component, an isometric style 'picture' is much easier to work with than plans and sections. Automobile manuals, for example invariably include exploded isometric views of the various components as they are discussed. They are much easier to relate to real objects. For the same reason, engineering firms often employ technical illustrators and model makers to create more recognizable 'images' of designs. With three-dimensional CAD we can see how our design looks, by using rendering or hidden line removal utilities.

There are other important advantages of 3D computer models over the traditional technical illustrations or scale models. As the design develops, or modifications are made, we can check for clashes with existing equipment for example. If it is not clear in wireframe form than we can use one of the rendering tools. Clearances also can be checked with the measuring tools that we have at our disposal.

We can produce up to date images of the model, simply. They can be produced overnight, if necessary, or at other non-productive times. Because they are produced from the design model, they may be considered a useful by-product of the design process, rather than a separate task.

With version 4 of MicroStation we can place our eye-point anywhere within the model, looking in any direction. We can produce successive images to simulate a 'walk-through', for example, or 'fly-around' of the model. Unlike most plastic models, we can show our clients views from any location.

Version 3.3 does not have the same facility for viewing the design from within the model itself. Still, we can produce work-arounds for many situations by using level control. That is, placing discrete parts of the model onto separate levels. Saved views can then be created for the required areas of the design.

Summary

Rendering and Hidden Line Removal are powerful tools that we can use to convert our wireframe models into 'real' objects. This chapter has been an introduction to rendering. The following two chapters, one each for version 3.3 and version 4, show you how to use these tools.

10 : Rendering version 3.3

As was mentioned in chapter 9 (Introduction to Rendering), there are major differences in rendering between versions 3.3 and 4 of MicroStation. This chapter deals specifically with rendering for version 3.3.

When working in a design file, we see the elements in wireframe form, where only the outlines of surfaces display, and the model can be seen through. With the RENDERING utilities we can produce more realistic views of our model with the hidden surfaces not displaying. Also, we can get high quality rendered or shaded images. Surfaces can be defined as being metallic, shiny, matte, etc. and light sources may be placed in the file for the rendering process. The rendered images may be photographed or saved to disk for use with other processes. As well, drawings can be produced, using the EDGES option. This is dealt with in chapter 12 (Drawing Production).

It is possible to vary the way the rendering process displays our model with the many options, or switches, available. You may find that the default settings are all that you require. Still look at the other options. There may be an occasion where a simple modification will greatly improve a presentation, or drawing.

Most of the options may be keyed in at the time of creating the hidden line removal or rendering. For some variables, however, you will require a text editor. This is true, in particular, for modifying the SURFACE DEFINITION FILES. These files are in ASCII format (i.e., standard text format).

Prerequisites

When installing MicroStation version 3.3 on a P.C., one question asked is whether or not we wish to install the RENDERING utilities. These must be installed before we can perform any of the hidden line removal or rendering processes.

The relevant program files, if they have been installed, are in the USTATION\RENDER directory. Also in this directory is a text file named RENDER.TXT. It is recommended that you obtain a printout of this file. You can use it as a complete manual for hidden line removal, rendering, edges design files etc. Don't be concerned if you can't get a printout immediately. Topics covered in this chapter provide enough information for most normal situations.

Hardware Requirements

To carry out the shading, or hidden line removal, the system performs many calculations. Unless your computer has expanded memory, this process can be very time consuming. If MicroStation finds that there is not enough memory, it writes the information to a temporary file on the disk. The time taken can be many times that of a computer with expanded memory. If you have extended memory with a RAM drive, then the problem can be alleviated by configuring MicroStation to use this for its temporary files. For rendering, the recommended amount of expanded memory required is 2 bytes per pixel of your graphics screen. For a 1024x768 resolution screen, this works out to be just over 1.5 Megabytes. If the rendering is in one quadrant of the screen then the amount of expanded memory required is one quarter of that for the entire screen.

Producing a rendered or shaded image also requires a graphics card capable of displaying 256 colors simultaneously. If this is not available then MicroStation performs a hidden line removal with panel fill. The outlines of the surface polygons are displayed in color 0 (usually white). Enclosed areas are filled with the color of the original element, giving a 'cartoon-like' effect.

Selecting from the Menu

While the paper or digitizer menu does not have the hidden line removal utilities on it, the standard sidebar menu, USTN.SBM, does. They are located in the UTILS sub-menu. We have a selection of 5 processes. They are:

ShadeS - Produces shaded images of a view.

ShadeE - Shades a selected element.

Hline - For hidden line removal of a view.

HlineE - Performs hidden line removal on a selected element.

Edges - Creates a design file of the hidden line removal.

Each of the above calls up a user command. If you are using the digitizer menu, an option available to you is to have the user commands called from a function key menu. The sidebar menu selections and their associated key-ins to call the user commands are as follows:

ShadeS - UC=MS_RENDER:SHADES
ShadeE - UC=MS_RENDER:SHADELEM
Hline - UC=MS_RENDER:HLINE
HlineE - UC=MS_RENDER:HLINELEM
Edges - UC=MS_RENDER:EDGES

Of these, all but Edges are raster based. That is, the hidden line removal, or shading, is performed at screen resolution. They appear on the screen only. We can save them to disk using a screen saving utility, such as SNAPSHOT or SAVERGB. Alternatively we can photograph them. We refer to Edges as a vector based hidden line removal. Hidden line removal with Edges is performed at design file resolution. Its output, is to a design file, which may be plotted like a normal file. We use Edges to produce drawings from our 3D models (refer chapter 12 - Drawing Production).

We can halt all the processes with a reset. If you have seen enough and don't want to wait for the entire view, then hit reset and the process will terminate. Updating the screen will return it to wireframe mode.

Rendering in a Fence

If a fence is present in the selected view, then only those elements contained in the fence will be processed. It is possible, by using fences, to have the screen contain a combination of wireframe form, hidden line and shading.

Shades, Hline and *Edges* allow us to enter options, sometimes called switches, which can affect the style and accuracy of the output.

Shades and Hline

The options for *Shades* and *Hline* come in 6 categories. They are:

-ct - Color Table Generation
-l - Light Sources
-o - Operating Mode
-pm - Polygon Meshing
-v - View Specifications
-z - Z buffer Parameters

When we select any of the Hidden Line Removal/Rendering utilities, we are prompted to 'Select View/Enter Options'. This is where our preferred values can be keyed in.

The file RENDER.TXT, mentioned earlier, has a full listing of the options or switches. The options that are generally 'played' with, and that we will look at, are those associated with the polygon meshing. These are the switches that affect the way that arcs, circles and curved surfaces appear when using hidden line removal or shading.

The Hidden Line Removal process represents arcs and circles with many short straight sections (i.e., chords). The number of chords is always between the minimum and maximum number specified. We can specify minimum and maximum values for the number of chords used. The minimum number allowed is 3, with 15 the default. The maximum number allowed, is 200, which is also the default. The process determines the number of chords used, based on the minimum and maximum values, and the size of the circle or arc.

To specify a minimum chord value we use the *-pmn* option. Entering '-pmn70' at the options prompt specifies that the minimum number of chords in any circle or arc is 70. Maximum chord value is entered with the *-pmx* option.

As well as the above, which affect the chord to arc tolerance we have a third variable. This is *-pmr* (polygon mesh resolution), which controls the way that curved surfaces display. These include surfaces of revolution, cylinders and B-spline surfaces. Values for *-pmr* are listed as between 200 and 10,000. The default is 1500. Values up to 32,000 for the -pmr switch can be entered, but the maximum stated in RENDER.TXT is 10,000.

Processing time is directly affected by these 3 options. The higher the figure used, to produce 'smoother' results, the longer the process takes. For general work the defaults are good enough. We can get an idea of how a view will look, quickly. We can then make any modifications that are required, to the view, before commencing the more time consuming final shading. With experimentation, you will find the combination of values that suits your application best.

Another switch that is very useful is *-vz*. This causes the screen to be cleared prior to hidden line removal or shading being performed. This switch should be used for the final shading at least. It ensures that there are no remnants of wireframe lines, remaining on the screen, showing in the rendering.

At the options prompt, keying in '-pmn50 -pmx100 -vz' sets the number of chords used to approximate arcs and circles to between 50 and 100. It also instructs the system to clear the screen first. We key the instructions in as shown, with a space between each option. There is no space between the option and the parameter (e.g., the *-pmx* and the *50*). In figures - 10.1, to 10.4 - on the following pages, the same model is used with different options and values.

Figure 10.1 is a standard Hline with the default settings for the options.

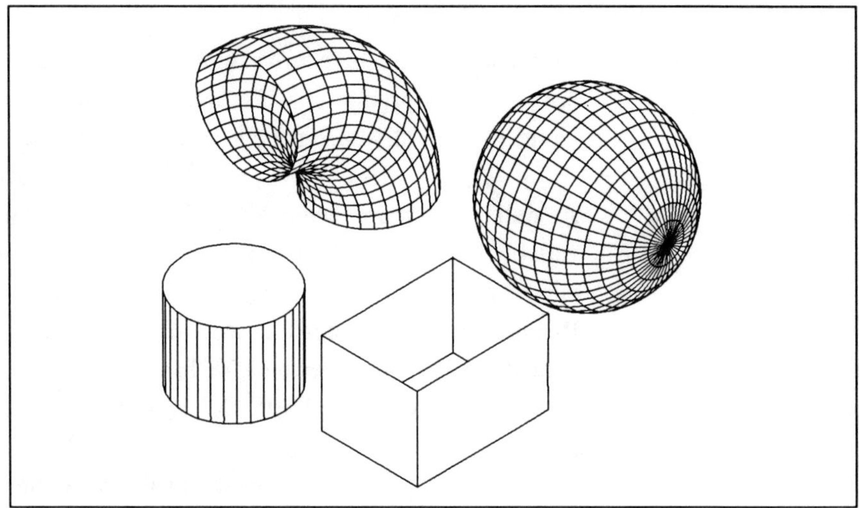

Figure 10.1 Standard Hline

Figure 10.2 has the maximum number of chords to arcs set to 20. That is, -pmx20. All other values are the default settings. This gives a much more 'coarse' representation of the curved surfaces.

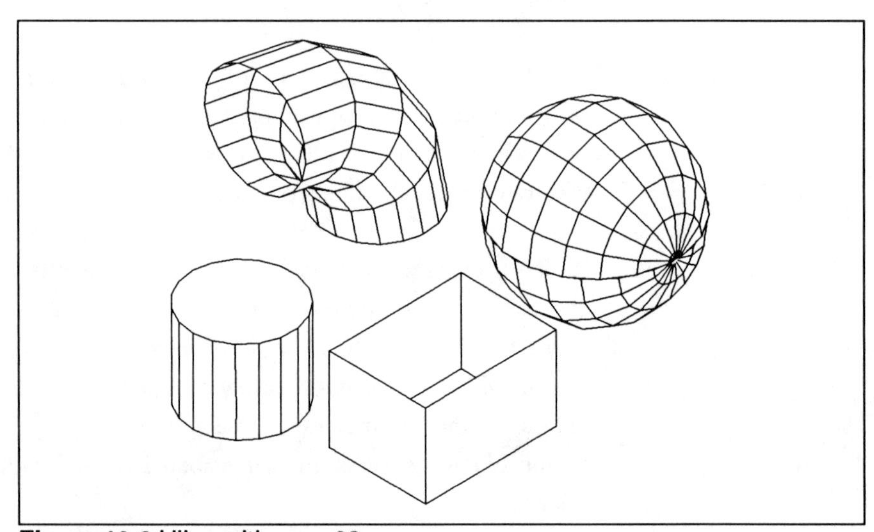

Figure 10.2 Hline with -pmx20

Figure 10.3 has the minimum number of chords to arcs set to 50, or -pmn50. Again, all other values are the default settings.

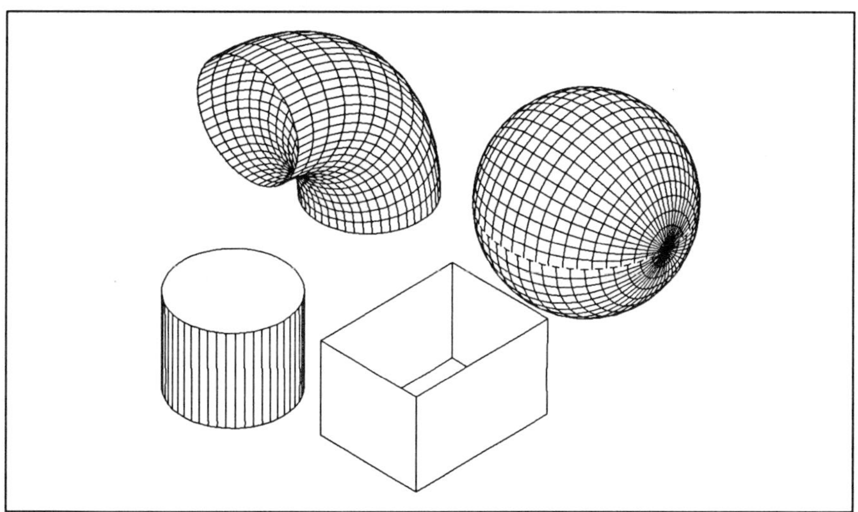

Figure 10.3 Hline with -pmn50

Figure 10.4 uses a -pmr value of 2500. All other options are at the default settings. Like figure 5.3 this also gives a finer resolution with the curved surfaces.

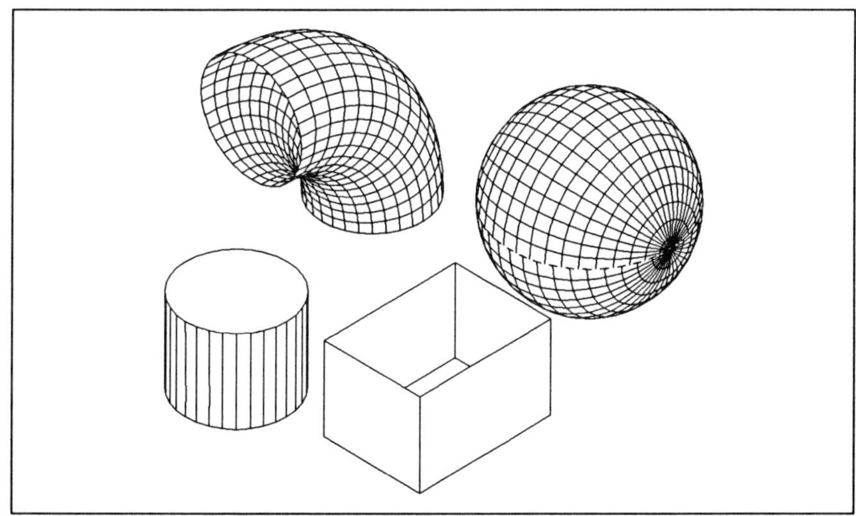

Figure 10.4 Hline with -pmr2500

Shades

As we saw in 'Hardware Requirements' the rendering or shaded-image facility of MicroStation requires a 256 color graphics card. Without a 256 color card we do not get a shaded image, but a hidden line removal with panel fill.

When we call up the SHADES user command MicroStation uses a Surface Definition File to create a temporary color table named RENDER.TBL. This is the color table that it uses to produce the shaded image. Surface Definition Files have the .SRF extension. Unless the rendering process finds a Surface Definition File with the same name as the model file (i.e., filename.SRF) the system uses a default file, SURFACE.SRF.

The maximum number of shades is 256 including the background and highlight colors. Therefore, the fewer separate colors that are used for rendering, the more shades of each color that are available. In other words, less colors can produce smoother shading.

Surface Definition Files

Surface Definition Files, having the '.SRF' extension, allow us to give different properties to elements of particular colors. We can have, for example, all the yellow elements shaded with a metallic sheen. Also, we can vary the number of colors used in the shading.

A design file may use more colors than are defined in the surface definition file. Here, the extra colors are assigned colors that have been defined. If you find that some elements are being shaded a different color to that expected, it is probably for this reason. For example, eight colors may have been defined in the surface definition file, and ten colors used in the design file.

Surface Definition Files are in ASCII format (i.e., normal text format). We may create new files or change the values in existing files with any text editor, such as EDLIN or EDWIN on PC's.

The following is the default surface definition file, SURFACE.SRF.

```
0   0    0    0                    background color
0   255  255  255  1.0  10  10     white
0   0    0    255  1.0  10  10     blue
0   0    255  0    1.0  10  10     green
0   255  0    0    1.0  10  10     red
0   255  255  0    1.0  10  10     yellow
0   255  0    255  1.0  10  10     magenta
0   255  128  0    1.0  10  10     orange
0   0    255  255  1.0  10  10     cyan
```

It consists of eight fields (columns). They are, in order -
OVER-RIDE, RED, GREEN, BLUE, DIFFUSE, REFLECTIVITY, ROUGHNESS, COMMENT.

The first line in a surface definition file defines the background color. From then on the lines describe elements created with the colors in numerical order. That is, line 2 describes elements with color 0, line 3 describes elements with color 1 and so on. Subtracting 2 from the number of the line in the surface definition file gives us the number of the color it is describing.

The maximum number of different surface types allowed is 32. So, the maximum number of lines allowed in a surface definition file is 33 (32 surfaces, plus background).The comment column can be used to note the color/effect that we are creating. We should also put the number of the color, in the design file, that it is describing. This makes it simple to make changes when necessary. If the default shading is good enough for your purposes then there is no need to create a new .SRF file for your design file. If you do want to modify the way some elements are shaded then the simplest way is to copy the file SURFACE.SRF to the same directory as your design file, giving it the same name as the design file except with a .SRF extension (i.e., filename.SRF). You may then modify this file with a text editor.

Briefly, the columns and their valid values are as follows:-

Column 1 - OVER-RIDE

0 - Ignore this line. Use corresponding line from SURFACE.SRF.
1 - Use this line. Ignore the corresponding line in SURFACE.SRF.
-1 - This line not used

Columns 2, 3 and 4 - RED, GREEN and BLUE respectively.

Setting each of these to values from 0 - 255 creates the various base colors.

Column 5 - DIFFUSE

The range for this column is 0.0 to 1.0
0 - Only show specular highlights.
0.1 - Glossy (or metallic) surfaces.
0.4 - Shiny surfaces.
1 - No specular highlights.

Column 6 - REFLECTIVITY

This column determines how much light is reflected from the surface. Common values are:-
1.0 - No light reflected, therefore surface appears grey or black.
10 - Non metallic (for normal matte surfaces).
1800 - Metallic (shiny).
Note: reflectivity is taken into account only when diffuse is set to a value less than one. That is, when we require some specular highlights.

Column 7 - ROUGHNESS

Must be a positive value.
0.3 - Smooth surfaces.
10 - Dull matte surfaces.

Column 8 - COMMENT

This column is useful for adding your comments such as the number of the color (in the design file) and what it will be when shaded.

For example, we may want a red element in our design file to be shaded with a metallic sheen. If our design file was named TRAIN3D.DGN, we would do the following:

- First, we create a surface definition file in the same directory as the design file. This can be done simply by copying SURFACE.SRF to be the same name as the design file, only with a .SRF extension.
- With DOS we would type 'COPY SURFACE.SRF TRAIN3D.SRF' from the DOS prompt.
- Using a text editor we would then alter the definition line for the color red. Red is color 3, so its description in the surface definition file is line 5. We modify this line to be:

  1 64 64 64 0.1 1800 0.3 Metallic (color 3)

  1, in the first column, tells the shading process to use this line and ignore the equivalent line in SURFACE.SRF.
  64 64 64 in columns 2,3 and 4 determines the color, grey.
  0.1 in column 5 creates a glossy surface.
  1800 in column 6 means a highly reflective surface.
  0.3 in column 7 tells the process that the surface is smooth.
  Metallic (color 3) in column 8 is a comment to remind us that the color we have changed is color 3 in the color table of the design file. The new color when rendered will be metallic.

The other lines would have 0 in the first column to indicate that the default file SURFACE.SRF should be used.

If we now use *ShadeS* on this file, any surfaces, that are colored red in the design file, will be rendered as shiny metallic grey. The colors will be taken from the color table RENDER.TBL which is created from the surface definition file. If you want to keep this color table for future use, just copy it to be another name. Each time ShadeS is used, a new version of RENDER.TBL is created. If your surface definition files are different from the default, then the color table, RENDER.TBL, will differ from the default color table.

Sample Surface Definition Files

MicroStation is delivered with sample Surface Definition Files. Apart from the default file SURFACE.SRF there are two others. They are SURFACE8.SRF and SURFAC16.SRF. They contain 8 and 16 surface definitions respectively. SURFACE8.SRF is the same as SURFACE.SRF.

With experience you should be able to create enough varying effects to give your models a much more lifelike look, when rendered.

Light Sources

As we discussed in chapter 9 (Introduction to Rendering), all rendering depends on the interaction of light with the various surfaces. Surface types are determined from a Surface Definition File. Light sources are cells that we place in our design file.

With version 3.3 we can place up to 16 light sources in our file. A light source is a cell named 'LIGHT'. MicroStation provides a standard cell library (LIGHT.CEL) containing a cell with this name. It is a normal cell, a sphere, and can be manipulated like any other cell. If you find it too big or too small, you can scale it up or down.

Light sources should be placed on a level that is separate from the model. This level can be disabled so that the light sources are not rendered along with the model. The rendering process searches the file for light sources, whether they are located on a displayed level or not.

Light emanating from each light source is in all directions. Distance from the model has no effect. Each light source is as bright as any other. To create a brighter light, we can place multiple light sources at the same location.

Once light sources have been placed in a file we can use Shades to render our selected view. Initially, the default values are all that is needed to get an idea of how the final image will look. We can move our light sources to change the shading on various elements, as required.

Often, we may want to move a light source in one direction (i.e., up or down, right or left). A simple way to do this is to SET BORESITE LOCK ON and use a data point to identify the light source. This allows us to move the light source without it 'jumping' to the ACTIVE DEPTH of the view.

Saving Shaded Images

We can save our rendered images to disk with a user commands such as SNAPSHOT. Chapter 13 (Utilities) has a section on this user command.

Edges

Whereas HLINE and SHADES produce output to the screen only, EDGES sends its output to a design file, which we refer to as an *Edges Design File*. The process creates a new design file into which all the visible elements are written. From this file, plots may be taken as from any design file. We can use EDGES to generate drawings from our 3D models. If a fence exists in the view to be processed, only elements within the fence will be considered. EDGES is a vector based process, meaning it works at design file resolution. Because of this, it is much more time consuming than the other rendering processes. We can run EDGES either from within MicroStation, or from the DOS prompt.

There are a number of options, or switches, available to us, which affect the generation of the edges design file. These may be keyed in at the 'Select view/Enter options' prompt when running EDGES interactively. They may also be included in an input file for running in batch mode from the DOS prompt. A third alternative is to include them (permanently) in the user command. Altering the user command should only be attempted by experienced programmers.

An advantage with running EDGES from DOS is that a batch file can be created to process more than one view, or file. This could be run overnight or at other times when the computer would be idle.

Options

When we create an edges design file we can control the generation of the file with the following options (switches). These options may be added to the EDGES.UCM user command, specified on the command line when running in standalone mode, or added to any line of an input command file.
Options available with the EDGES process are:

-ED2, -ED3

Selects either 2D or 3D edges design file

-EF

To specify the name of the edges design file. The default is the same name as the design file with a '.HLN' extension.

-HL

If positive, the hidden lines will be placed on this level. Default is -1 and no hidden edge elements are generated.

-HC, -HS, -HW

Color and weight of the hidden edge elements can be specified with -HC and -HW respectively. The default is for the elements to have the same color and weight as the design file elements.

The line style of the hidden edge elements can be specified with -HS. Here, the default is style 1 (dashed).

-V

specifies the view number to be processed. Default is view 1.

-P, -N

Projected surfaces and surfaces of revolution are approximated by a number of rectangles. The default (-P) is to display profiles only (i.e., the outside edges of the surfaces). With the -N option, we can specify that the rectangles are displayed also.

Figure 10.5 shows an edges design file that was created with the default setting. Only the profiles of the objects are shown. With the -N option, the generated file would look similar to that in figure 10.1, with the rectangles displayed.

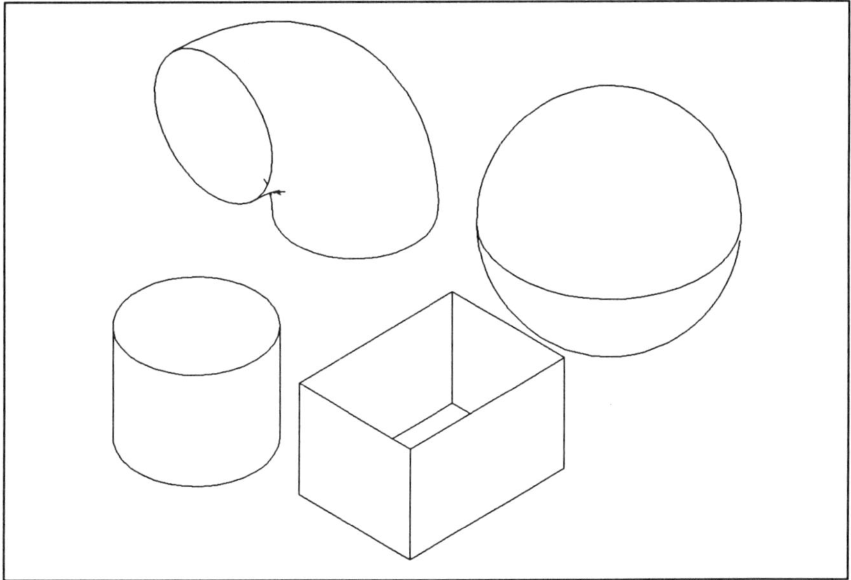

Figure 10.5 Standard Edges Design File

-M, -L

Arcs and circles are approximated by a number of straight chords. This number is based on the size of the circle and will be between the minimum and maximum values.

Maximum value is specified with -M. The default is 200 and the maximum 300.

Minimum value is specified with -L. The default is 15 and the minimum 3. Increasing the value of the minimum value gives smoother rounded surfaces at the cost of longer processing time.

-U

must be specified when the hidden line removal task is invoked from a user command.

Options are entered in the form: -V1 -L50 -EFfilename

Each option is preceded by a hyphen, and there is no space between the option and its parameter (e.g., no gap between -L and the value 50).

Creating Edges Design Files

Edges design files may be generated from within Microstation by selecting EDGES in the sidebar menu. Alternatively, we can key in UC=MS_RENDER:EDGES to run the user command. We are then prompted to 'Select view/Enter options'. It is here that we can key in one or more options prior to selecting the view for processing.

Edges processing may be aborted with a RESET from the tablet or mouse, or CONTROL C from the keyboard.

Batch Mode

Edges design files may optionally be generated from outside Microstation. The primary graphics screen displays the visible elements of the file as they are processed, giving a visual indication of progress.

Microstation must have been run prior to using hidden line removal, to load the resident drivers.

We can create an edges design file from view 1 of a file TEST.DGN, by keying in the following: \USTATION\RENDER\HIDDEN TEST.DGN

To the above key-in the various options or switches may be added for different conditions.

Running Edges outside MicroStation, it is possible to process several views of one file, or several files, sequentially through an input command file.

We create the input command file using a text editor. For example, if we had two files - BUILDING.DGN and BLDMODS.DGN, the following input command file (e.g. INPUT.CMD) could be created:

```
BUILDING.DGN -V1 -EFBLD1.HLN
BUILDING.DGN -V2 -EFBLD2.HLN
BLDMODS.DGN -V1 -EFMODS1.HLN
```

Providing that MicroStation had been run previously, we would then start the process by entering :-
\USTATION\RENDER\HIDDEN @INPUT.CMD

This would produce three edges design files:
BLD1.HLN containing visible edges of view 1 from BUILDING.DGN
BLD2.HLN containing visible edges of view 2 from BUILDING .DGN
MODS1.HLN containing visible edges of view 1 from BLDMODS.DGN

Processing of the files could be completed overnight, without supervision.

Display Parameters

Before running any of the rendering utilities, the view to be processed should be FITted. This will set the display depth to the minimum required to display all the elements. We can then window area to, or place a fence around the part of the file that we want to process.

While we can window area, or zoom in and out, this version of the rendering utilities does not let us take sections through our models. That is, we can set the display depth such that we are inside the model. The process will attempt to proceed but the results will not be reliable. If this presents a problem, the best solution is to separate your file out with level control. Discrete sections of the model can be placed on separate levels. The other option is to upgrade to version 4 which does allow us to move anywhere within the model.

Set Perspective

A viewing parameter that we have not looked at yet is that specified with the SET PERSPECTIVE key-in. Our views are normally displayed with parallel projection. That is, objects with the same dimensions, display on the screen with the same dimensions. This is irrespective of their relative position in the viewing volume of the view. In real life objects that are further away would appear smaller than those that are closer to us.

With this key-in we specify our eye-point to be a particular distance in front of the display volume of the view. The display is then updated to show the same volume of the design, but with perspective. To illustrate this we will use the box that we created in the first part of the tutorial section of the book. This is a 500x600x400 rectangular box with text describing each face.

When we look at the box from the FRONT all we see is a shape and the text describing the front and back faces (figure 10.6). The words are on top of each other. MicroStation shows the view with parallel projection. The back face is directly behind the front face and the same dimension. We can't see readily that there is another face behind the front face. If we looked in the other views we would be able to see the other faces.

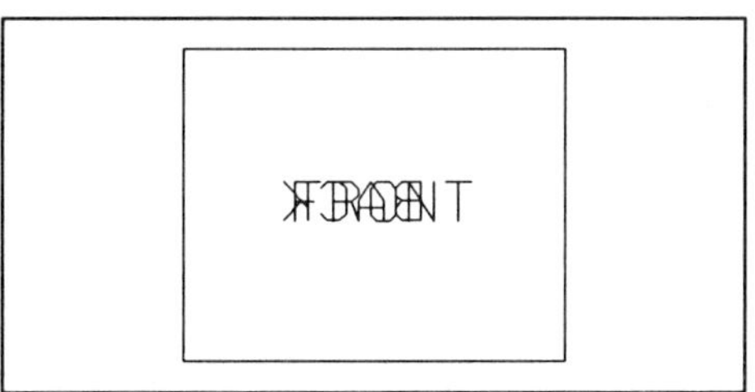

Figure 10.6 Front View - Perspective 0

If we key in SET PERSPECTIVE 750 and place a data point in the FRONT view, when prompted, the display changes. We can now see the other faces of the box (figure 10.7).

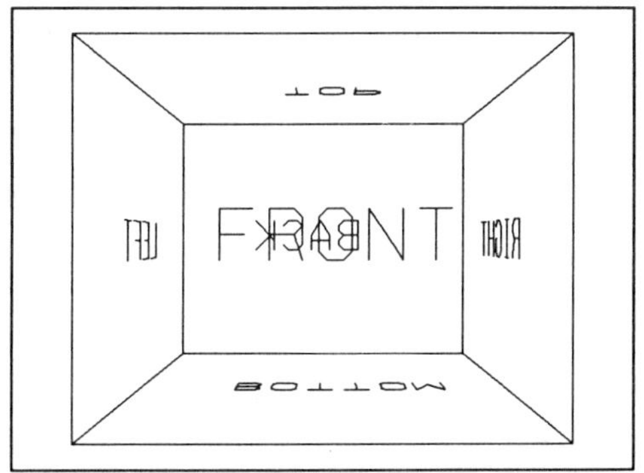

Figure 10.7 Front View - Perspective 750

We now see the box with the faces that are further away from our eye-point appearing smaller. This is a display parameter only, we cannot tentative to elements in this view. We can, however, still work in the other views. As we work in the other views, this view (with perspective set) updates like a normal view.

We can set perspective in any view. For example, rotating the FRONT view 45 degrees about the Y-axis would give something similar to figure 10.8

Figure 10.8 Rotated Front View

If we set perspective to 750 in this view then we get a much better idea of what the model looks like, as in figure 10.9. By using different views as we work in 3D we can form a picture in our mind of the model. Setting perspective is a quick way to see the model in a more normal aspect.

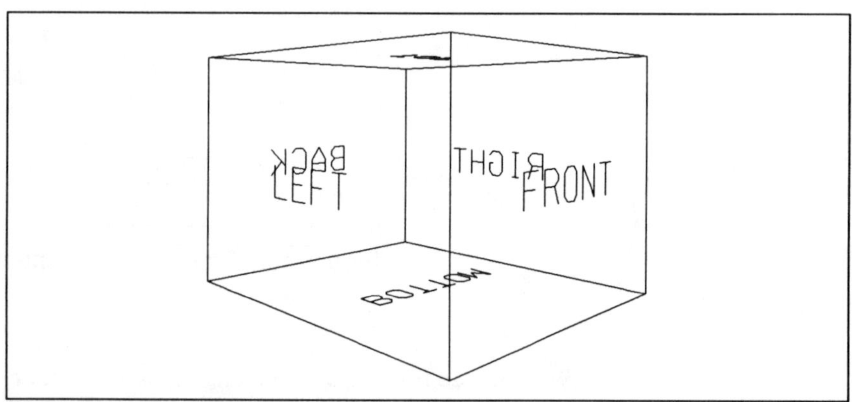

Figure 10.9 Rotated Front View - Perspective 750

How Perspective is Set

Figure 10.10 is a view from the top of a front view in which we have set the perspective to be 750. The model is a box similar to the one we made in the tutorial. The lines emanating from the eye-point are the projection lines. These determine the relative size of the various objects in our view.

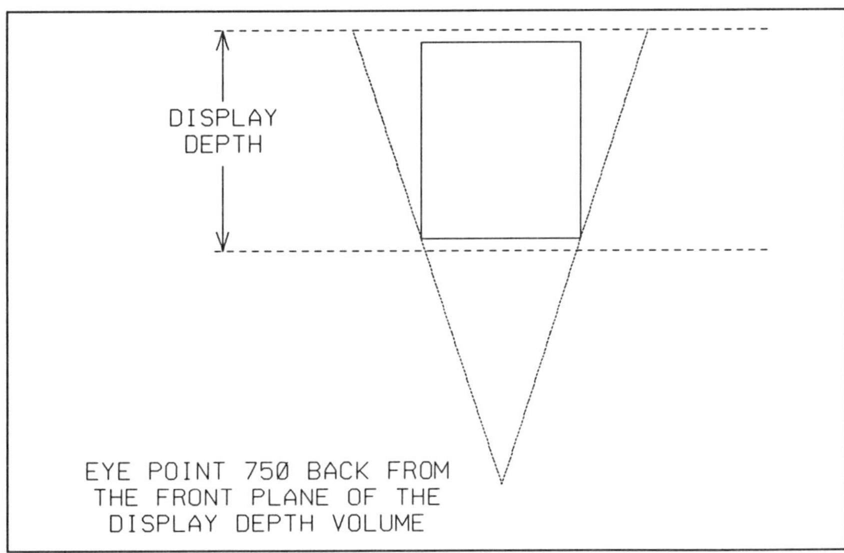

Figure 10.10 Setting Perspective

All objects are, in effect, projected onto the rear plane of the display depth of the view. The angle of the projection lines is determined by the distance of the eye-point from the front plane of the display depth. If we set our eye-point very close to this plane, then the perspective will be very pronounced. By experimenting with different values, we can usually arrive at a suitable perspective view for our purposes.

In figure 10.11 we can see clearly that setting perspective is similar to projecting the model back onto a common plane. Sizes of the elements are determined by their distance from the eye-point. More distant elements don't have to be projected as far and therefore do not increase in size as much.

To set the view back to normal parallel projection, we key in SET PERSPECTIVE 0 followed by a data point in the view.

We can use the rendering utilities, including EDGES, on views with perspective. In this way we can produce more realistic 'pictures' of the design showing perspective.

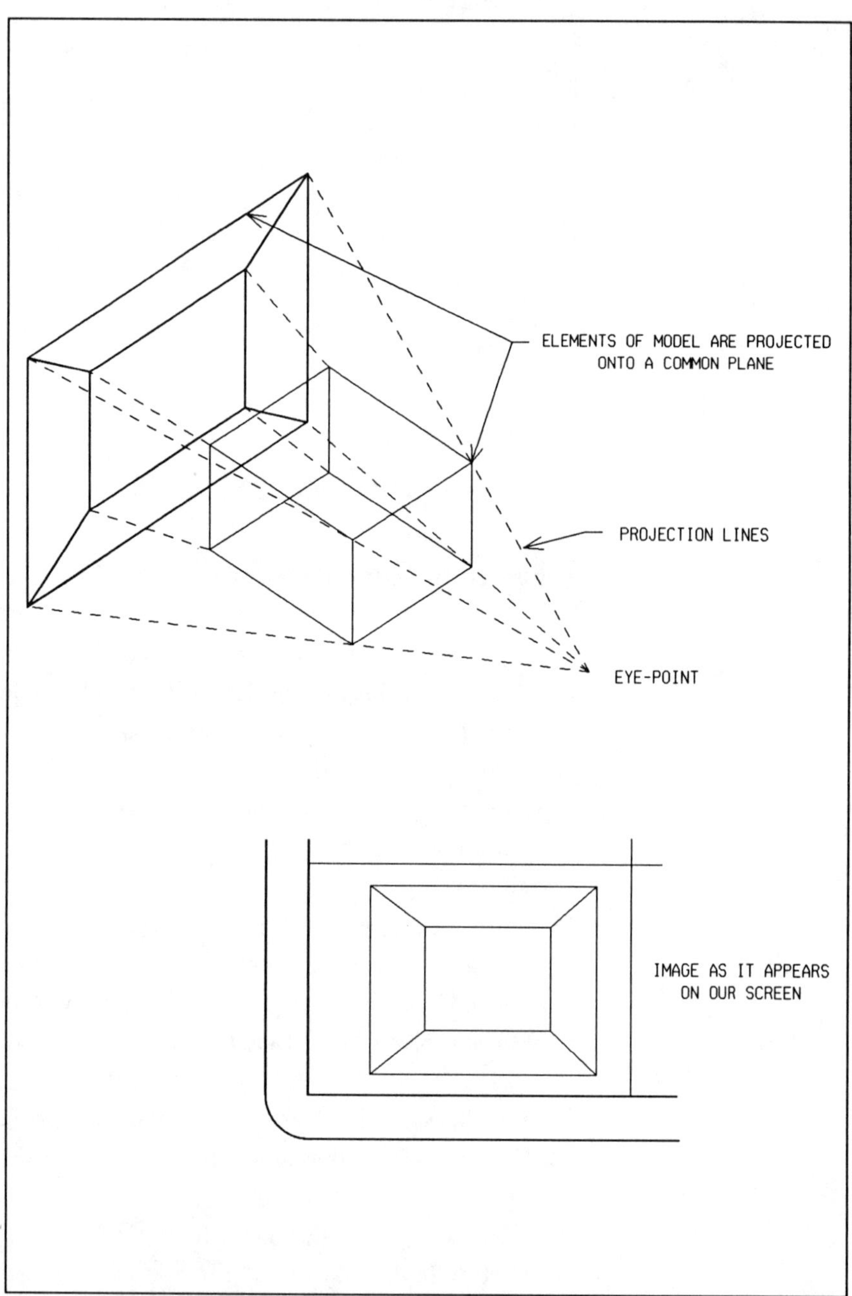

Figure 10.11 Set Perspective Explained

11 : Rendering version 4

In chapter 9 (Introduction to Rendering) we learnt the different terms associated with rendering. We also looked at how rendering works. In this chapter we will discuss the various tools that we can use with version 4, for hidden line removal, rendering and edges design files.

Render Options

When we select 'Render' from the 'View' pull-down menu, we are shown a choice of eight options (figure 11.1).

After choosing an option, we are prompted to select a view. The selected view is then rendered according to the chosen option.

This is a temporary display only. Updating the view returns it to its default display. We can change this default display also, with the View Attributes settings box. First we will see what the various options mean.

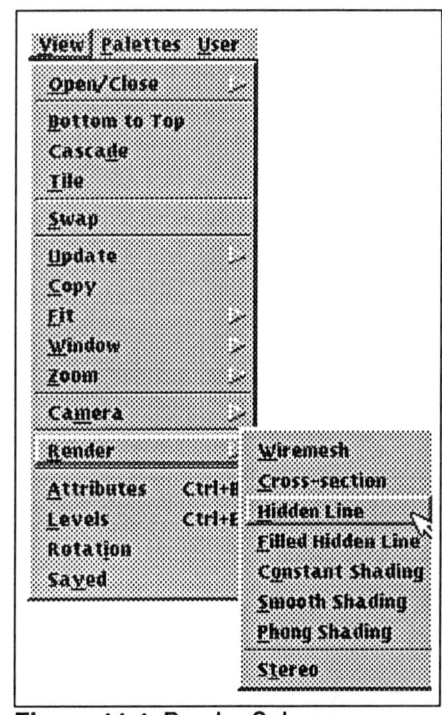

Figure 11.1 Render Sub-menu

To illustrate the various display options, we will use an ISO view of the first model used in chapter 8 (B-Spline Constructions). Figure 11.2 shows the wireframe view of this model.

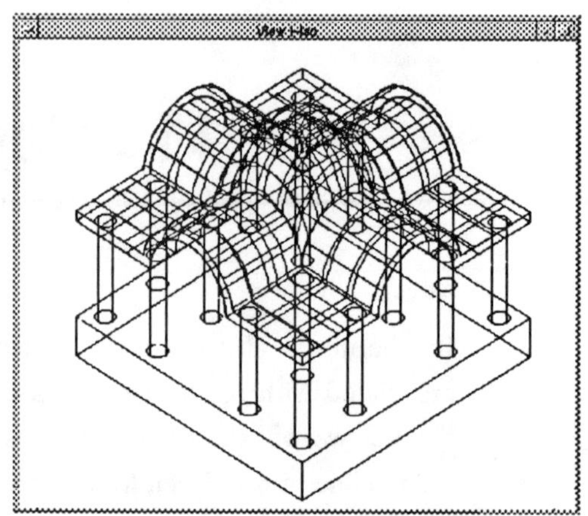

Figure 11.3 Default Wireframe Display

Wiremesh

In figure 11.3 we see the same model displayed in WIREMESH mode. This is similar to wireframe display, in that we can still see through the model. Where it differs, is that curved surfaces are shown in more detail. In our model this is illustrated with the b-spline surface 'roof' and the cylindrical columns.

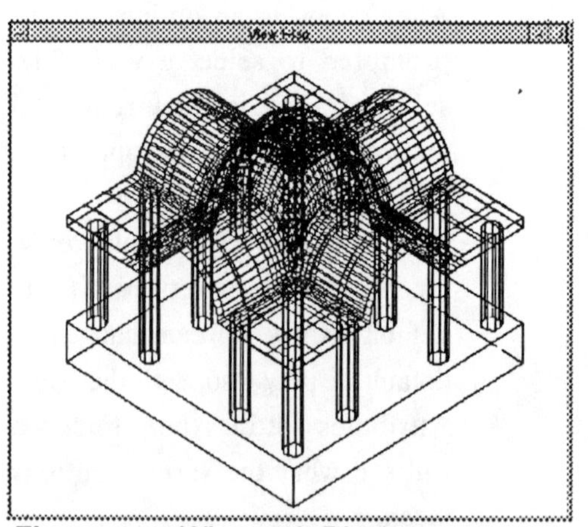

Figure 11.2 Wiremesh Display

Cross-section

Cross-section, as the name suggests, displays a section through the model, in the selected view. The section is taken at the ACTIVE DEPTH of the view. In this instance (figure 11.4), a RIGHT view has been used. ACTIVE DEPTH for the view was set at the center line of the first row of columns in the view.

We can vary the location of the section by changing the ACTIVE DEPTH of the view. As with all of these 'Render' options, this is for display only. Updating the view returns it to its default setting.

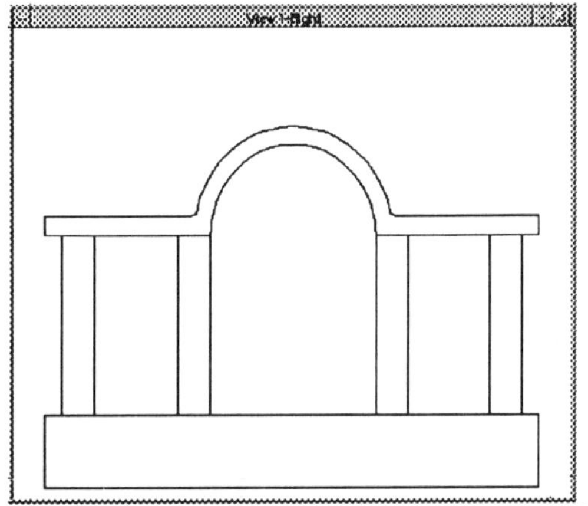

Figure 11.4 Cross-section taken at Active Depth

Where an auxiliary coordinate system (ACS) is active, the section is taken in the XY plane of the ACS. For the illustration, an ACS was set up along the center line of the columns along the right side. Figure 11.5 shows an ISO view of the model with the triad for the ACS visible.

Figure 11.6 shows the result of selecting 'Cross-section' display in this view. We see that, though we have an ISO view, the section has been taken in the XY plane of the ACS. It has not been taken parallel to the view, at the ACTIVE DEPTH.

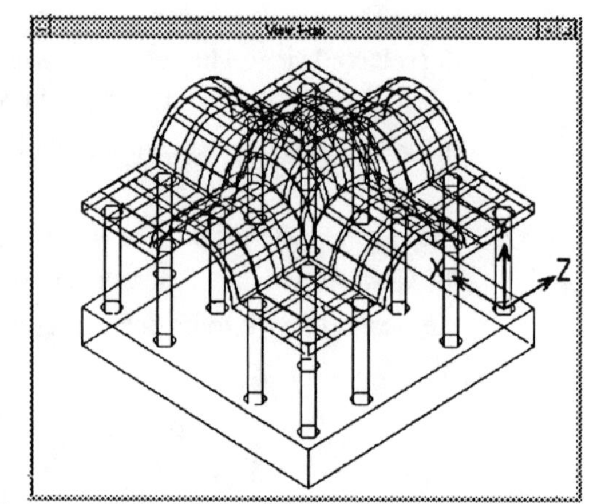

Figure 11.5 ISO view with ACS triad visible

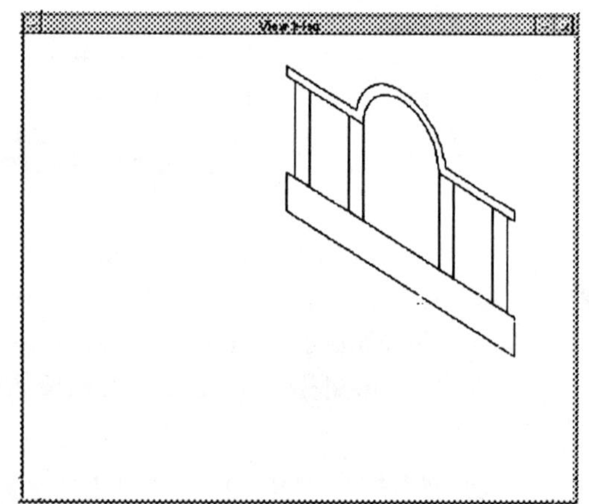

Figure 11.6 Cross-section with the ACS active

Hidden Line

With 'Hidden Line' we see a view of our model with the hidden lines removed. This results in a display similar to 'Wiremesh' except that we don't see through the model. Hidden lines/surfaces are not displayed.

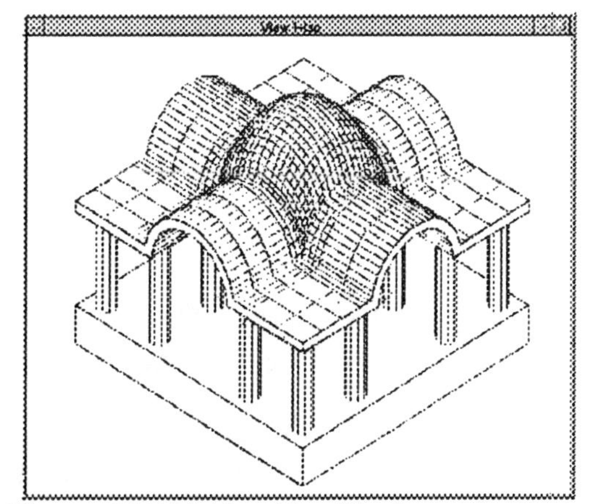

Figure 11.7 Hidden Line Display

Filled Hidden Line

Like 'Hidden Line' but polygons filled with colors of the elements (figure 11.8).

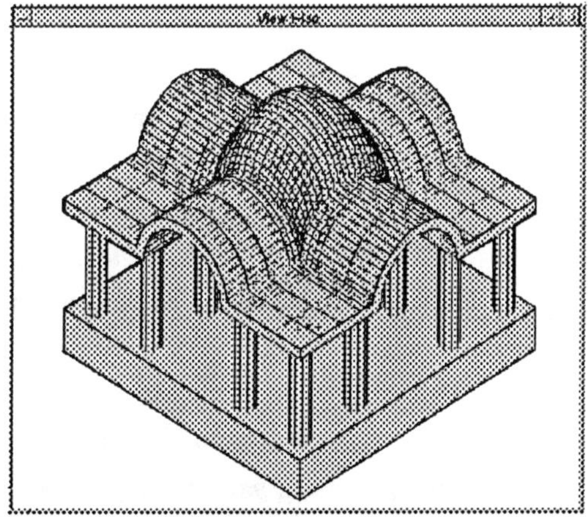

Figure 11.8 Filled Hidden Line Display

Constant, Smooth and Phong Shading

In Chapter 9 (Introduction to Rendering), we discussed each of these forms of shading or rendering. Figure 11.9 shows our model after the ISO view has been shaded. With shading we have more variables to consider. To change the way the model displays we can place light sources in the file. We can specify also, the surface characteristics of elements.

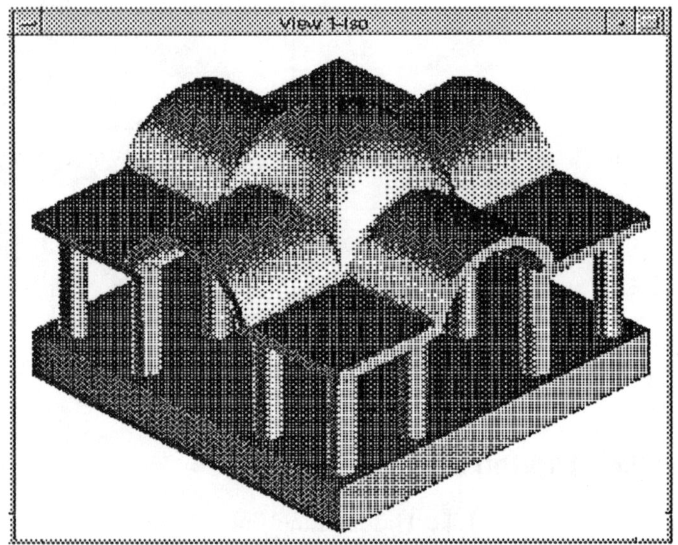

Figure 11.9 A Shaded Image of the Model

Stereo

We can set a view/s to display with a stereo effect. To see the stereo effect requires special 3D glasses to be worn. With this setting the view is rendered twice, once for the view as seen from the left eye, and once as seen from the right eye. These views are superimposed on each other. With the 3D glasses, the right eye's view is filtered out by the lens for the left eye, and vice versa.

This display requires the camera for the view to be on. If it is not on, then MicroStation turns it on before performing stereo rendering. Information on setting camera parameters follows later in this chapter.

View Attributes

Along with temporary view displays, we can set a view to be continuously rendered through the View Attributes settings box. When we click on the 'Display Mode' field we get an option menu allowing us to select from eight display modes. Stereo is not included here, but 'Wireframe' is. To see the new display for the view we have to update it. Display modes selected here remain in force, as the view default, till changed in the setting box or with the alternative key-in.

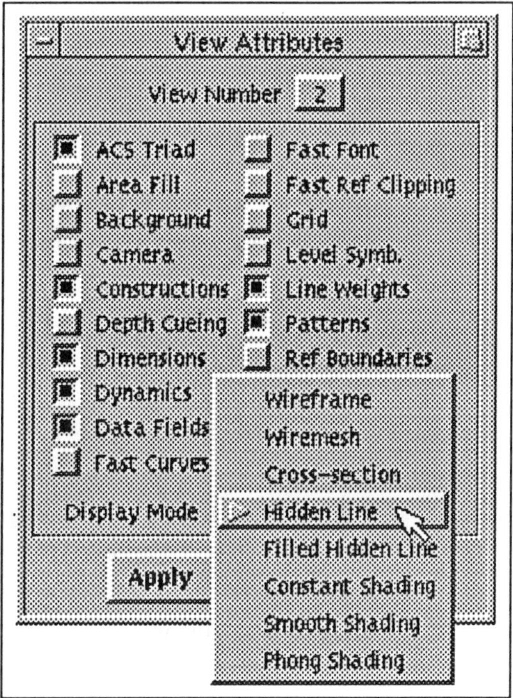

Figure 11.10 Setting Display Mode

Depth Cueing

Included in the view attributes is Depth Cueing. This setting can be used with shaded images. When set, elements that are furthest from the viewing point are dimmed as a visual aid, giving a greater feeling of depth to the shaded image.

Alternative Key-ins

We can set the view display via the dialog/settings boxes, or we can use key-ins. For one-off, or temporary rendering, we key-in *'RENDER VIEW parameter'*, where the parameter is one of:
WIREMESH, SECTION, HIDDEN, FILLED, CONSTANT, SMOOTH, PHONG or STEREO.

Updating, or working in the view returns it to its default display. To render a view as CONSTANT shading, for example, we key in:

RENDER VIEW CONSTANT

To set a view to be continuously rendered we use *'SET VIEW parameter'*, where the parameter is one of:

WIREFRAME, WIREMESH, SECTION, HIDDEN, FILLED, CONSTANT, SMOOTH or PHONG.

After choosing the view, we have to update it to see the new display.

To set a view to continuously display as CONSTANT shading, we key in:

SET VIEW CONSTANT

Setting a view to continuously display as rendered slows down screen updates, tentative points etc. appreciably. For most cases we work in wireframe mode, and when required, render a view using the 'RENDER VIEW' key-in or selection from the menu.

Views and Fences

We can render part of the contents of a view by using a fence. After the fence is placed in the view, we key-in *'RENDER FENCE parameter'* to complete the operation.

Where a view is set to be continuously rendered, we can use the 'UPDATE FENCE' key-ins. 'UPDATE FENCE', or 'UPDATE FENCE INSIDE' will update (i.e. render) the contents of the fence only. 'UPDATE FENCE OUTSIDE' will update the rest of the view, leaving the fence contents as is.

It is possible to have a view where different sections of it have been rendered in different manners as figure 11.11 shows. Here a fence was placed around the left part of the model and 'RENDER FENCE FILLED' keyed in. This procedure was repeated for the other two sections using 'SMOOTH' and 'HIDDEN' respectively.

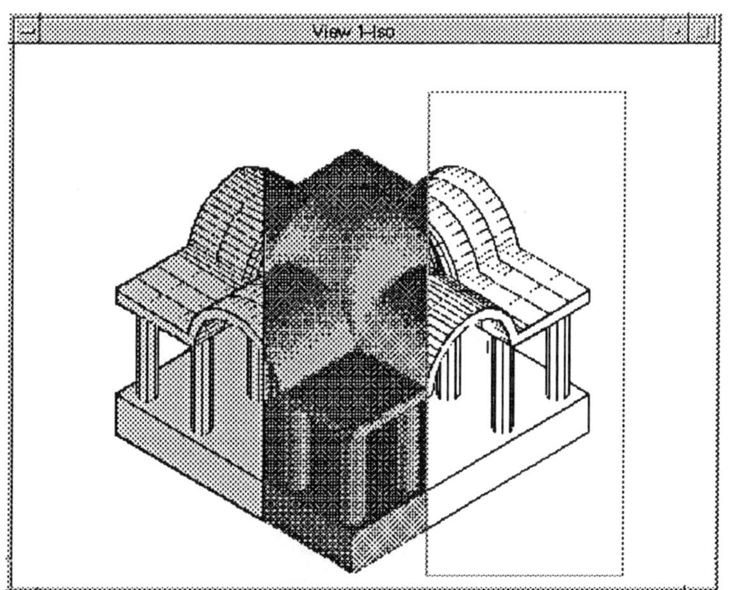

Figure 11.11 Rendering Contents of Fence

Rendering Settings Box

We can control the way that rendered images are displayed with the settings in the Rendering settings box. This is accessed by selecting 'Rendering' in the 'Settings' pull-down menu. While full explanations of the various settings are given in the MicroStation documentation, we will briefly discuss them here.

Figure 11.12 shows the settings box. You will see that it is divided into two sections, 'Polygon Generation' and 'Lighting'

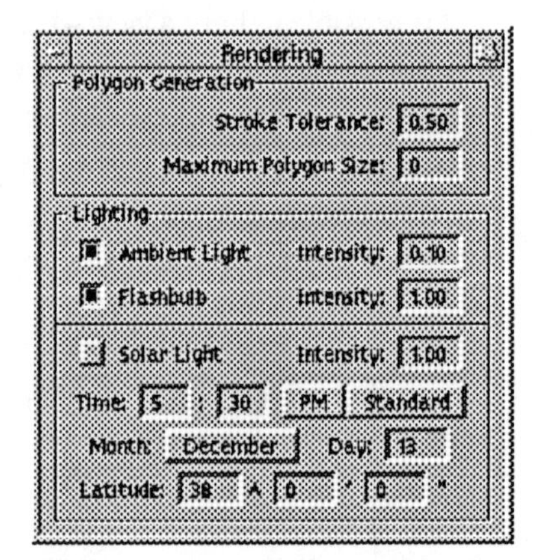

Figure 11.12 Render Settings Box

Polygon Generation

When a view is rendered, the model is broken down into smaller polygons. Generation of the polygons is controlled by two factors - Stroke Tolerance and Maximum Polygon Size.

Stroke Tolerance

Curved surfaces are rendered by breaking them into polygons. The Stroke Tolerance value determines the size of these polygons. A small value produces a large number of small polygons, forming a fine mesh, which follows the original surface closely. Larger values produce correspondingly larger polygons. A mesh formed with large polygons does not match the original surface as closely, but display times are much less. Figure 11.13 shows a view after being rendered HIDDEN, with the default Stroke Tolerance (0.50).

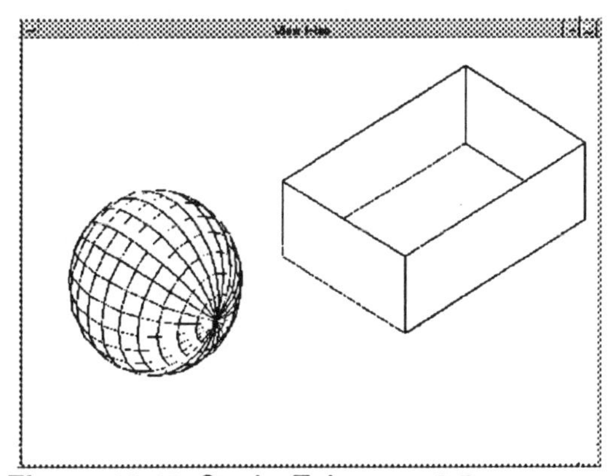

Figure 11.13 Stroke Tolerance 0.5

Figure 11.14 shows the same view, after rendering (Hidden) with a Stroke Tolerance of 0.10. The polygon mesh forming the sphere is much finer than in the previous figure. There is no change in the open rectangular box.

Stroke Tolerance affects the treatment of curved surfaces only.

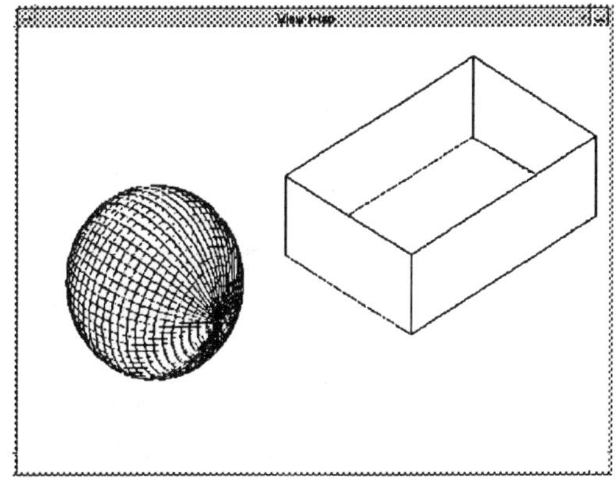

Figure 11.14 Stroke Tolerance 0.1

Maximum Polygon Size

Treatment of polygons (i.e., flat surfaces) by the rendering process is controlled by the Maximum Polygon Size setting. While this has no significance when we render a view as HIDDEN, it becomes important with shading. With the default setting of zero, there is no breaking of polygons into smaller polygons. The result is as for the box shown in figures 11.13 and 11.14.

Figure 11.15 shows the effects of changing the Maximum Polygon Size. Here, it has been set to 50. We can see that the rendering process has broken the flat surfaces (polygons) into smaller polygons. Maximum Polygon Size does not affect the treatment of the curved surfaces.

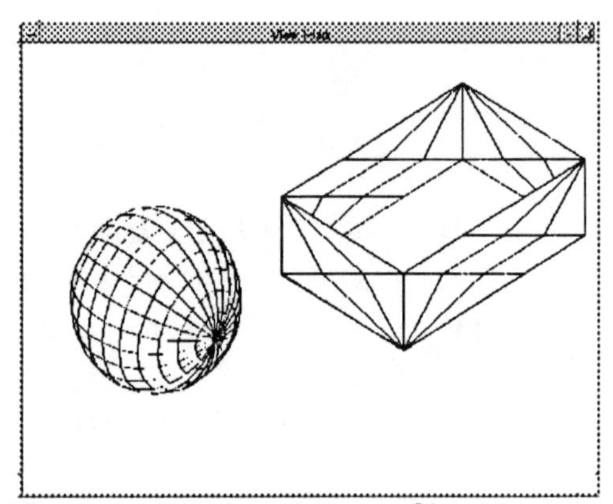

Figure 11.15 Max. Polygon Size 50

Modifying the Default Settings

With 'Constant' rendering, each polygon is assigned a color depending on its orientation and the location of light sources. Figure 11.16 shows the view after being rendered (CONSTANT) with the default settings. Notice that the sphere has a slightly tiled look, and each side of the box is shaded a single color.

In figure 11.17 the same view has been rendered, with the Maximum Polygon Size set to 50, and the Stroke Tolerance set to 0.1. Here, the sphere is much smoother. Also, the sides of the box are no longer colored uniformly. The rendering process has broken the flat surfaces into smaller polygons and assigned a color to each of them. This allows for a variation across the area of the flat sides reflecting the location of a light source. The sphere also has been broken into smaller polygons than with the default setting. This creates a much smoother effect. Both these modifications create a more realistic effect at the cost of longer processing time.

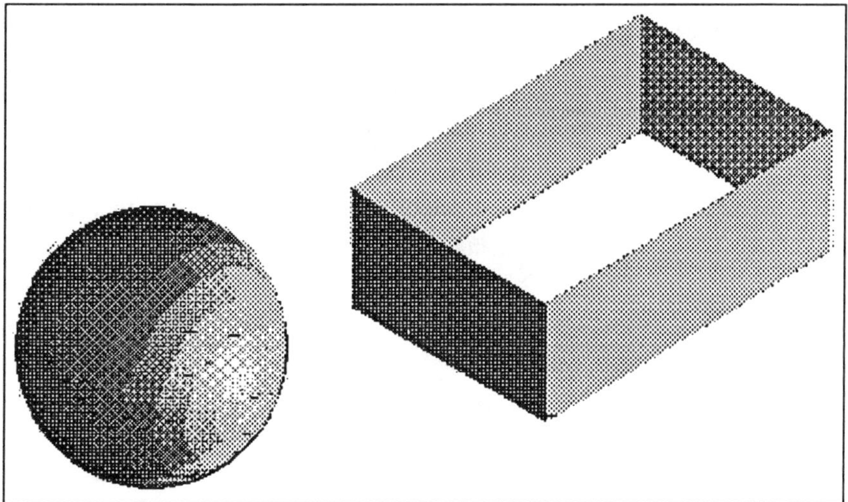

Figure 11.17 Constant Rendering with Defaults

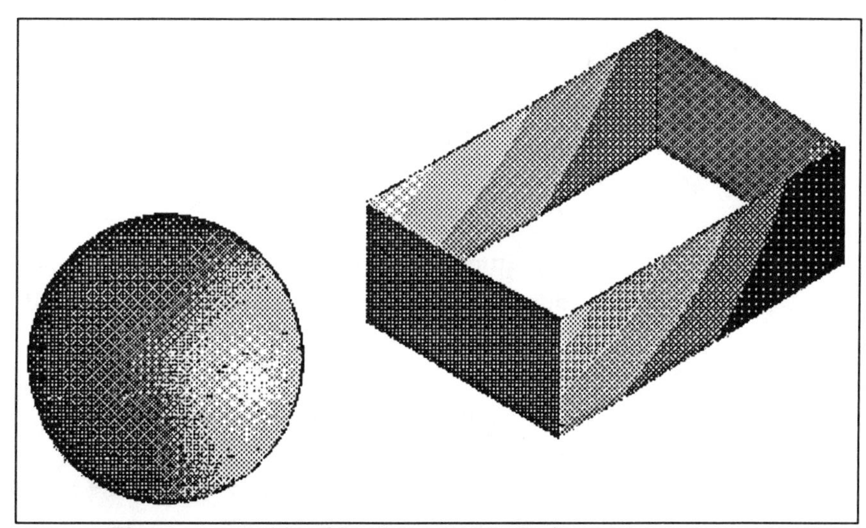

Figure 11.16 After Modifying the Defaults

Lighting

In the Lighting section of the Rendering settings box, we have 3 types of lighting at our disposal - Ambient, Flashbulb and Solar. Each can be set to be on or off. As well, the Intensity for each can be set to a value in the 0.0 to 1.0 range.

Ambient Light

This can be considered as background lighting. All elements are illuminated an equal amount by ambient lighting. A value of 0 specifies no ambient light, while a value of 1.0 illuminates all objects totally. A default of 0.1 is set. Increasing this value will bring out detail in objects that would not normally receive light (i.e., facing away from the light source).

Flashbulb

Flashbulb lighting is directional and originates from the viewing or camera location of the view. This is similar to a flash on a camera, hence the name. We can vary the intensity of the flash from 0.0 to 1. By default the flashbulb is on with an intensity of 1.

Solar Light

This lighting source simulates light from the sun. We can set the time, date and latitude. Where the time specified is during daylight hours then lighting is provided from the calculated direction of the sun. If the sun is calculated to be below the horizon, then solar lighting is not provided.

Light settings only have an effect when we use one of the shading options with rendering. That is, CONSTANT, SMOOTH, GOURAUD, PHONG or STEREO.

As well as these lighting settings the rendering process takes into account any light source cells placed in the design file.

Adding Realism to Shaded Images

So far we have discussed the mechanics of displaying our model as a rendered image. We have also looked at settings that can change the appearance of our rendered images. Now, we will look at how to create different effects with the rendering processes (i.e., shading). We will be using light sources that we place in the file, and Material Tables that tell the rendering process what type of surface an element is.

From chapter 9 (Introduction to Rendering) we know that shading/rendering depends on the reflection of light off surfaces. The surfaces may be anything from dull matte through to highly reflective. Also, reflections range from diffuse to mirror-like.

We create our models using colored elements on various levels. To get a realistic rendered image we can link these design file colors and/or levels to particular surface characteristics with Material Tables. As well as the characteristics used by MicroStation rendering, we can include items which are used by other software packages - for example, ModelView and RenderMan. Here, we will be dealing with those characteristics which are recognized by MicroStation. First, we will discuss light sources.

Light Sources

With version 4 we have a choice of light sources. These are cells with special names, contained in a cell library (LIGHTING.CEL) that comes standard with MicroStation. We can select from three types of light source, by using the relative cell. Cell names and their corresponding light source types are:

PNTLT - point light
DISTLT - distant light
SPOTLT - spot light

Figure 11.18 shows each of the cells. Enter data fields with each cell allow for specifying light color, intensity, beam distribution etc. Information about light sources is read from the file when a view is first rendered.

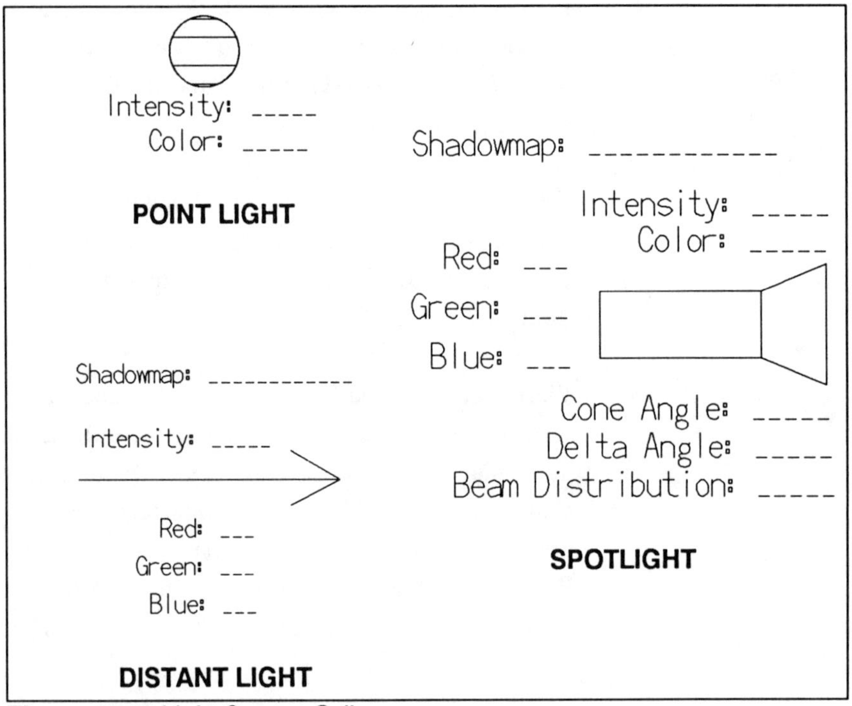

Intensity: -----
Color: -----

POINT LIGHT

Shadowmap: ------------

Intensity: -----

Red: ---
Green: ---
Blue: ---

DISTANT LIGHT

Shadowmap: ------------

Intensity: -----
Color: -----

Red: ---
Green: ---
Blue: ---

Cone Angle: -----
Delta Angle: -----
Beam Distribution: -----

SPOTLIGHT

Figure 11.18 Light Source Cells

Often, we make modifications to the light sources, after rendering has been performed. When this occurs, we must use the key in 'DEFINE LIGHTS' prior to performing further rendering. This ensures that the new lighting information is read into the file before we render the view.

Light cells are created as construction elements. Their display can be disabled with the 'SET CONSTRUCTION OFF' key-in. They may be manipulated like normal cells. Scaling up or down has no effect on their operation when shading a view.

Each light source has different characteristics.

POINT LIGHT sources radiate light, in all directions, from the origin of the cell.

DISTANT LIGHT sources cast parallel rays in the direction of the arrow placed in the cell.

SPOT LIGHT sources cast light in a conical beam from the center of the cell. The direction of the light is towards the wide end of the spotlight.

To show the variation in the respective lights we will use a simple model. For these exercises we will accept the default values for the light sources. Later, you can try modifying them by placing values in the enter data fields. Much of the work with rendering is trial and error to get the right effect. The following exercises will show how the different light sources illuminate the model.

To start, create a new 3D design file named RENDER1.DGN

- Enter the file and place a 1500x750 block in a TOP view using 0,0,0 and 1500,750,0 for the two points.
- Place 750x1000 blocks to form the left and right sides.
- Place a 1500x1000 block to form the back side
- Attach the cell library LIGHTING.CEL to the design file.

We will look at the POINT LIGHT first

- Place a point light cell (AC=PNTLT) at XY=750,375,500. Any view can be used for this operation because a point light source radiates light in all directions.

Figure 11.19 shows a view of the model with the point light in place. The light source in the figure was placed in a FRONT view.

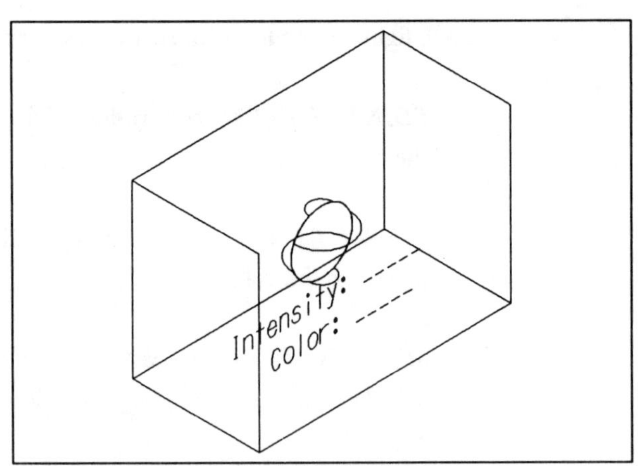

Figure 11.19 Model with PNTLT cell

For the rendering, use the ISO view.

- Select 'Rendering' in the 'Settings' pull-down menu, to access the Rendering settings box (refer figure 11.12).
- Set the Flashbulb and Ambient lights to off.
- Set the value of the Maximum Polygon Size to 10.
- Put the rendering settings box behind the ISO view window by clicking in the top right corner of the settings box.
- Key-in 'RENDER VIEW SMOOTH' and select the required view.

Point Light (AC=PNTLT)

In our exercise, we turned off the other light sources, Flashbulb and Ambient. The POINT LIGHT is the only lighting in the design file, located in the center of the open sided box.

You will see that all the interior surfaces are have been illuminated while the surfaces that face away from the light source are not. Figure 11.20 shows, the model after rendering. We can see that light radiates in all directions from the origin of the point light cell.

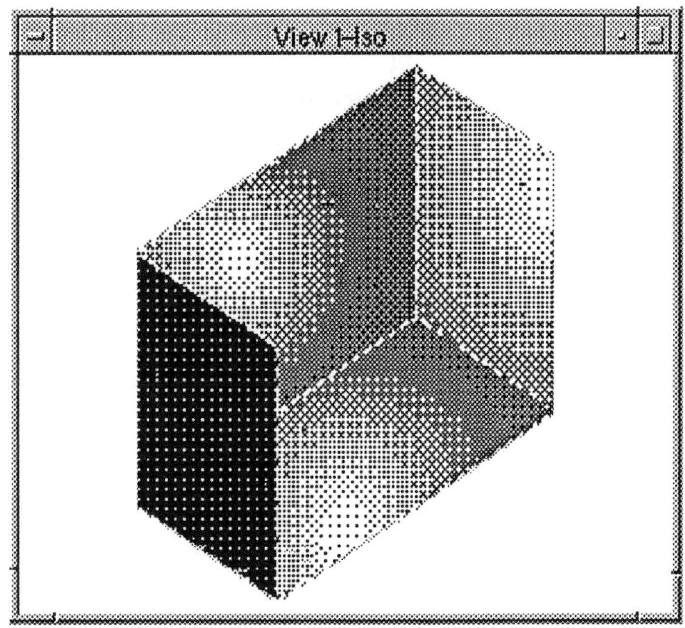

Figure 11.20 Effects of a Point Light

Try rotating the view various amounts and then rendering it again. Because the only light source in the file is the cell we placed, only the inside surfaces will be illuminated. This is true, no matter which view or view rotation we render.

If the Flashbulb and/or Ambient lights in the Rendering settings box were set to 'on', rendering would take these into account also.

Note:
There is no need to key in 'DEFINE LIGHTS' after rotating or manipulating a view. This applies also, when changes are made to the settings in the Rendering settings box. The 'DEFINE LIGHTS' key-in is necessary only when a modification has been made to the cell light sources. Modifications include moving the light sources or changing a value in the enter data fields, as well as adding or deleting a light source.

We will use the same exercise to look at the other types of light source.

In this exercise we will use a DISTANT LIGHT source.

° Delete the POINT LIGHT source cell.
° Check that the active angle is 0.
° Using a FRONT view, place a DISTANT LIGHT source (AC = DISTLT) at XY = 750,375,500. We have to use a specific view because this form of lighting is directional. A FRONT view gives us the orientation that we want. The direction of the light is indicated by the arrow in the light source cell.

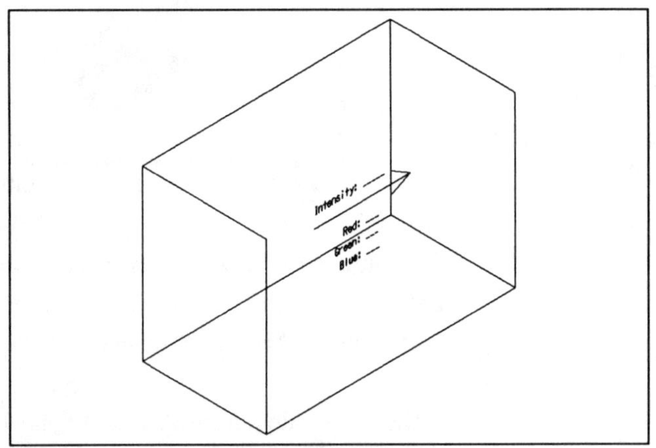

Figure 11.21 Model with DISTLT cell

Check that the Rendering settings are unchanged from the previous exercise. That is, Flashbulb and Ambient lights should be set to 'off'. We have changed the light source information since we performed the previous rendering so the lighting information has to be re-defined.

To complete the exercise:

° Key in 'DEFINE LIGHTS' to read in the new or modified lighting information
° Key in 'RENDER VIEW SMOOTH' and select the required view.

Distant Light (AC = DISTLT)

You will see that the surfaces facing the direction of the light source are illuminated (figure 11.22). This occurs whether the surface is behind or in front of the light source. In our model, the left face is behind the light source. Surfaces lying parallel to the rays are not illuminated (here, the bottom and back surfaces of the open box).

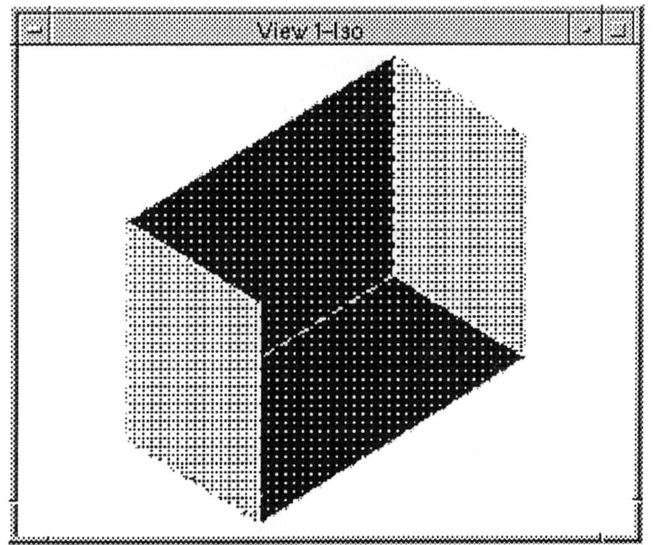

Figure 11.22 Effects of a Distant Light

Distant Lights are directional light sources, with parallel light rays. No matter where in the design file this form of light source is placed, surfaces that are facing it are evenly illuminated. Rays are cast in the direction of the X-axis of the cell, which is indicated by the arrow.

Because they provide even illumination, Distant Lights can be used to provide general lighting from a particular direction. They differ from the ambient light setting in the Rendering settings box. This is not directional. It affects all surfaces evenly, no matter which direction they are facing.

We will repeat the previous exercise, this time using the third type of light source, a SPOT LIGHT. We will first place the cell for the SPOT LIGHT in the center of the model, as we did with the other light sources. We will then move the light source to a point outside the model and repeat the rendering process.

- Delete the DISTANT LIGHT source cell.
- Again make sure that active angle is 0.
- Because SPOT LIGHTS are directional place the cell (AC=SPOTLT) in a front view, at XY=750,375,500. This will give the light the orientation that we require (figure 11.23).

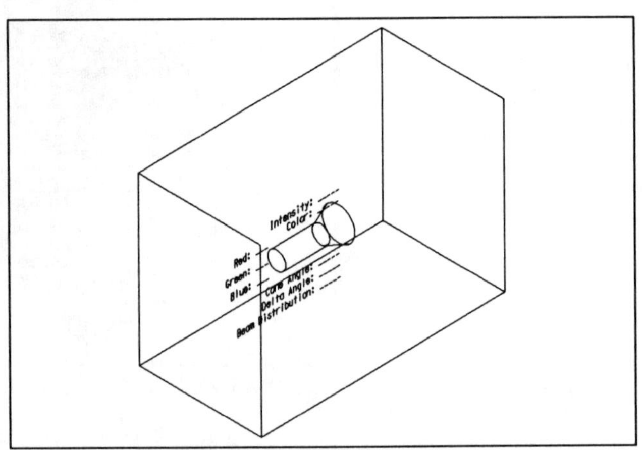

Figure 11.23 Model with SPOTLT cell

To complete this part of the exercise:

- Key in 'DEFINE LIGHTS' to read in the new lighting information.
- Key in 'RENDER VIEW SMOOTH' and select the required view.

Your rendered view should resemble figure 11.24. Part of the surface that is facing the spot light is illuminated.

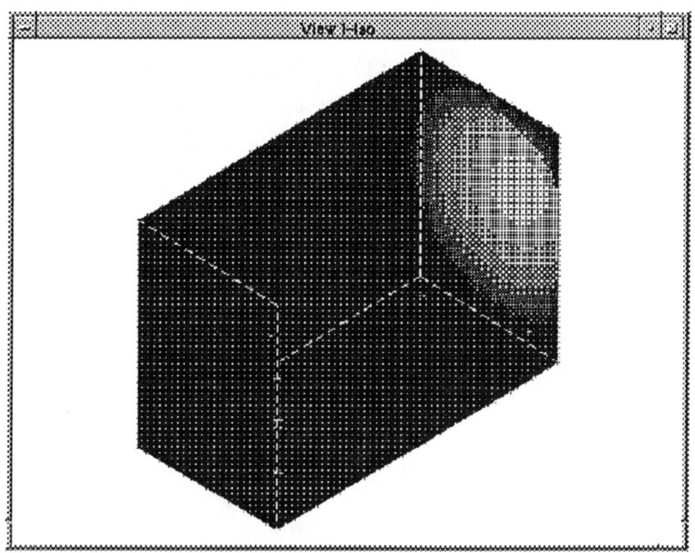

Figure 11.24 Effects of a Spot Light

Spot Light (AC=SPOTLT)

Spot lights are directional with the light being cast in a conical beam, similar to a torch or flashlight. Light is cast in the direction of the wide end of the spotlight, and emanates from the center of the cell. Enter data fields are provided for Cone Angle, Delta Angle and Beam Distribution, as well as for color and intensity.

These allow us to widen or narrow the beam of light. We can also vary the spread of the beam. Intensity of light ranges from maximum intensity to zero across the cone delta angle. Beam distribution refers to the concentration of light with respect to the center line of the light. Higher values concentrate the beam more in the central region.

To further demonstrate the conical beam effect of the SPOT LIGHT, we will move the cell outside the model and render the view again.

Move the light cell to a position outside the model as follows:

- Select the Move Element tool.
- Identify the light source cell.
- Key in DL=-1000 for the distance to move.

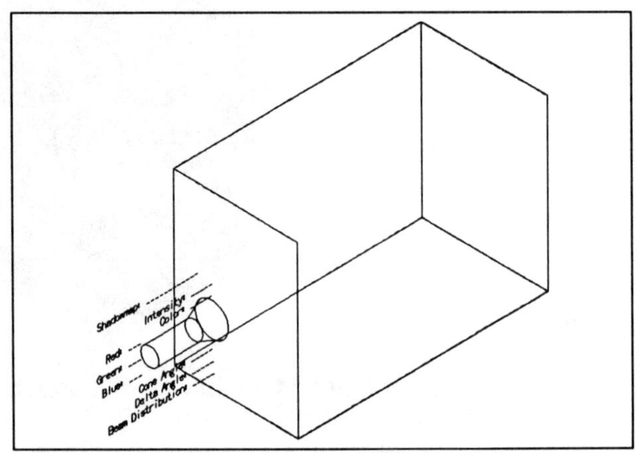

Figure 11.25 SPOTLT cell outside model

- Key in 'DEFINE LIGHTS'. We have moved the light source. This instructs the system to read the new information before we render the view.
- Key in 'RENDER VIEW SMOOTH'

Our rendered view (figure 11.26) shows clearly that the light beam is spreading. While the radius of the beam is quite small at the left side of our model, it illuminates the whole right side. As well, we can see that it has illuminated parts of the back and bottom sides.

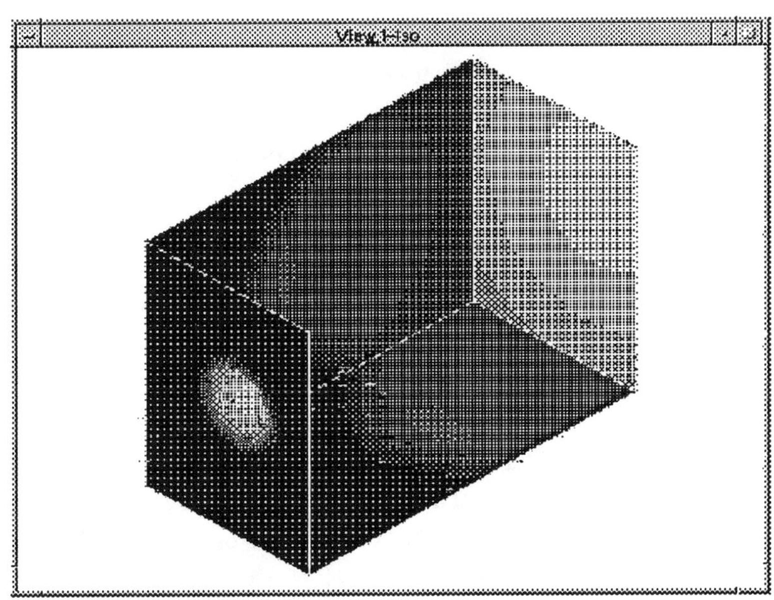

Figure 11.26 Spotlight Beam

You will also notice that we do not have shadows. The light from the light source does not take into account surfaces that are between the light source and any given surface. There are other software packages that can generate shadows. MicroStation files can be used with Intergraph ModelView, and there is an interface to RenderMan.

Placement of Lights

Design files can contain any number of light sources. These can be placed to highlight specific areas, or to simulate actual lighting conditions.

We have seen that the positioning of lights such as SPOT and POINT lights is important. With SPOT lights this is important on two counts - the distance from the surfaces, and the direction they are facing. POINT lights radiate light in all directions.

DISTANT lights may be placed anywhere in a file, but the direction they are facing is relevant.

We know that we can provide lighting for our model, using the lighting settings in the Rendering settings box. We can also place special cells in our design file. These cells, along with the lights set in the rendering settings box are used to determine lighting conditions.

We can also specify surface characteristics for objects in our model. We can specify properties for elements of a particular color, or on a particular level, or a combination of both of these. We do this with MATERIAL TABLES.

Material Tables

Material Tables are ASCII files (i.e., standard text files). Figure 11.27 shows a listing of the default table that comes with MicroStation. This file, named DEFAULTS.TBL, contains macros describing the standard colors, as well as a range of other colors and various finishes. A macro is merely a grouping (between the braces) of characteristics under the one name.

Macros take the form:
Macro_name { List_of_characteristics}

For example the macro describing the color red is:
redd { color = (1.0,0.0,0.0); }

This macro is named 'redd' and it consists of the general purpose color definition 'color = (r,g,b);' where r,g,b are values for red green and blue respectively. These values range from 0 to 1. In this instance, the value for red is 1 and for the others, 0.

A simple way to learn how to write macros is to follow the format of those included in DEFAULTS.TBL. As well as being able to define colors we can define surface finishes. Values for each characteristic also range from 0 to 1.

Once a macro has been defined, we can refer to it by its name alone.

```
white       { color = (1.0,1.0,1.0); }
redd        { color = (1.0,0.0,0.0); }
grn         { color = (0.0,1.0,0.0); }
blu         { color = (0.0,0.0,1.0); }
mtblue      { color = (0.05,0.17,1.0); }
yellow      { color = (1.0,1.0,0.0); }
purple      { color = (1.0,0.0,1.0); }
turq        { color = (0.0,1.0,1.0); }
black       { color = (0.0,0.0,0.0); }
gold        { color = (0.8,0.5,0.1); }
ltgold      { color = (0.02,0.012,0.002); }
grass       { color = (0.15,0.25,0.04); }
offwhite    { color = (0.85,0.85,0.85); }
stucco      { color = (0.85,0.85,0.70); }
ltgrey      { color = (0.7,0.7,0.7); }
mdgrey      { color = (0.5,0.5,0.5); }
dkgrey      { color = (0.2,0.2,0.2); }
tan         { color = (0.4,0.3,0.25); }
dkbrown     { color = (0.3,0.15,0.05); }
rust        { color = (0.50,0.15,0.05); }
mars        { color = (0.6,0.2,0.0); }
glossy      { finish = 0.99; reflect = 0.70; transmit = 0.00; }
shiny       { finish = 1.00; reflect = 0.45; transmit = 0.00; }
shiny2      { finish = 0.95; reflect = 0.15; transmit = 0.00; }
shiny3      { finish = 0.60; reflect = 0.05; transmit = 0.00; }
crystal     { finish = 0.99; reflect = 0.60; transmit = 0.80; }
flat        { finish = 0.00; reflect = 0.00; transmit = 0.00; }
trans       { finish = 0.00; reflect = 0.00; transmit = 1.00; }
chrome
 {
  finish = 0.9;
  reflect = 0.6;
  transmit = 0.0;
  color = (0.6,0.6,0.7);
 }
flatmetal
 {
  finish = 0.6;
  reflect = 0.3;
  transmit = 0.0;
  color = (0.6,0.6,0.8);
 }
metalic     { finish = 0.6; reflect = 0.20; transmit = 0.00; }
window      { finish = 1.0; reflect = 0.70; transmit = 0.30; }
window2     { finish = 1.0; reflect = 0.10; transmit = 1.00; }
window_nite   { finish = 1.0; reflect = 0.30; transmit = 0.70; }
water       { finish = 0.8; reflect = 0.70; transmit = 0.10; }
water_nite    { finish = 0.8; reflect = 0.70; transmit = 0.10; }
fountain      { finish = 0.0; reflect = 0.20; transmit = 0.50; }
mirror      { finish = 1.0; reflect = 1.00; transmit = 0.00; }
```

Figure 11.27 DEFAULTS.TBL Material Table

Linkages are made in the material table by associating a group of defined characteristics with a range of levels and colors. Linkages have the form:

for level LRANGE color CRANGE { CHARACTERISTICS }

where:
LRANGE is a range of levels.
CRANGE is a range of colors.
CHARACTERISTICS is a list of characteristics.

Your MicroStation documentation has a full description of how to use material tables, but we will go through a simple exercise to demonstrate them in action.

To complete this exercise you will need to create a text file using an ASCII text editor. That is, the text file has to be in ASCII (normal text) form. The file may be created with any word processor or text editor that can produce ASCII format files.

- First create a new 3D design file named RENDMAT1.DGN.
- Using the ISO view in figure 11.28 as a guide, place 4 white spheres in the file. Place the spheres on separate levels (i.e. levels 1 to 4).
- Place 3 DISTANT LIGHT sources (AC=DISTLT), as shown. That is, one facing in the positive X direction, one the positive Y direction, and the third the negative Z direction. They can be placed anywhere in the file.
- Using your text editor/word processor, create a text file with the same name as the design file and the extension 'M'. For this exercise the text file would be called RENDMAT1.M. Place it in the same directory as the design file. Into this file, type the following:

```
@defaults.tbl
for level 1 {flat; gold;}
for level 2 {shiny; gold;}
for level 3 {glossy; gold;}
```

- Now, enter the design file and, after setting the ambient and flashbulb lights to off, render the iso view ('RENDER VIEW SMOOTH').

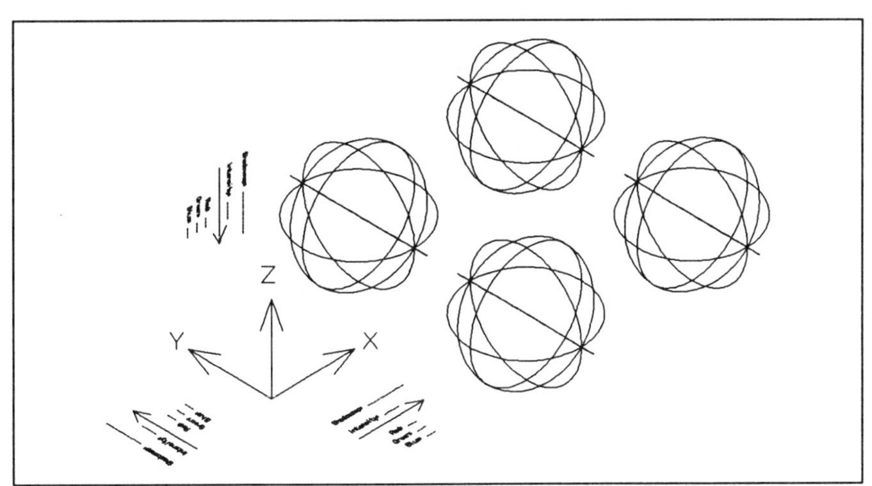

Figure 11.28 Spheres & Light Sources

Figure 11.29 shows the four spheres after rendering. The topmost sphere is on level 1, the left sphere on level 2, the bottom sphere on level 3. Each sphere looks different, after the rendering process despite them being the same color, when placed in the file.

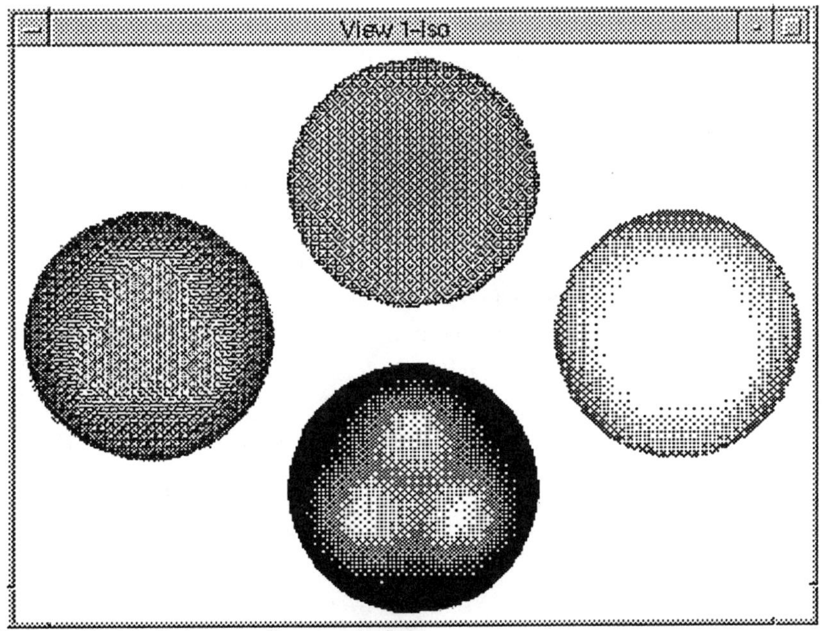

Figure 11.29 Spheres after rendering

Here, the rendering process found a file with the same name as the design file, but with a '.M' extension. It used the information in this file to determine the properties of the elements. Where no information was given for the elements on level 4, a system default was used. Our instructions were:

- Line 1. '*@defaults.tbl*'
  Here we instructed the rendering process to include information from the file 'defaults.tbl'. From here we can refer to information in this file by the macro name alone.

- Line 2. '*for level 1 {flat; gold;}*'
  This instructed the renderer that all elements on level 1 had the properties of being 'flat' and colored 'gold'. Neither of these macros are contained in our file, so the other included file 'defaults.tbl' was used.

  Because no color was specified, elements of all colors on level 1 would be treated identically.

- Line 3. '*for level 2 {shiny; gold;}*'
  As for line 2, except that the surface properties are 'shiny'. Again this macro is found in the file 'defaults.tbl'.

- Line 4. '*for level 3 {glossy; gold;}*'
  As for lines 2 and 3, with the surface properties being 'glossy'.

As you become more experienced with rendering, you may develop your own default color and surface property tables. These could be used exclusively, or along with the default table. We can include more than one file in a materials table. Each included file must be preceded by the '@' symbol, and be on a separate line. Path names may be included in the file specification.

Macros may be referenced more than once in a materials table. For example, you may have *{glossy; gold;}* along with *{glossy; redd;}*.

As with light source cells, information from materials tables is read when a view is rendered initially. If we modify the materials table for a file, while MicroStation is still active, we must key in 'DEFINE MATERIALS', to cause the modified information to be read.

Saving Shaded Images

From the previous exercises, it is clear that we have many options for creating realistic shaded images. Production of these images can be quite time consuming. When the image is exactly how you want it, it can be saved to disk for future recall. We can save shaded images in various formats, namely 'TARGA, TIFF, RGB or PICT'.

We access the 'Save Image' settings box by selecting 'Save Image As...' in the 'View' pull-down menu. This settings box (figure 11.30) allows us to:

– set the view to be saved
– choose the format and resolution of the shading
– select the type of shading and whether it is to be Stereo or not.

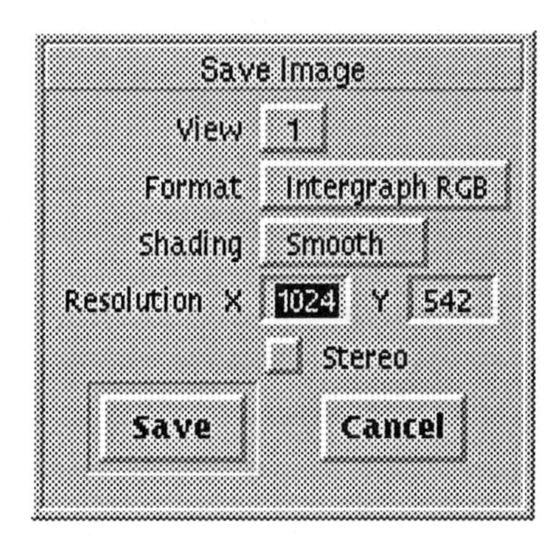

Figure 11.30 Save Image settings box

When we select View, Format or Shading we are given a selection of options. For the views, this is 1 to 8. For Format and Shading, our options are as shown in figure 11.31.

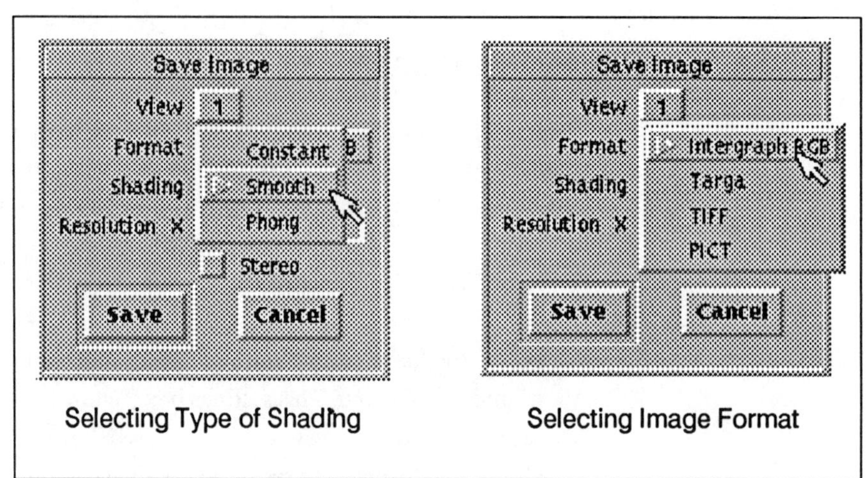

Selecting Type of Shading Selecting Image Format

Figure 11.31 Saving an Image

When we click on the 'Save' button we are prompted for a name and directory for the saved image. It is not necessary to create a shaded image prior to saving it. When 'Save' is chosen the process saves the selected view as a shaded image, whether or not it is currently shaded.

Note:

No matter what resolution screen you are using to save the image, it is saved at the highest resolution. This allows for work to be completed on a workstation having a 'cheaper' graphics card installed. A workstation with a high quality graphics card then can be used for final presentations, or for photographs to be taken of the image.

Displaying Saved Images

Saved shaded image files may be displayed with software that can read the particular file formats. We can display the saved images from within MicroStation also.

We select 'Image' from the 'Display' option of the 'File' pull-down menu as figure 11.31 shows. From here we are prompted for the file to display. We can do this from any design file, whether 3D or 2D.

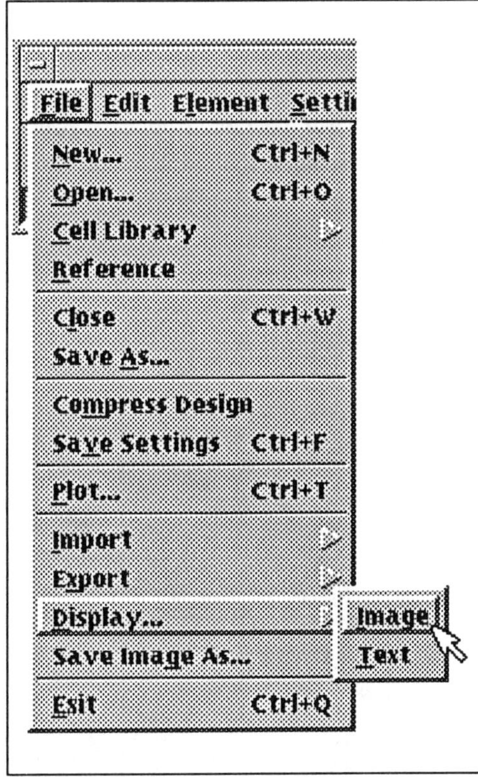

Figure 11.32 Selecting Display Image

Using View Cameras

While we are creating our three-dimensional models we normally work in views with parallel projection. That is, there is no perspective associated with objects. Parallel lines appear parallel on the screen and distant objects appear at the same scale as near ones of the same dimensions. In real life this is not how we see them. The more distant objects appear smaller to us.

With version 3.3 of MicroStation, we use a key-in 'SET PERSPECTIVE'. With version 4 we have a 'camera' associated with each view. We can turn the camera on in any view. We also have various settings that allow us to vary the type of lens, the position of the camera and where the camera is aimed.

We can place the camera anywhere in the design file, looking in any direction. By moving the camera and its target point progressively through a model, saving the (rendered) views each time, we can produce a 'walk-through' presentation of our design.

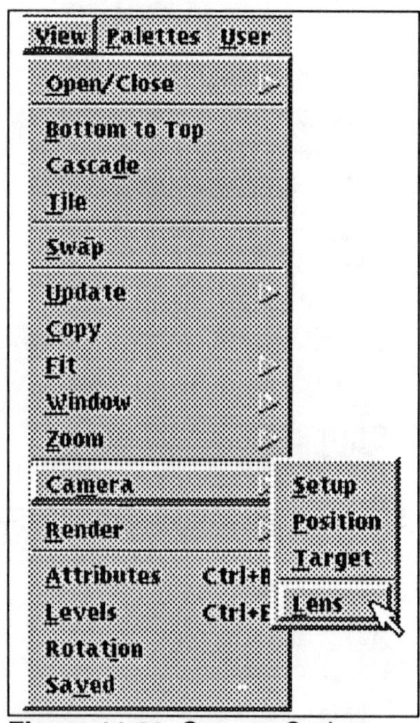

Figure 11.33 Camera Options

When we select 'Camera' in the 'View' pull-down menu, we are given four options (refer figure 11.33).

Selecting 'Lens' opens the Camera Lens settings box (figure 11.34). Here we can specify the view to use and the lens Angle and Focal Length. We also have a choice of Standard lenses which we can use.

If a Standard Lens is chosen, the lens Angle and Focal Length are automatically entered.

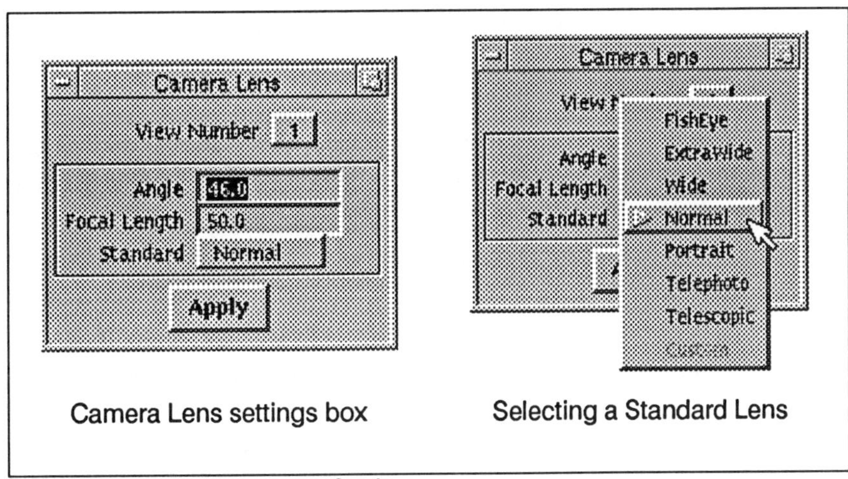

Camera Lens settings box

Selecting a Standard Lens

Figure 11.34 Camera Lens Settings

When we have completed setting up the camera parameters, clicking on 'Apply' causes the chosen view to be updated showing the effect of the lens setting. This uses the orientation of the view as the default for setting the camera position and target point.

In practice, we would be required to specify the camera location and target point. As an illustration we will consider a simple model, consisting of a slab (2000x1000x500) with 3 cylinders (150 radius x 300 high) inside. The origin of the slab is at 0,0,0 as shown in the diagram below.

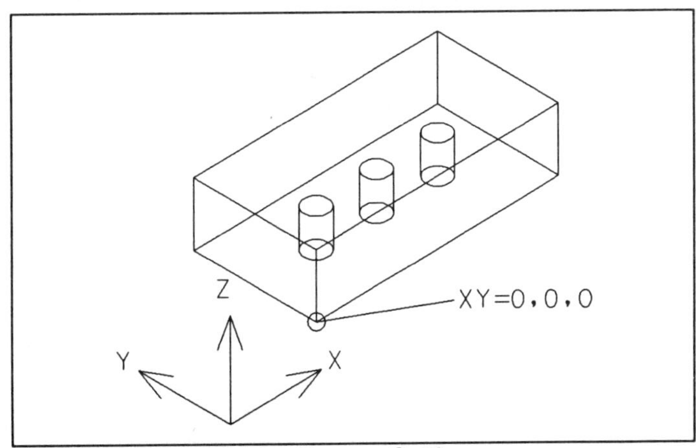

Figure 11.35 Model for Camera Settings

To set up a view with camera position and target point, we go through the following steps:

- Select 'Setup' from the 'Camera' options (refer figure 11.33). We are prompted first for a view to be used as the camera view. For this example, the RIGHT view was selected. At this point we are shown a viewing 'pyramid', and as we move our cursor, the pyramid moves. This is a display of the viewing parameters that we are setting for our chosen view. Note that it is restricted by Display Depth. If the Display Depth of a view is not large enough, only part of the viewing pyramid will be displayed.
- We are then prompted for the camera target point. This can be a data point or a key-in. Here, a key-in was used, XY=-1000,500,250. This is a point on the center line of the slab and 1000 left of the left face.
- Next, we are prompted for the camera position. Again, a key-in was used here, XY=3500,500,250. This point is on the center line of the slab and 1500 to the right of the right side.
- Our next prompts are for the front and back clipping planes respectively. Figure 11.36 shows the screen as the back clipping plane was being set. When this has been completed the chosen view is updated to reflect the new viewing parameters with the camera turned on. Figure 11.37 shows the RIGHT view after it updated to reflect the new viewing parameters.

Whereas the original RIGHT view gave us no real idea of what we were looking at, the view with perspective shows the model more clearly. We can see the side walls diminishing in the distance. Also, because we are looking at the model in wireframe mode, we can see the cylinders behind one another. Models displayed with perspective are easier to visualize, even in wireframe mode.

In the above illustration we used points outside the model. We can set the camera position and target points anywhere in the design file. Also the camera position and target points can be set individually. This allows the camera to be moved around a target point, and vice versa.

To set a view back to parallel projection we can turn the camera off. This is done by with the key-in 'SET CAMERA OFF' followed by a data point in the view.

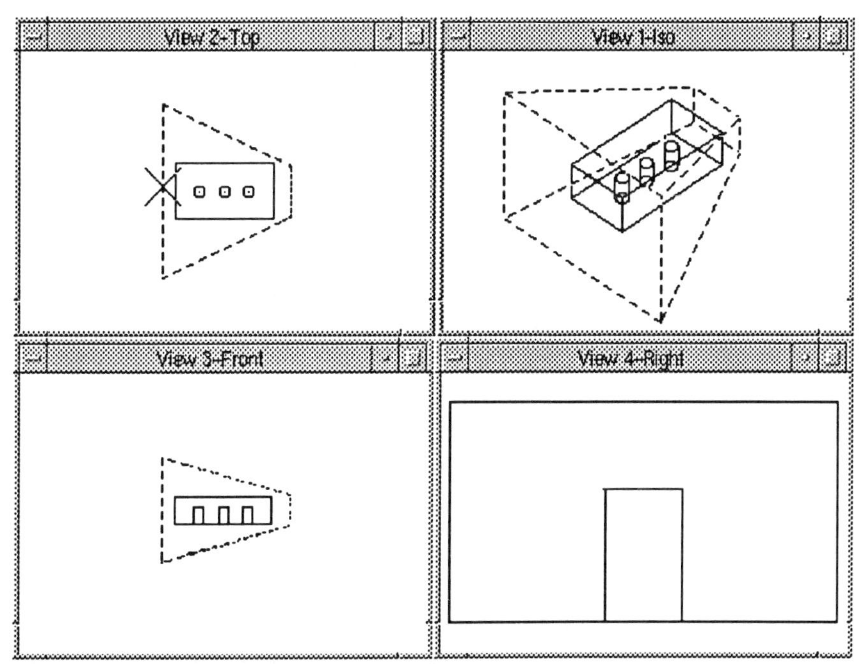

Figure 11.36 Setting Clipping Planes for Camera

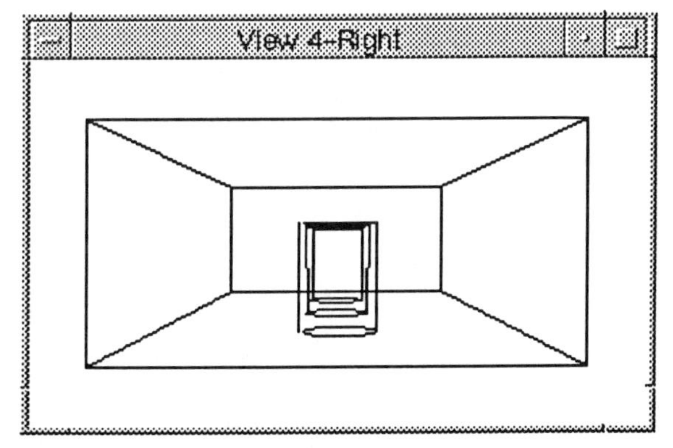

Figure 11.37 Right view after updating

Active Background

Normally the background to our design file is a black screen, or some other color if we change the color table. For normal design work this is fine. With version 4, we can use a saved image as a background. Updates to views with active backgrounds are very slow, but for presentation purposes this feature can be useful. For example, we can render a model, with an active background displayed, and save the results in yet another file.

To display a saved shaded image as a background we first key in:
'ACTIVE BACKGROUND filename'
to specify which file is to be used. Next we key in:
'SET BACKGROUND ON'
and select a view. When the view is updated, instead of the normal background, we will have the shaded image that we selected.

Edges Design Files

To this point, the displays we have discussed have been raster displays. That is, they are screen displays only and are at the resolution of the screen. To get plots or printouts, special plotters/printers are required. Where we need scale drawings from our three-dimensional model we need to generate design files displaying the visible edges only.

To do this we select 'Visible Edges' from the 'Export' option in the 'File' pull-down menu. This opens the Visible Edge File Generation Dialog box (figure 11.38), which is divided into three sections, 'Output Format', 'Hidden Edges' and 'Mesh Resolution'.

To illustrate some of the options, we will use the model, shown in ISO and LEFT views in figure 11.39. This was constructed from simple three-dimensional elements (sphere, slab, cone and cylinders) and a surface of revolution.

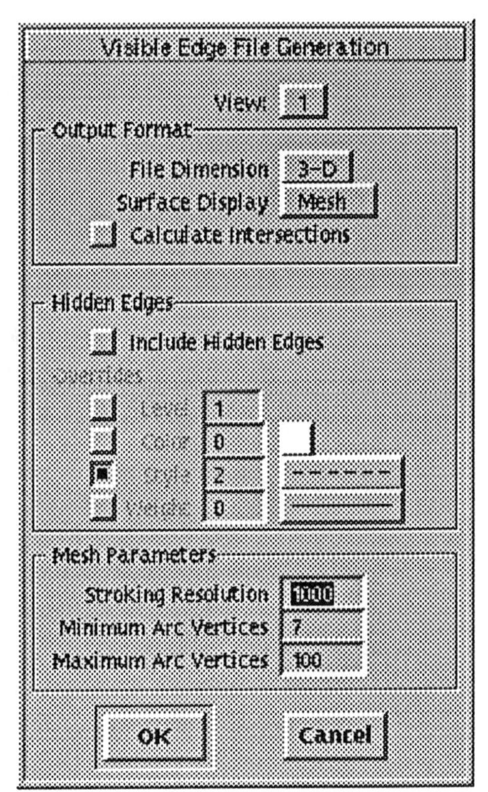

Figure 11.38 Visible Edges Dialog Box

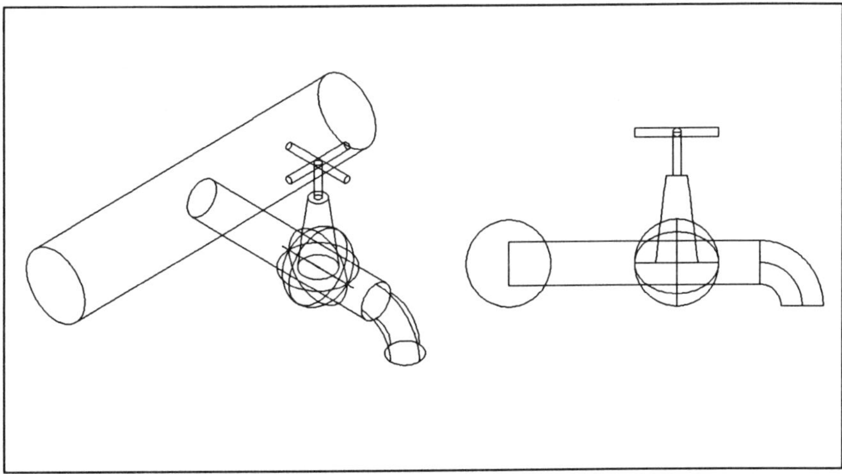

Figure 11.39 Visible Edges Model

Output Format

File Dimension

This allows us to specify the output file dimension as either '2D' or '3D'.

Note:

Output files in 3D can be viewed correctly from one viewpoint only. Visible edges will vary from view to view. If a non-standard view is used, the rotation of the view should be documented. When the visible edges file is used, the view has to be set to the same rotation as when it was generated from the model. This does not apply to 2D output files.

Surface Display

We can have surfaces displayed as a number of small rectangles (Mesh), or as the outer boundaries only (Profile).

Calculate Intersections

When this option is on, intersection lines between elements are calculated and placed in the output file. Calculation of intersections increases processing time significantly.

Hidden Edges

With this option we can specify that the hidden edges be included in the generated file by setting 'Include Hidden Edges' to on. We can set the symbology and level for the hidden edges also.

Mesh Parameters

Stroking Resolution

Determines the accuracy of curved surfaces. Values range between 100-32,000. The default is 1,000 which produces a reasonable mesh with moderate processing time. Increasing the Stroking Resolution can increase processing time considerably.

Minimum/Maximum Arc Vertices

Linestrings are used in visible edges design files to approximate arcs. The number of vertices in these linestrings is related to the size of the arc. It must be a figure within the range specified by the minimum and maximum settings. The minimum value must be 3 or greater, and the maximum 200 or less.

When all settings are as required, clicking on the 'Save' button opens the Save Visible Edges Design File dialog box. This is similar to the Save Design File dialog box. By default, visible edges design files are given the same name as the design file, but with a '.HLN' extension. This can be changed as required.

Figure 11.40 shows examples of visible edges design files of our model with various settings. Except where noted, default settings were used.

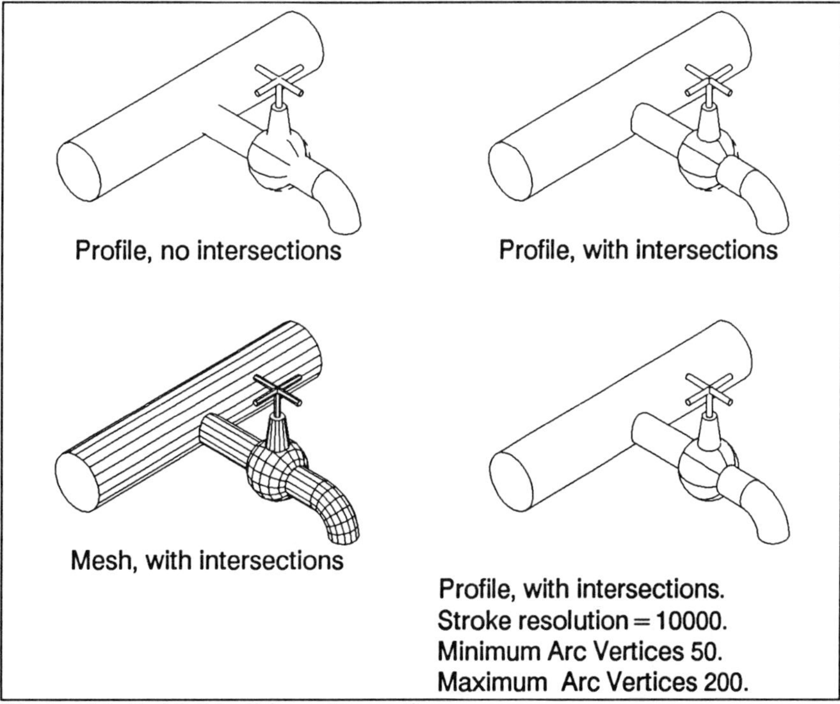

Figure 11.40 Visible Edges Alternatives

12 : Drawing Production

For the most part, when we work in 2D design files we are thinking of producing drawings. We draw the plans, elevations, details etc. as separate items. Whenever any modifications are made, we have to check through all our 'drawings' and make the necessary corrections. We can reduce the number of changes by using reference files, where appropriate.

Our thinking is different with 3D. We create a three-dimensional computer model of our design. This can consist of one file or many files referenced to each other. Once we have produced a 3D model, we can extract drawings from it using the EDGES option of the HIDDEN LINE REMOVAL utilities. EDGES sends its output to a design file, which may be plotted like a normal file.

There are many ways to create drawings, both 2D and 3D from the model file/s. Here, to get you started, we will go through two methods. For the example we will use the model produced in the first part of chapter 7 (Advanced Techniques). For both methods our aim is to produce a drawing showing PLAN, FRONT ELEVATION, RIGHT ELEVATION and ISOMETRIC view. Both methods described in the following pages have their strengths and weaknesses.

Method 1 allows the drawing to be seen in wireframe form, and we add all dimensioning using the 3D model file. There is a link with the model right up to the when we produce the Edges file. The disadvantage is that the whole edges design file has to be re-created each time there is a change to the model.

Method 2 breaks from the model prior to adding the dimensioning. Each drawing view has its own file, which requires care with naming conventions. This is offset by being able to create parts of the drawing 'on the fly' and separately. Modifications may be made to separate parts of the drawing without having to re-create the whole drawing with the Edges process.

Method 1

This method involves 5 steps.

- 1. Create a new 3D 'drawing' file.
- 2. Reference the model file/s to the drawing file as often as necessary to create the required views for the final drawing.
- 3. Move, rotate, clip and scale the reference files, where necessary, to form the drawing in wireframe mode.
- 4. Add dimensioning and any notes.
- 5. Run the Edges utility to create a 3D edges design file to use for plotting the drawing.

With this method we can view the finished drawing, in wireframe mode, at any time. We don't have to disturb the model files, or the view set-ups in these files, to do this. We set the drawing up in another file, having the model file/s referenced to it. Also, we have the advantage of being able to produce preliminary plots as we work on the model.

Often, a model consists of several files referenced to one another. This can make the initial setting up of the drawing file laborious. Offset against this, is the fact that the drawing file is still 'attached' to the model file/s. Any changes to the model will be reflected immediately in the wireframe drawing file. Running the Edges process will produce the latest version of the drawing each time.

A limiting factor for this method, is the number of reference files we can have attached. With version 3.3, we can have a maximum of 32 reference files. Take, for example, a model consisting of five files referenced together. Our drawing file in this instance could have a maximum of six views and/or details (i.e., 6x5 = 30 reference files). There is much more flexibility with version 4 because we can have up to 255 reference files.

To create the drawing, do the following:

Step 1.

Create a new 3D design file in which to produce the wireframe form of the drawing.

Step 2.

For our exercise the model consists of one file. As we have 4 views to create for the drawing, we need to reference the model file to the drawing file 4 times.

We can differentiate between the reference files by using meaningful logical names. Use PLAN for the plan view, FELEV for the front elevation, RELEV for the right elevation and ISOM for the isometric.

Step 3.

Turn off all other views on your screen but one and make this a TOP view. Using a TOP view allows us to use DX or DL for moving the reference files to set up our drawing. Make sure that reference file snap is on for all reference files.

Where the reference files appear initially, we can use as the PLAN view for the drawing. So, we need to move the reference files for the other views away from here.
For this exercise:

- Move FELEV -500 units in the Y direction (DX=,-500).
- Move RELEV 650 in the X and -500 in the Y direction (DX=650,-500).
- Move ISOM 800 in the X direction (DX=800).

Figure 12.1 shows how your file screen should look after moving the four reference files, ready for the next step.

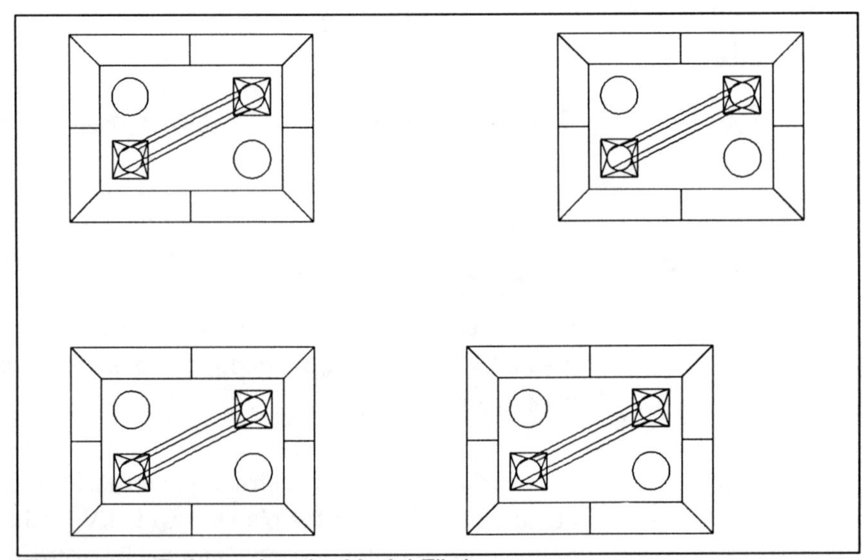

Figure 12.1 Referencing the Model File/s

We now have to rotate 3 of the reference files to create the view orientations for the drawing. Remember to snap to a point in the reference file you are rotating, for the axis of rotation. This ensures that at least part of the reference file stays within the display depth of the view. Also, by selecting the points for the axis of rotation wisely, the views will remain aligned with one another. This will reduce the amount of manipulation required to make the drawing look balanced.

Using figure 12.2 as a guide for the rotation points, rotate the 3 reference files as follows:

- FELEV: -90 (90 degrees backward, about the X axis) - point 1.
- RELEV: -90,-90 (90 degrees backward, about the X axis and 90 degrees clockwise about the Y axis) - point 2.
- ISOM: ,,45 (45 degrees anti-clockwise about the Z axis) - point 3.
  then -55 (55 degrees backward about the X axis) - point 3.

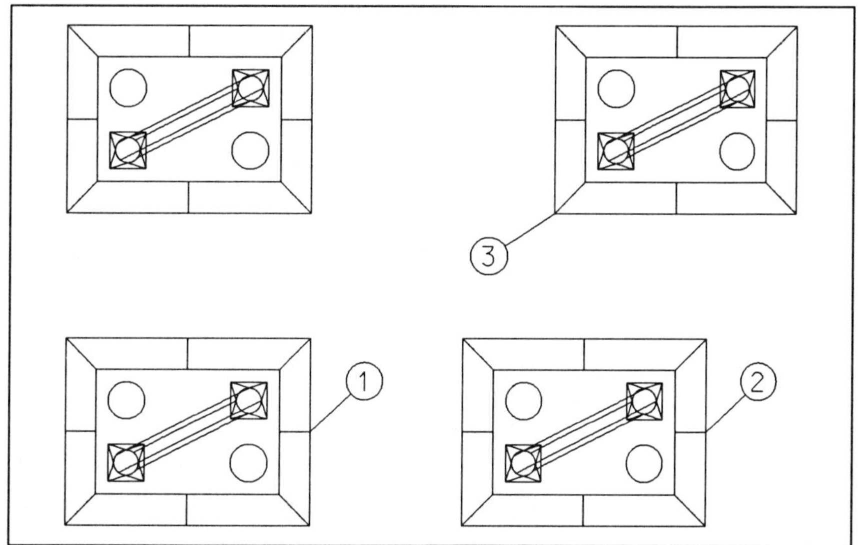

Figure 12.2 Rotation Points for Reference Files

Your TOP view should now look like figure 12.3. It is a wireframe picture of our drawing, prior to text being added. Here we can make any alterations to the layout, if necessary, to make the drawing look better.

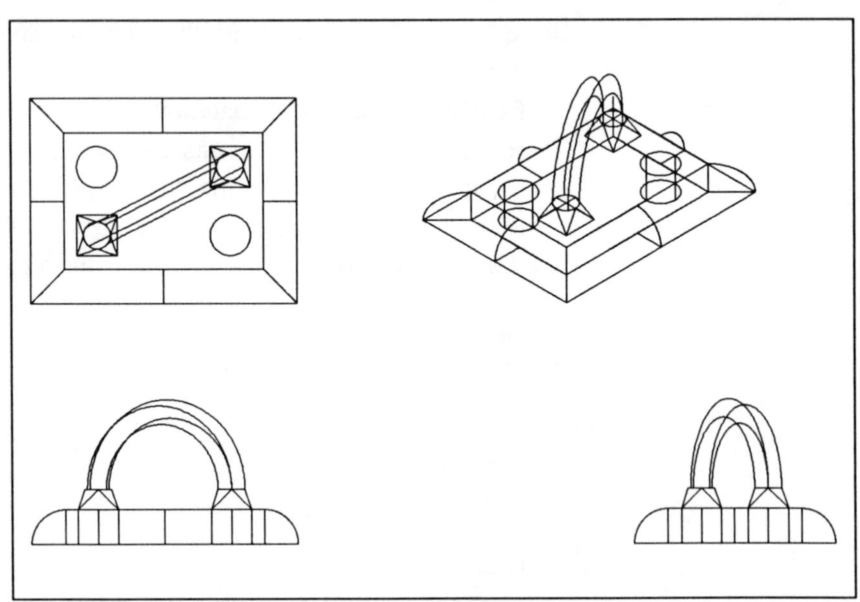

Figure 12.3 After Rotating the Reference Files

Step 4.

We are ready to add the dimensioning and any notes. The dimensioning should be placed on a level by itself. As this file, so far, consists entirely of reference files, the dimensioning can be put on any level.

Placing the dimensioning in this file ensures that we are using the model for the information, thus avoiding possible errors. We can reference the title block to this file also, to see how the final drawing will look. The display of the title block should be turned off when we run the Edges process, to save time. The title block can be referenced again to the Edges file, for plotting.

If you do decide to leave the title block displayed, and run the Edges process, make sure that the title block is behind the model. Title blocks often contain shapes in them and when you run hidden line removal, the shape is a surface. If it is in front of any part of the model, then it will 'hide' that part of the model.

Step 5.

We can now run the Edges process to create the edges design file. Remember that Edges works like a file fence. If there is another file of the same name as the edges design file, then it will be over-written and lost.

There are two alternatives for creating the Edges file:

- We can leave the text and dimensioning turned on and it will be processed with the elements. Text is converted to linestrings. The edges file can be either 2D or 3D.
- A more efficient way is to reference the drawing file to the Edges file. The drawing file has all the text and dimensioning in it. This reduces processing time and allows us to modify the text as text. The edges file should be a 3D file in this case.

Figure 12.5 (next page) shows graphically the system of referencing files to create the drawing file from which the Edges design file is generated.

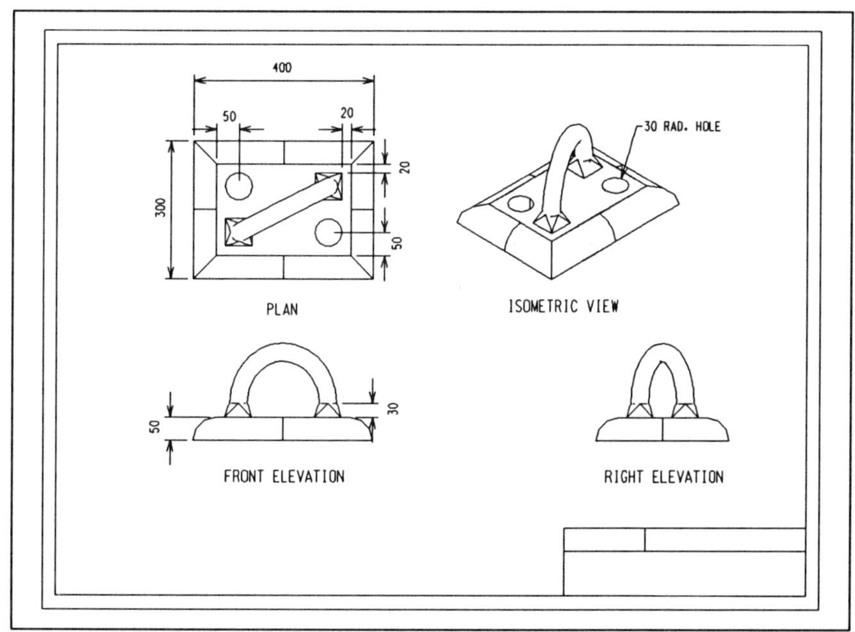

Figure 12.4 Edges File with Text and Title Block

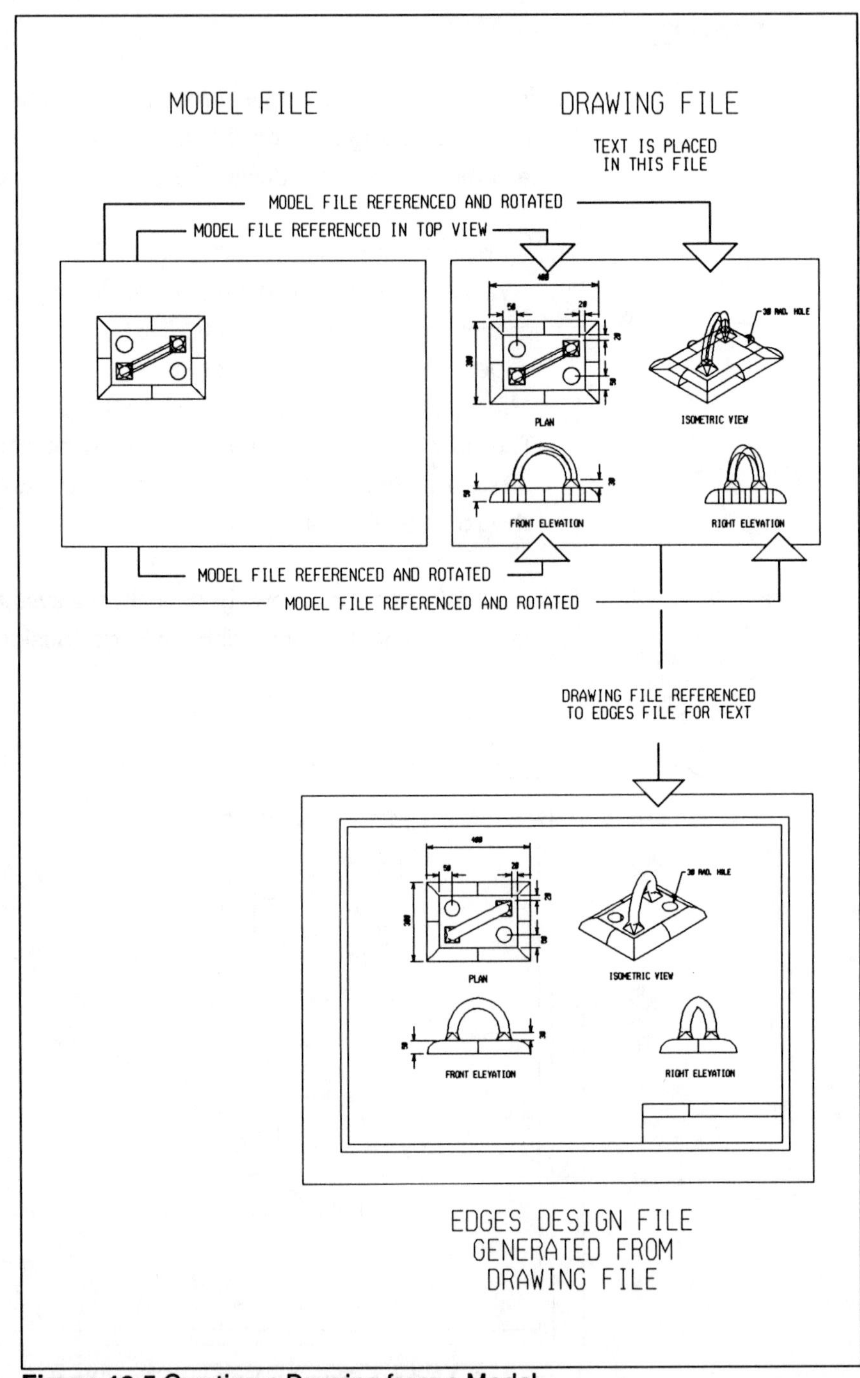

Figure 12.5 Creating a Drawing from a Model

Method 2

The steps involved in this method are as follows:

– 1. Make SAVED VIEWS of each aspect of the model that we require for the drawing.
– 2. Create 2D edges design files of each saved view.
– 3. Create a new 2D drawing or plot file.
– 4. Reference the edges design files and the title block to the new drawing file.
– 5. Add dimensioning and notes to the drawing or plot file.

The main point to watch with this procedure is to ensure that each edges design file has a unique name. Remember that the Edges process works like a file fence. If we give an edges file the same name as an existing file then the existing information is over-written. The edges design files are all being created from the same design file, so we will use the naming option to give them different names. The default is for them to have the same name as the originating design file, but with '.HLN' as the extension. Once we have set up the saved views, we can run the Edges process on each view when appropriate.

By creating 2D edges files, we don't have to rotate reference files to create the final drawing. We simply move them to the required place.

To create the drawing, using this method, do the following:

Step 1.

Make saved views of each part of the drawing. We are making a drawing with a PLAN, FRONT ELEVATION, RIGHT ELEVATION and ISOMETRIC. The saved views we will create are PLAN, FELEV, RELEV and ISOM. These all align with standard MicroStation views, which makes our task simple.

° Use VI=TOP for PLAN, VI=FRONT for FELEV, VI=RIGHT for RELEV and VI=ISO for ISOM. Remember to FIT the views before saving them with the SV=viewname key-in.

Step 2.

We can run Edges for each saved view individually, with the following:

- Create a 2 dimensional edges file.
- Give each file a unique name.

This will give us four edges files from which to create the drawing file.

Version 3.3

A much more efficient way is to run the process in batch mode. We can let the process run during a period when the computer is not normally used. As we only have four views for the drawing, we can set our screen up with four quadrants. We will have a saved view in each.

Set the screen up so that:

- View 1 has the saved view ISOM, (VI=ISOM).
- View 2 has the saved view PLAN, (VI=PLAN).
- View 3 has the saved view FELEV, (VI=FELEV).
- View 4 has the saved view RELEV, (VI=RELEV).

We now have to create an input file for the Edges process. We use a text editor to do this. For the example, we will call the input file EDFTEST.INP. If the model file name is MODEL3D.DGN, the input file would be as follows:

```
MODEL3D.DGN -ed2 -v1 -efISOM
MODEL3D.DGN -ed2 -v2 -efPLAN
MODEL3D.DGN -ed2 -v3 -efFELEV
MODEL3D.DGN -ed2 -v4 -efRELEV
```

Each line in the input file is telling the process to use the design file MODEL3D.DGN, create a 2 dimensional edges file, use the view specified (-v), and name the new file as specified (-ef). To run the process we type, from the DOS prompt:

HIDDEN @EDFTEST.INP

remembering that MicroStation must have been run previously.

The result will be four files:

ISOM.HLN from view 1
PLAN.HLN from view 2
FELEV.HLN from view 3
RELEV.HLN from view 4

For further information on Edges file creation, refer to chapter 10 (Rendering version 3.3).

Version 4

We can set the required parameters in the 'Visible Edges File Generation' dialog box. Further information on Edges file creation can be found in chapter 11 (Rendering version 4)

Step 3.

Create a new 2D design file, to be the drawing file.

We will reference the Edges design files to it to produce the drawing.

Step 4.

Reference the Edges files to the drawing file. Move and scale them, if necessary, to form the drawing. Also, reference the title block to this file.

Step 5.

Add dimensioning and notes.

Where you have various scales, use REFERENCE FILE DIMENSIONING. This will give the correct dimensions, even when the reference files are scaled. The 'drawing' is now ready for plotting.

Both methods give the same result. For simple models, the first method would suffice. For more complex models, where several files are used to create the model, the second method is probably more viable.

You may prefer a combination of both, where the model is still able to be compared directly with the edges file/s. This can be done by putting dimensioning in the model file used for creating the Edges files. Dimensioning would have to be on separate levels for each view. This avoids conflicts with dimensioning in a 'front' view being visible in a 'right' view of the same part of the model.

With the two methods described as a basis, you should be able to come up with a system for creating drawings from your 3D model.

13 : Tips & Tricks

This chapter contains a collection of ideas, tips and information that you may find useful.

RTP/RTM

This is the most basic tip and applies for all software packages, particularly when you are using a new tool or procedure. When anything goes wrong or you cannot understand why a particular tool, or key-in is not functioning correctly. Before calling for help, the first action that you should take is to:

RTP - *Read The Prompts*.

Prompts, and error messages are there to assist. If this doesn't solve the problem, the next line of attack is to:

RTM - *Read The Manual.*

Again, the manuals often have very helpful information.

This is not to say that occasionally there are genuine 'bugs' in software, but in a majority of cases, it is operator error. If a tool has worked properly previously, it is very unlikely that it will stop working.

Menus

MicroStation allows us to use various styles of menus. We have a choice of:

- Digitizer
- Sidebar
- Function key
- Palettes, settings and dialog boxes (version 4)

While it is difficult to have a standard menu that satisfies all users, we can create our own menus for particular projects or disciplines.

Function Key Menus

Function key menus are very simple to make, from inside MicroStation, using the SET FUNCTION key-in. Alternatively, with version 4, choose 'Function Keys' in the 'User' pop-down menu.

You may have a number of function key menus that are accessed from a starting menu. At least one entry on each of these other menus should recall the starting menu. In this fashion you can go between your various function key menus without having to use an 'AM = filename,FK' key-in.

Sidebar Menus

We need an ASCII text editor to create or modify sidebar menus. The standard (version 3.3) sidebar menu, USTN.SBM, is comprehensive. However, you will find that many of the 3D tools are several sub-menus deep. This can be annoying if you are using a number of these tools regularly. While it is not as much of a problem with version 4, the various tools still are spread out in different palettes.

Figure 14.1 shows examples of sidebar menus containing commonly used 3D tools. On the left is a menu for version 3.3, while on the right is a similar menu for version 4. These menus have been configured for a standard VGA screen.

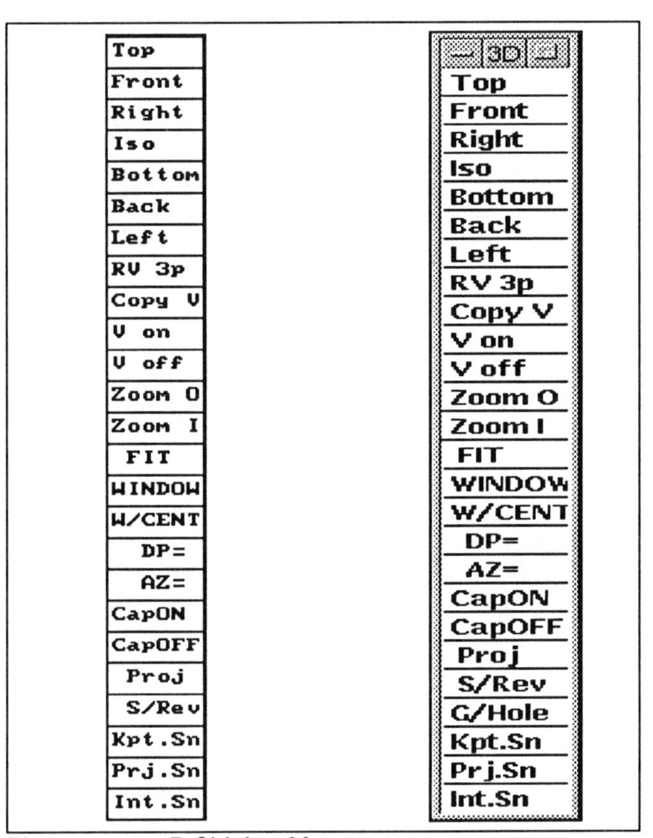

Figure 13.1 3D Sidebar Menus

These are single level menus. That is, there are no sub-menus and each tool is accessed directly.

On both menus, the options for setting display depth (DP=) and active depth (AZ=) are the version 3.3 style. That is, even with version 4, the tools work as they do in version 3.3. We do not get the graphics in version 4, that are associated with these tools when chosen in the 3D palette.

You will see in the menu listing that this is done by using the key-ins '/DDEPTH' and '/ADEPTH' respectively for the set display and active depth menu options. These are alternative key-ins to 'DEPTH DISPLAY' and 'DEPTH ACTIVE'.

If you want to use these menus, create an ASCII text file containing the information from one of the following listings. As indicated, both menu files are present on the accompanying disk that can be purchased with this book.

This first listing is for version 3.3 users. The file should be placed in the MicroStation DATA directory. Depending on your screen configuration, you may be able to place more entries in your menu.

Note:
Here, due to space limitations, the first line has been separated into two. Make sure, if you are creating this file, that you combine the first two lines into one.

\V33\3DMENUV3.SBM

```
MAIN_MENU title=3d, color=(9,4), width=6, height=1, rows=25,
column=1, border, by_row, vline
   'Top','vi=top', line /nosave
   'Front','vi=front', line /nosave
   'Right','vi=right', line /nosave
   'Iso','vi=iso', line /nosave
   'Bottom','vi=bottom', line /nosave
   'Back','vi=back', line /nosave
   'Left','vi=left', line /nosave
   'RV 3p','p,viewpl', line /nosave /color=(9,3)
   'Copy V','copy view', line /nosave /color=(9,3)
   'V on','vi on', line /nosave /color=(9,3)
   'V off','vi off', line /nosave /color=(9,3)
   'Zoom O','zoom out', line /nosave /color=(9,7)
   'Zoom I','zoom in', line /nosave /color=(9,7)
   ' FIT','fit', line /nosave/color=(9,7)
   'WINDOW','window', line /nosave /color=(9,7)
   'W/CENT','window center', line /nosave /color=(9,7)
   '  DP=  ','/ddepth', line /nosave /color=(9,5)
   '  AZ=  ','/adepth', line /nosave /color=(9,5)
   'CapON','ACT CAP ON',line /nosave /color=(9,8)
   'CapOFF','ACT CAP OFF',line /nosave /color=(9,8)
   ' Proj','p,prjele', line /color=(9,8)
   ' S/Rev','p,surrev', line /color=(9,8)
   'Kpt.Sn','p,keysnp', line /nosave
   'Prj.Sn','p,snaplk', line /nosave
   'Int.Sn','inte', line /nosave
```

Next is the listing for the version 4 users. Again this file should be placed in the MicroStation DATA directory.

Note:
Here, due to space limitations, the first line has been separated into two. Make sure, if you are creating this file, that you combine the first two lines into one.

\V4\3DMENUV4.SBM

```
MAIN_MENUtitle=3d,color=(9,4),width=6,height=1,rows=26,
column=1,border,by_row,vline
   'Top','vi=top', line /nosave
   'Front','vi=front', line /nosave
   'Right','vi=right', line /nosave
   'Iso','vi=iso', line /nosave
   'Bottom','vi=bottom', line /nosave
   'Back','vi=back', line /nosave
   'Left','vi=left', line /nosave
   'RV 3p','p,viewpl', line /nosave /color=(9,3)
   'Copy V','copy view', line /nosave /color=(9,3)
   'V on','vi on', line /nosave /color=(9,3)
   'V off','vi off', line /nosave /color=(9,3)
   'Zoom O','zoom out', line /nosave /color=(9,7)
   'Zoom I','zoom in', line /nosave /color=(9,7)
   ' FIT','fit', line /nosave/color=(9,7)
   'WINDOW','window', line /nosave /color=(9,7)
   'W/CENT','window center', line /nosave /color=(9,7)
   '  DP= ','/ddepth', line /nosave /color=(9,5)
   '  AZ= ','/adepth', line /nosave /color=(9,5)
   'CapON','ACT CAP ON',line /nosave /color=(9,8)
   'CapOFF','ACT CAP OFF',line /nosave /color=(9,8)
   ' Proj','p,prjele', line /color=(9,8)
   ' S/Rev','p,surrev', line /color=(9,8)
   'G/Hole','group hole', line /color=(9,8)
   'Kpt.Sn','p,keysnp', line /nosave
   'Prj.Sn','p,snaplk', line /nosave
   'Int.Sn','lock snap intersection', line /nosave
```

General

Under this heading is grouped a mixture of items, some relevant to version 3.3 only, some for version 4, and others for all versions.

Using Data Points

- With 3D work it is sometimes necessary to work in a view with elements, both in front of and behind the required element. This can lead to confusion when placing a tentative point, as it may be hard to see which element has highlighted, because of the cursor. Rather than using a tentative, use a data point. The cursor can be moved away from the element and reset used until the correct element highlights.

- When Boresite Lock is enabled, elements identified with a data point do not change their view Z-value to the Active Depth. This is a particularly useful feature for placement or manipulation of light sources for rendering.

Views

- It is a good idea to have orthogonal views set up in at least two quadrants, or windows of your screen. This simplifies the task of setting ACTIVE DEPTH and DISPLAY DEPTH, graphically.

- FIT in version 3.3, does not fit reference files . A 'trick' is to place 2 active points in the active file. The two points should be on diagonally opposite corners of a rectangular or cubic volume in the active file. This volume should include the volume which contains the reference files. A FIT in the active file will then include the active points, which in turn embrace the volume occupied by the reference files.

Cell Libraries

3D files can grow in size rapidly, thereby slowing screen updates, tentative points etc. Often, in the design stage, we don't need the fully detailed information that is required for the final presentation.

For example, much mechanical equipment can be represented by simple shapes in the initial stages. Figure 13.2 shows an example of a small motor, both as simple shapes, and in its final form. The cells are shown in wireframe form and after hidden line removal.

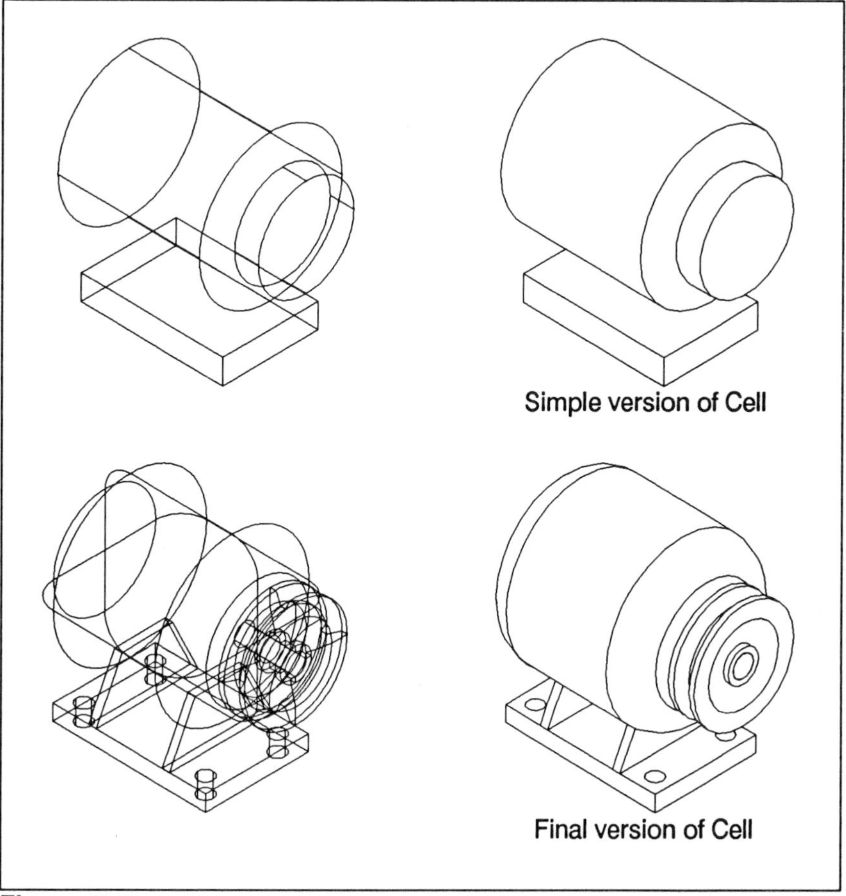

Figure 13.2

Stereo Pairs

With version 4, we can create stereo rendered images on screen. To see the image in 3D, or stereo, requires special glasses to be worn. The screen has two images, superimposed on each other. One image is the view as the left eye would see it 'in real life'. The other image is the view from the view point of the right eye. The glasses filter the right image from the left eye and vice-versa.

Stereo pairs also can be created by producing two images of the model. The two images should be rotated a small angle, relative to each other, about the Y-axis. This angle should be about 2 degrees. That is, set up the view for the left eye and render it. Save this image. Next, rotate the view minus two degrees about the Y-axis, then render and save it.

When the stereo pairs of the model have been created, they can be photographed, then viewed through a normal stereo viewer.

Saved Images

With both version 3.3 and version 4 of MicroStation we can produce rendered images which we may want to archive. We can save them to disk, but they are usually large files. A saved image using SNAPSHOT with version 3.3 and a 1280x1024 resolution screen, for example, takes about 1.3 Mb.

One way to overcome the problem is to use a compression program like PKZIP, which is readily available from most bulletin boards. Saved image files can be compressed by up to 98% in many cases. With these compression factors, we can fit up to 10 shaded images on a single 1.2 Mb floppy disk, for archiving.

Utilities

MicroStation is delivered with various utility programs. Included with these programs are translators, screen savers and memory mappers (for system managers/analysts). We will discuss some of these utilities available with MicroStation, that are relevant to 3D work.

Version 4

With version 4 we have utilities that can be accessed from within MicroStation. We can import or export DXF files, save 3D files as 2D files and save rendered images in various formats. These operations are performed with dialog and settings boxes which are accessed from the 'File' pull-down menu.

The 'Export' selection of this menu gives us a number of options.

DXF

With the DXF option we can create a DXF file of a MicroStation design file. DXF is the standard protocol for transferring CAD files between different systems and software packages. As with any translation program of this type, the results may not be one hundred percent compatible with all CAD software packages.

3D/2D Translations

This option is '2D' if the active file is 3D and vice versa. With it we can save our 3D design file as a 2D file. We select a view to be used to convert to 2D. We can nominate a view number, or a standard view such as TOP, FRONT, RIGHT etc. A 2D file is then generated from the nominated view. There is a check box also, where we can specify whether or not attached reference files are to be translated . If we are in a 2D design file we can save this as a 3D file with this option. We can save the elements at a specified Z value.

Another option is to translate an attached 3D cell library into 2D.

Other Options

Other menu selections allow us to generate Visible Edges files and save rendered images (refer to chapter 11, Rendering version 4).

Version 3.3

Most of the utilities available with version 3.3 of MicroStation are accessed from the Utilities Menu of MCE. As with version 4, we can create 2D design files from our 3D files and vice versa. We can also create DXF format files, for transferring between software packages.

To save rendered images with version 3.3, we can call up a user command such as SNAPSHOT.

Snapshot

Once files (or views) have been rendered they may be saved to disk with the SNAPSHOT user command which is invoked with the UC=MS_SUPL:SNAPSHOT key-in.

The user command prompts the user to select a view to be saved and options for the saved image.

Options

All options are preceded by a hyphen (-) and are passed to the snapshot program (SNAPSHOT.EXE) as command line arguments. The following options are available:

*-U*required when SNAPSHOT is activated from a user command.

*-T*Specifies that a transparent background should be used. By default a non-transparent background is used.

-CRtable
*-CLtable*Specifies a color table for the right (or left) screen be placed into the output file. This ensures that the correct color table is used when the shaded image is recalled at a later date. Whenever 'Shades' is run a new color table is automatically generated and stored in RENDER.TBL.

It is usually desirable to store this color table in the output file when a rendered image is saved. This could be accomplished with one of the following:
-CRrender.tbl if the rendering is done on the right screen or
-CLrender.tbl for the left screen.

*-Iinputfile*Specifies the input file. This is the currently active design file when SNAPSHOT is called.

*-Ooutputfile*Specifies the output file. By default a new file is created with the same name as the input file with the extension '.SAV'.

*-Vviewnumber*Specifies the view to be saved (1-8).

*-Llevel*Specifies the level (1-63) for the raster element (i.e., the image). This is placed, by default, on level 1.

*-3*Specifies that a three-dimensional output file should be created.

*-B*Specifies that a border should be placed around the raster element. By default, no border is placed.

*-Fformat*format for the raster element may be specified :
1: Binary (bit mapped)
2: Byte (color) - - - DEFAULT
9: Run length encoded binary

To use snapshot, you must have one of the graphics cards that it supports. These are listed in a text file 'SNAPSHOT.DOC' which should be located in the same directory as the user command. This is the MicroStation SUPL directory.

Note:
If you do not have this user command, it is available from the MicroStation Bulletin Board.

When you have saved your shaded images, you may retrieve them just as you do a design file. By default they have the extension '.SAV'. You may have a number of these files referenced to one another, but the last file to be read in will 'paint' over the previous screen display.

This can be used to create a presentation, where different views of a model can appear on the screen one after another.

Images saved with snapshot can be displayed only through MicroStation. Another user command, SAVERGB.UCM, allows us to view the saved images from outside MicroStation. We do this with a display program MSSLIDE.EXE. Again, there is a text file SAVERGB.DOC that comes with the user command. This explains how to use both the user command and the display program. As with SNAPSHOT, this user command also, is available on the MicroStation Bulletin Board.

14 : Practice Examples

Exercises given on the following pages are for practicing the concepts and tools as you learn them. After each group of examples you will find instructions for at least one method of completing each model. If you use different tools and methods to those shown, don't be concerned. With CAD, there is usually more than one way to construct a model. In fact, you will find that the instructions given to complete the various models are not consistent in their approach. This has been done deliberately to show different methods.

Create a new 3D file for the practice models. This will keep them separate from the exercises used in the tutorial section of the book.

Try to complete each 'model' before looking at the instructions following. You should find that the hints given at the beginning are enough to get you started.

Chapter 1 Examples

Exercises shown in figures 14.1 to 14.3 are to provide practice in placing shapes in the various views.

- Use the most convenient view to place your shapes.
- Where shapes are the same as others, copy them rather than creating each shape individually.
- Use precision inputs where appropriate. This includes using a tentative to an existing element followed by a DX= or DL= key-in.

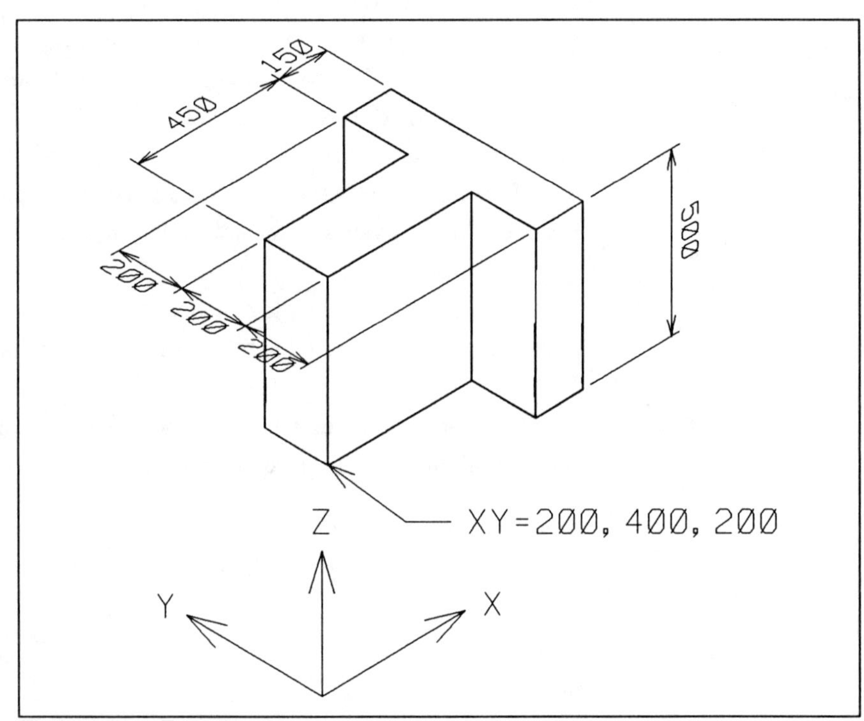

Figure 14.1 Exercise 1

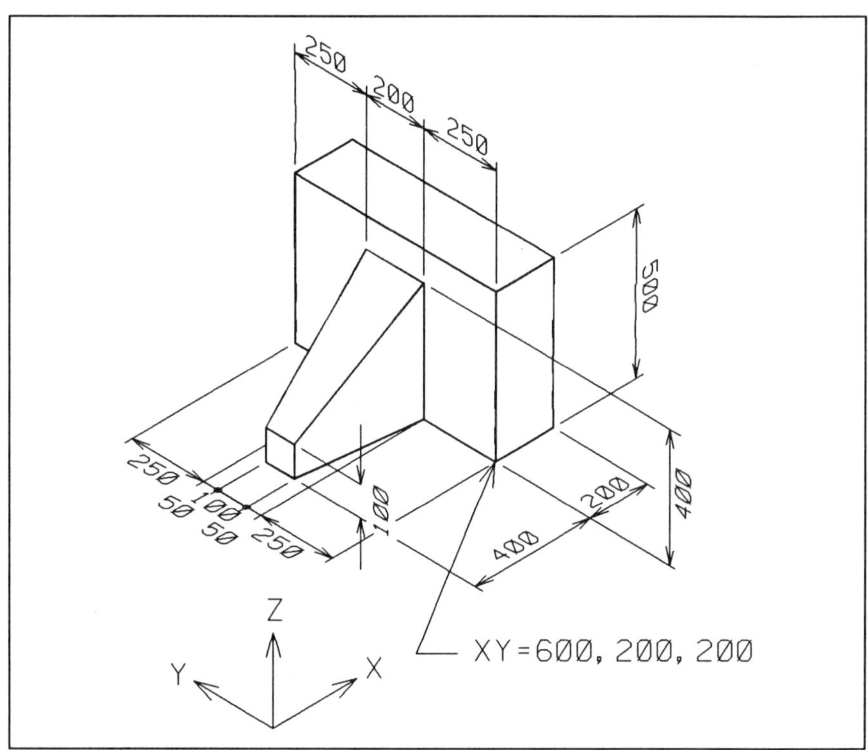

Figure 14.2 Exercise 2

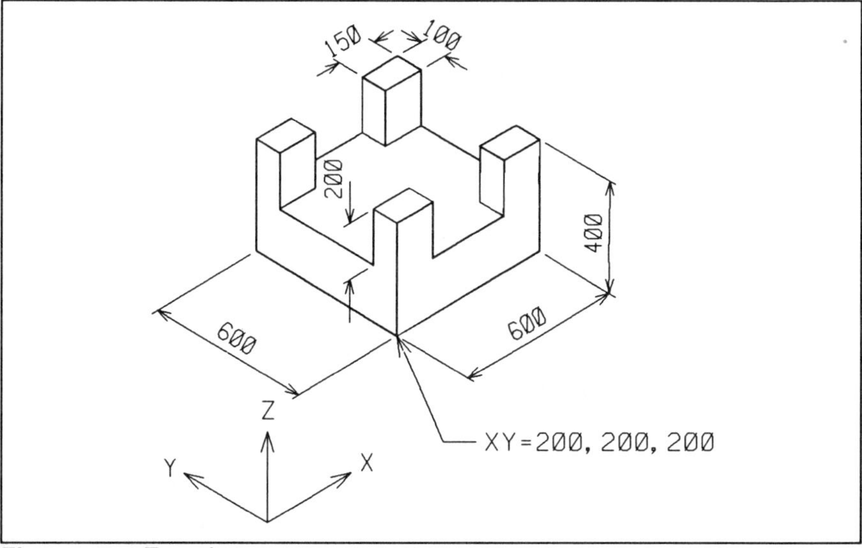

Figure 14.3 Exercise 3

Exercise 1

A. Create the base shape in the TOP view. This can be done by placing a shape, or placing lines and chaining them together later. Another way is to place 2 blocks, forming the 'tee'. The 2 blocks could be used, as a guide, to create the overall shape of the base.

To place the shape, the following procedure could be used:

- Make the TOP view 'current'.
- Select the *Place Shape* tool from the menu.
- Key in XY=200,400,200 for the first point. Then key in the following to specify the other points of the shape:
  'DX=450', 'DX=,-200', 'DX=150', 'DX=,600', 'DX=-150', 'DX=,-200', 'DX=-450', and finally 'DX=,-200'.

We could have used DL equally as well because we are working in a TOP view.

B. Copy the base shape vertically 500 units. Use the *Copy Element* tool and identify the shape in any view. Key in 'DL=,,500' for the distance to copy the element

C. Place one of each unique shape. That is, only place shapes which have not been placed previously. If a shape with the same dimensions and orientation has been placed before then we can use copy.

D. The shape shown hatched can be copied to the other 2 indicated locations. All 3 shapes (or blocks) have the same dimensions. Copying is quicker than having to create the shape.

E. Similar to D (above).

F. Similar to D and E (above).

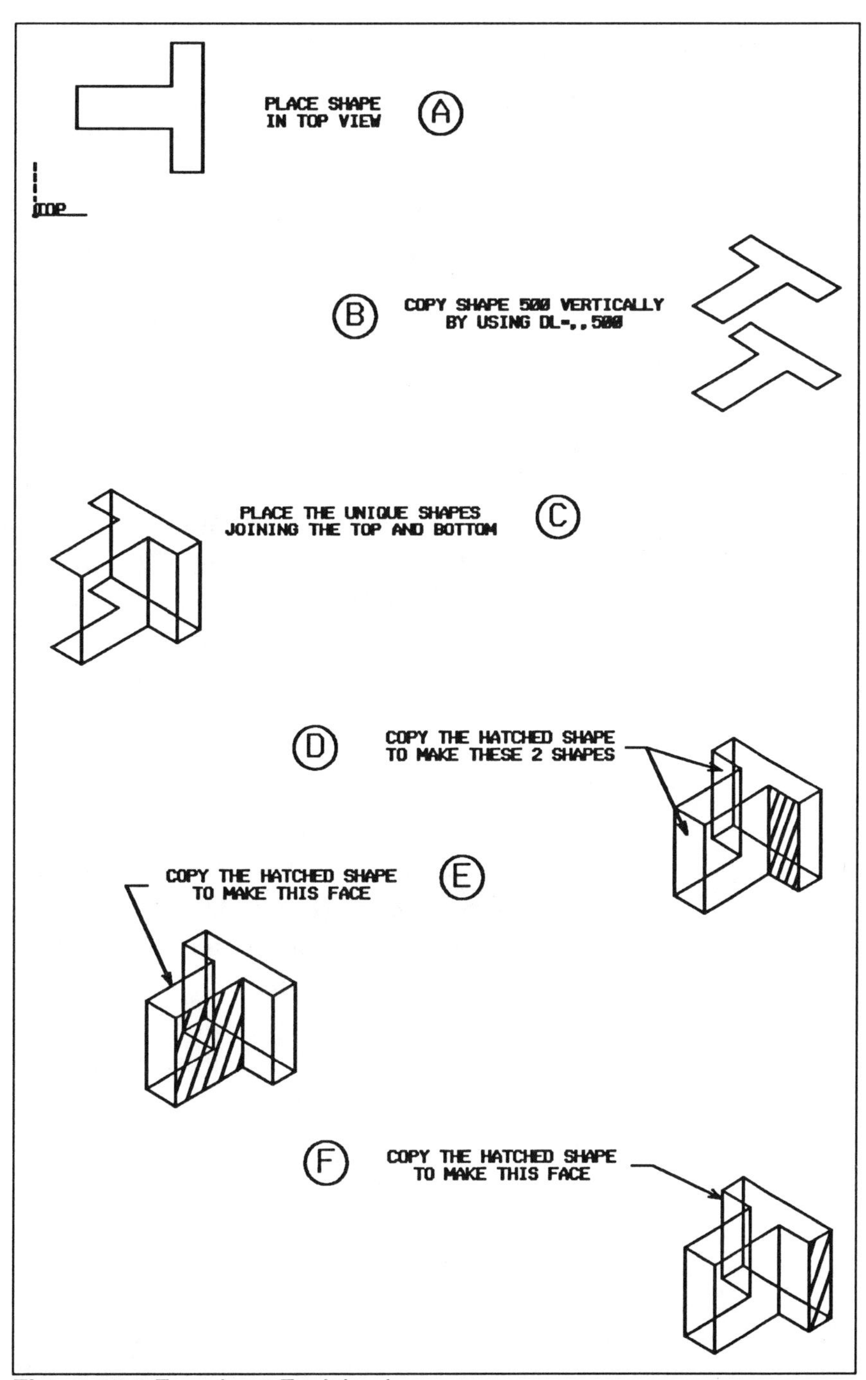

Figure 14.4 Exercise 1 Explained

Exercise 2

A. Create the shape of the base in a TOP view. This can be done by directly placing a shape using precision key-ins as follows:

- Make the TOP view current.
- Select the *Place Shape* tool from the menu.
- Key in XY=600,200,.200 for the first point followed by: 'DX=200', 'DX=,600', 'DX=-200', 'DX=,-200', 'DX=-400,-50', 'DX=,-100', 'DX=400,-50', 'DX=,-200', 'DX=200'.

B. Still with the *Place Shape* tool, use the ISO view to place a shape as follows:

- Identify POINT 1 as the first point followed by POINT 2.
- Tentative to POINT 2 then key in 'DL=,,100' for POINT 3.
- Tentative to POINT 1 then key in 'DL=,,400 for POINT 4.
- Identify POINT 1 for the closing of the shape.
- The same method can be used for creating the other similar shape. It could also be created with the mirror element tool in a TOP view.

C. Place the sloping top surface and the end shape by snapping to the shapes already placed.

D. Using a RIGHT view the back face can be created using the *Place Block* tool. It does not have to be created in its correct location, necessarily, as it is a simple matter to move it to the correct location later. This block can then be copied to form the other face.

E. A similar method to that of D above, can be used for creating the side face. We would use a FRONT view this time. The block can then be copied to the other side.

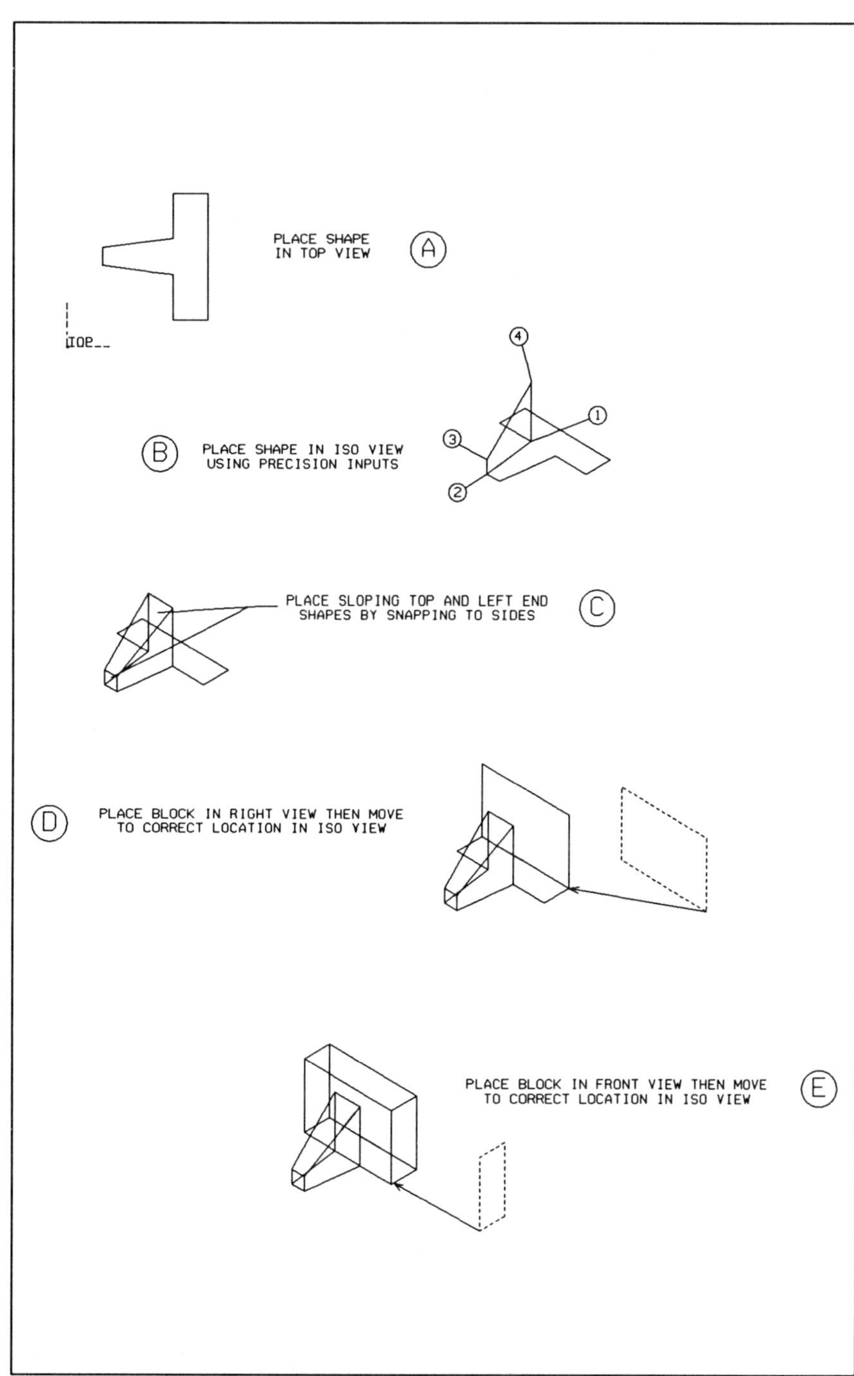

Figure 14.5 Exercise 2 Explained

Exercise 3

To simplify this model, consider it to be in two main parts, the base and the four 'lugs'.

A. In a TOP view, place a block for the base. Copy this block vertically 200, using DL=,,200 to specify the distance.

B. Again, in a TOP view, place a block for one of the four lugs.

- As displayed in the figure opposite this can be placed with a tentative to a corner of an existing block followed by keying in DX=-150,100.
- Copy the block 200 above (i.e., the block placed in B). These two blocks can be used to place the vertical sides of the lug, later.
- Copy the uppermost block to create the top surface of the other 3 lugs, using DL to specify the distance. Use the other views to check that you are placing them in the correct location.

C. Place two of the vertical faces on the base, using the *Place Shape* tool in the ISO view. Similarly, place two of the vertical faces on the lug presently having the two blocks. These vertical faces can then be copied to the other positions to finish the model.

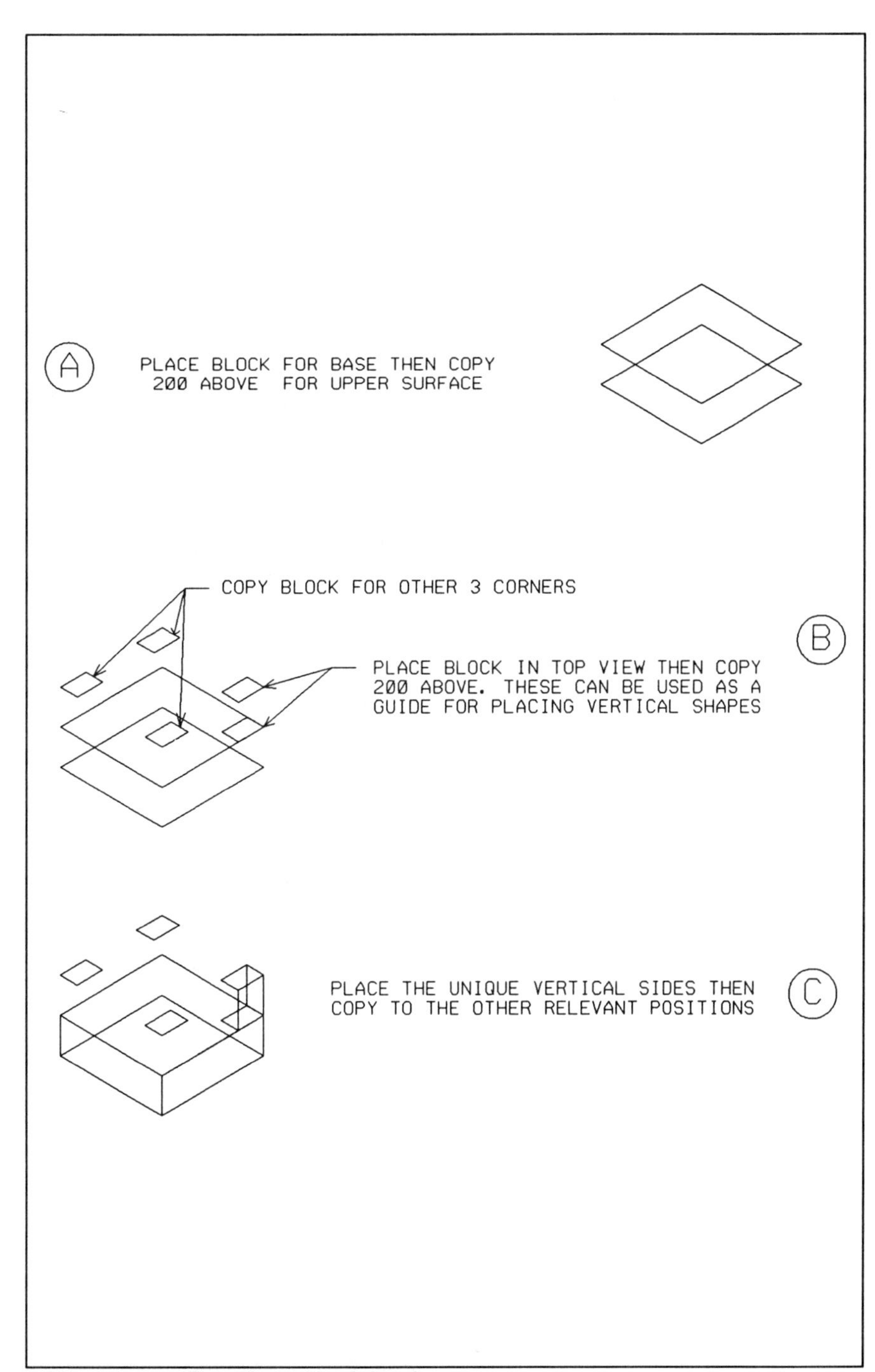

Figure 14.6 Exercise 3 Explained

Chapter 3 Examples

In chapter 3, we discussed projection and surfaces of revolution. Examples in this section can be constructed using those techniques.

Projection

Exercises shown in figures 14.7 and 14.8 are to provide practice in making models using projected shapes. To start, complex models should be separated into less complicated components.

Remember that all of the tools you had at your disposal with 2D can be used in 3D. With some, the orientation of the view has to be considered. Where possible, copy existing elements. Copying is quicker than creating a new version of an existing part. Also, with copying existing elements, we don't have to worry about which view is being used. Elements retain their existing orientation when being moved and copied.

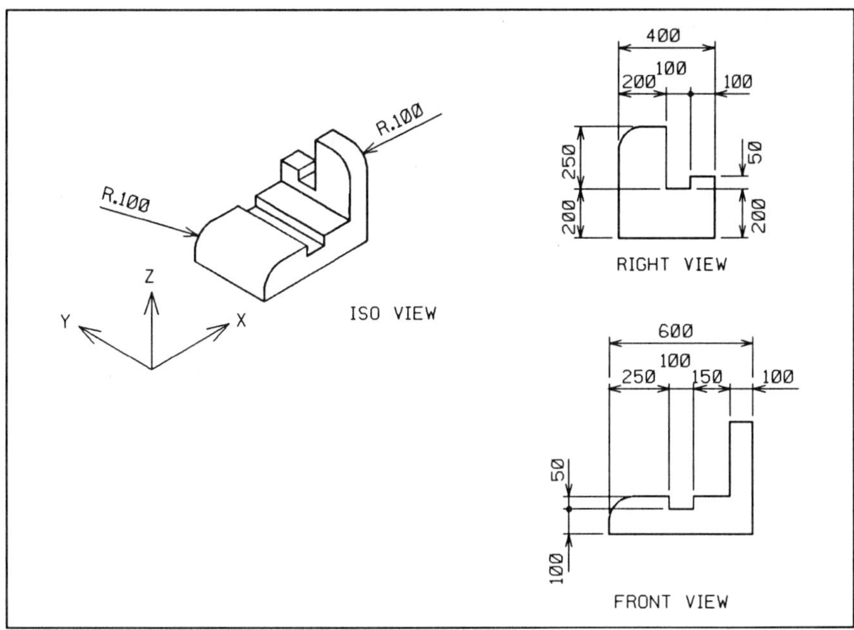

Figure 14.7 Exercise 4

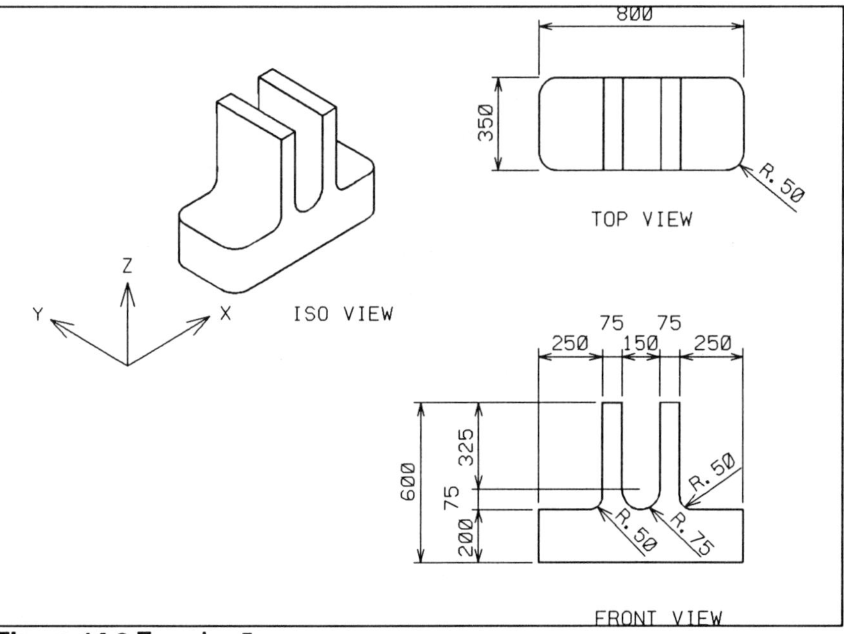

Figure 14.8 Exercise 5

Exercise 4

This model can be simplified by breaking it into two parts. Each of these parts can be formed by projecting a shape. Placing the shapes in the correctly orientated view, initially, makes the task simple.

A. Using a FRONT view, create the shape for the main body. The shape can be created initially as lines. The arc can be added as a fillet construction with the appropriate tool. When it is correct, the lines and arc can be chained together to form a shape.

B. A similar method to that outlined in A above can be employed to create this shape. It should be placed in the RIGHT view for the correct orientation.

C. Once the two shapes have been completed, they should be located correctly, relative to each other. This is only necessary where the two shapes weren't created in the correct locations in the first instance.

D. Check that ACTIVE ANGLE is set to 0 and ACTIVE SCALE is set to 1.

- Select ACTIVE CAPMODE ON.
- The shape of the main body can be projected using a precision input to specify the distance (DL=,400).
- The end section of the model could also be used to specify the distance.

E. With the same settings as for D above project the end section shape, using DL=100 as the precision input to specify the distance.

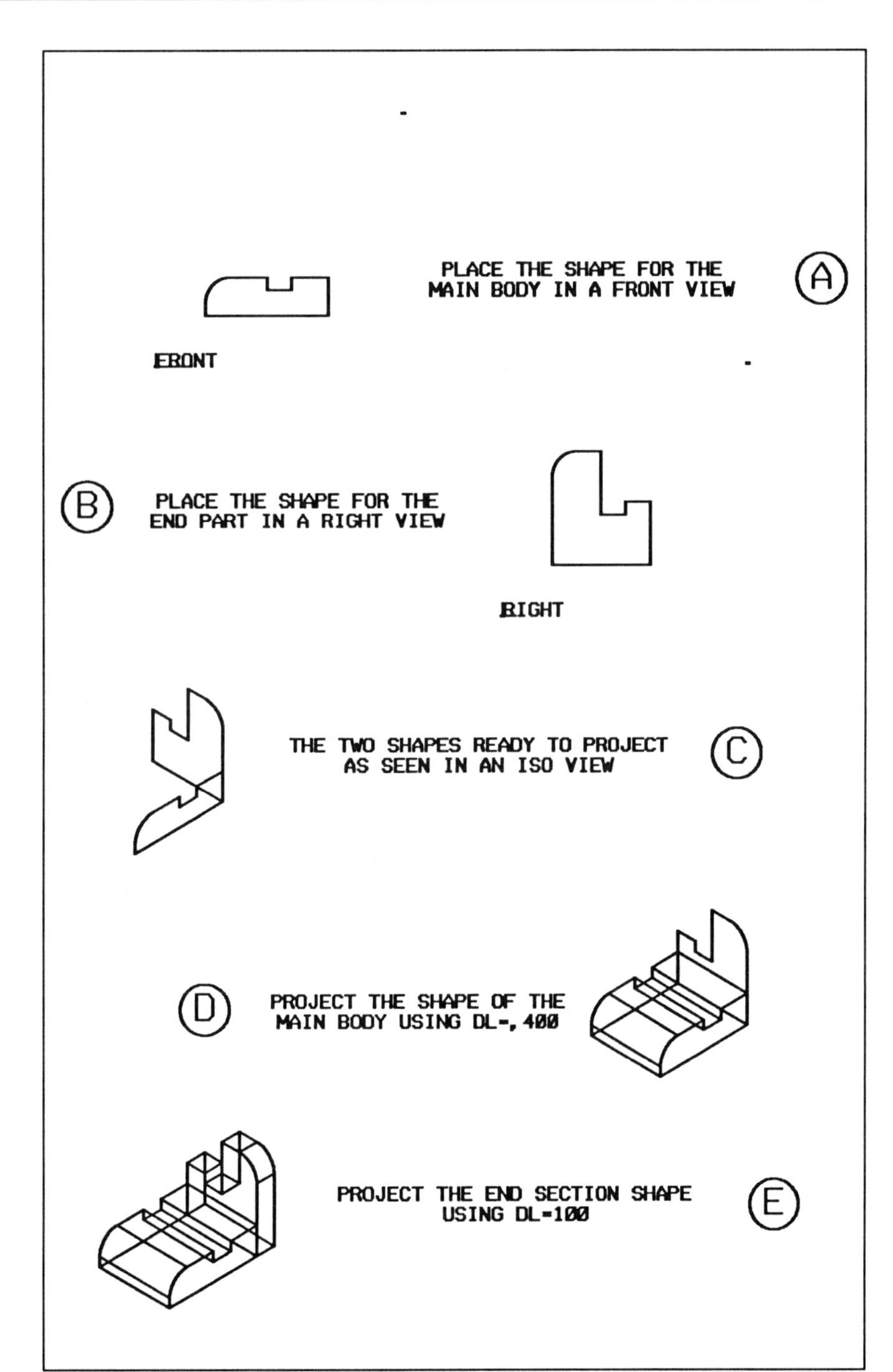

Figure 14.9 Exercise 4 Explained

Exercise 5

We can separate this model into three parts, the central section and the two end pieces. Creating the shapes in the correct views, simplifies the overall job.

A. Create the shape for the central section in a FRONT view.

B. Use a TOP view to make one of the end sections.

C. Make sure that the two shapes are in the correct location relative to on another.

D. Ensure that ACTIVE ANGLE is 0 and ACTIVE SCALE is set to 1.
Select ACTIVE CAPMODE ON.
The end section can now be projected

E. The opposite end section can be made by MIRROR COPYING the projected surface of the other. This can be done in a FRONT view.

F. Again, using the same settings as for D (AA=0, AS=1 and CAPPED SURFACES), project the central part of the model. One or the other of the end sections can be used to specify the distance.

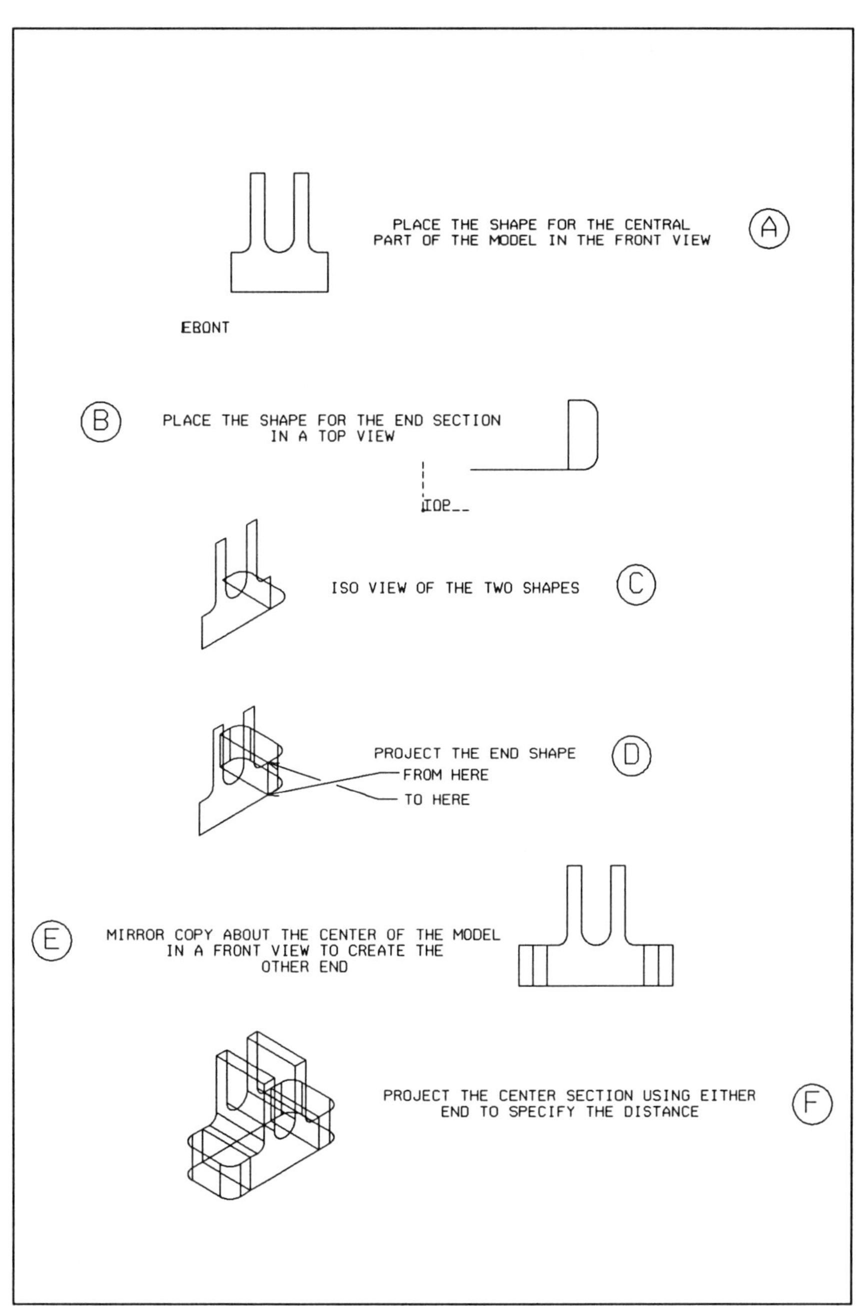

Figure 14.10 Exercise 5 Explained

Surface of Revolution

We use surfaces of revolution to create models that are circular in nature, with a constant section. We construct these models in two steps. We create the typical cross-section first. We then use the *Construct Surface/Solid of Revolution* tool to complete the object.

When creating a surface of revolution, care must be taken with defining the axis of revolution. A view must be used that has its Z-axis aligned with the required axis of revolution.

Figure 14.11 shows a model that can be created simply using surfaces of revolution. The circular o-rings have a diameter (thickness) of 3.8 and are located centrally in the grooves. Use a radius of 5 for both curved sections on the main body of the model.

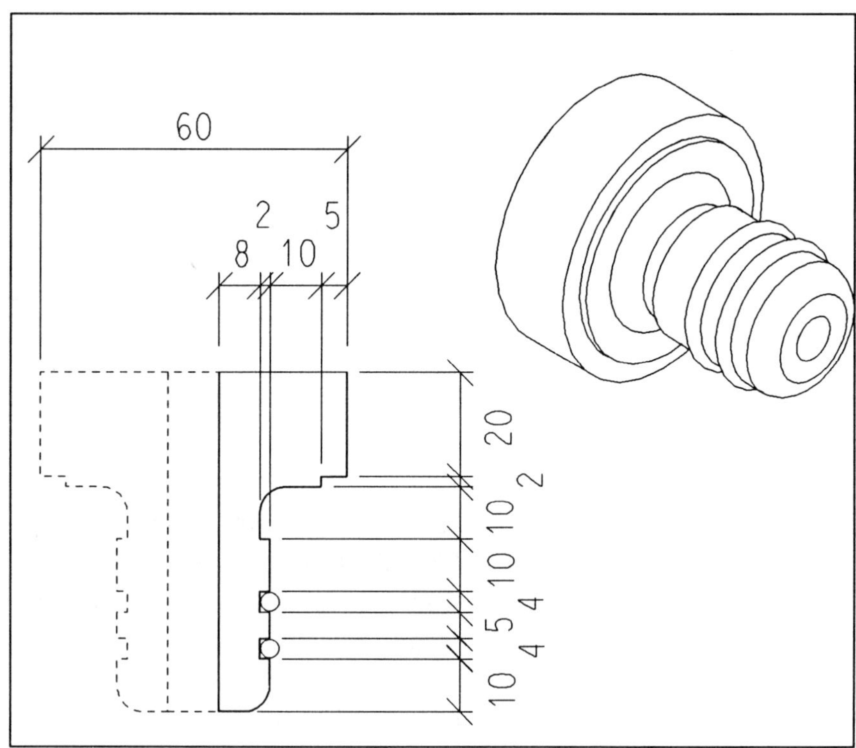

Figure 14.11 Exercise 6

Exercise 6

Using a TOP view, construct the cross-section of the main body. This can be done using standard 2D techniques.

Chain together, as a complex shape, the elements forming the main body.

Place the two circles to represent the cross-sections of the o-rings.

Place a center line, which can be used to specify the axis of revolution. This step is not mandatory, but it makes life easier when we are creating the surface of revolution. An active point would be as effective.

To create the surface of revolution, first set up your screen with a TOP view and a FRONT or BACK view of the model.

- Select the *Construct Surface/Solid of Revolution* tool.
- Set the angle of revolution to 90 degrees. Whether the angle should be plus or minus will depend on the direction that the section is to be rotated. For a FRONT view it will be plus and for a BACK view it will be minus.
- Identify the main body section in the FRONT view for simplicity.
- Define the axis of revolution in the FRONT or BACK view by identifying the center line.
- To form the other three sections key-in 'DL=0' three times.
- Repeat a similar process for the two O-rings.

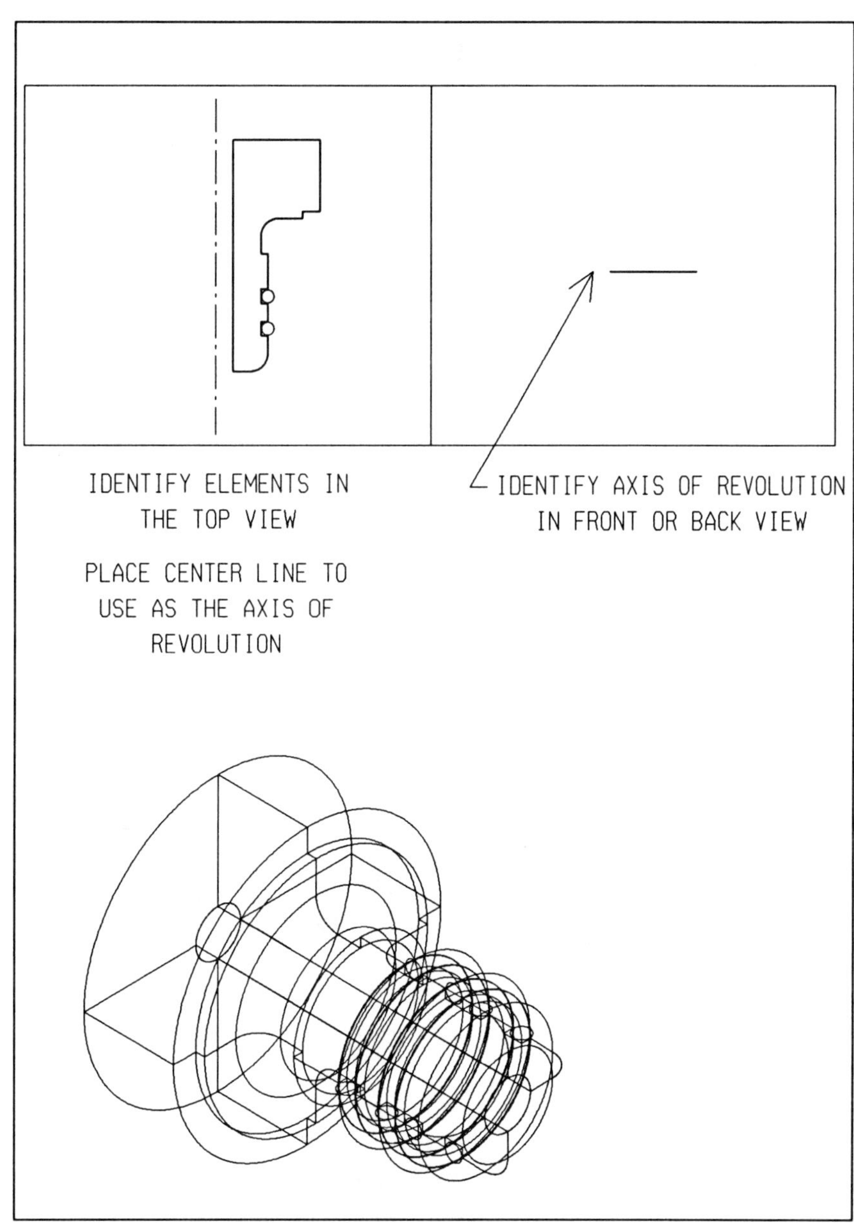

Figure 14.12 Exercise 6 Explained

Chapter 8 Example (version 4)

This model, shown in figures 14.13 and 14.14, incorporates the use of a B-spline surface construction. As with other examples we have looked at, the model should be separated first into its basic parts.

You can use a slab for the base.

A fence projection can be used to create the columns in one operation.

To create the domed roof, use the *Construct B-spline Surface by Edges* tool.

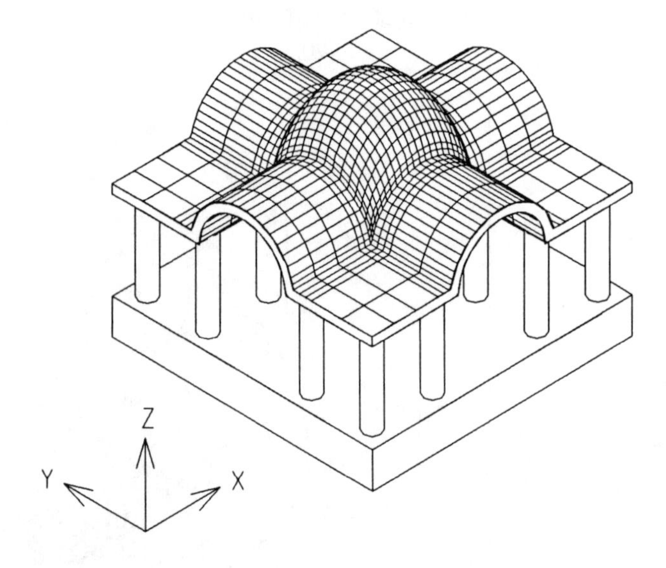

Figure 14.13 B-spline Surface Exercise

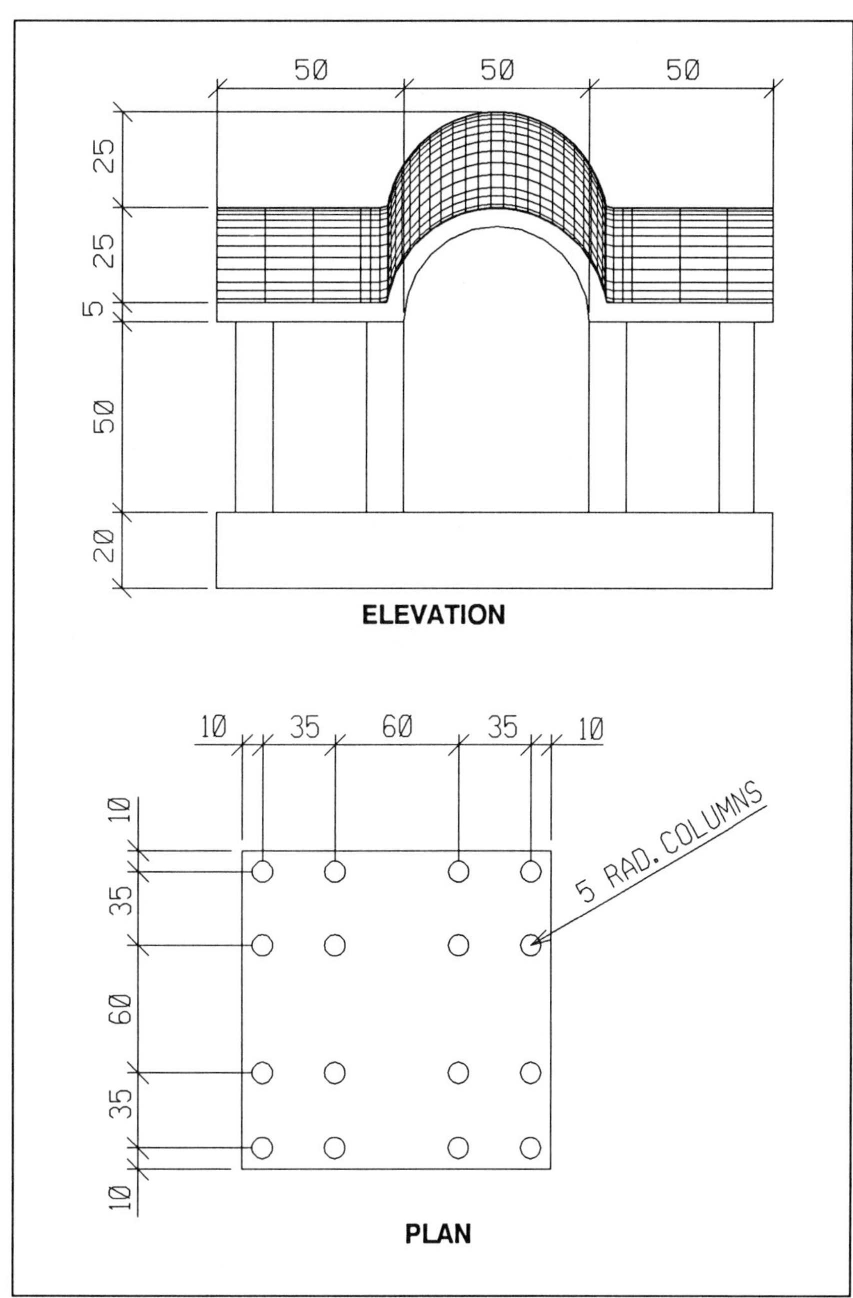

Figure 14.14 Exercise 7, Elevation and Plan

Exercise 7

This model can be considered to consist of three separate parts - the base, the columns, and the domed roof.

A. We can place the base as a slab. Circles for the columns can then be placed, in the plane of the top surface of the slab, in a TOP view. Following this, one of the outer roof edges can be created as a complex shape. Elements forming the shape (arcs and lines) can be placed in a FRONT view in their correct position, 50 above the base.

B. Set the active angle to 90 degrees. Using the *Rotate Element by Active Angle (Copy)* tool, create the left roof edges. Copy the front edge to the back and the left edge to the right to complete the roof edges.

C. In a TOP view place a fence around the circles representing the column sections. Use the 'FENCE SURFACE PROJECTION' key-in and project the columns vertically by 50. Before projecting, make sure that the active angle and active scale are 0 an 1 respectively.

D. Using a similar method to that used to create the roof edges, create the edge limits for the top skin of the roof.

E. With the edges just created, the roof can be created with the *Construct B-spline Surface by Edges* tool.

Use the same method to create the inner skin of the roof.

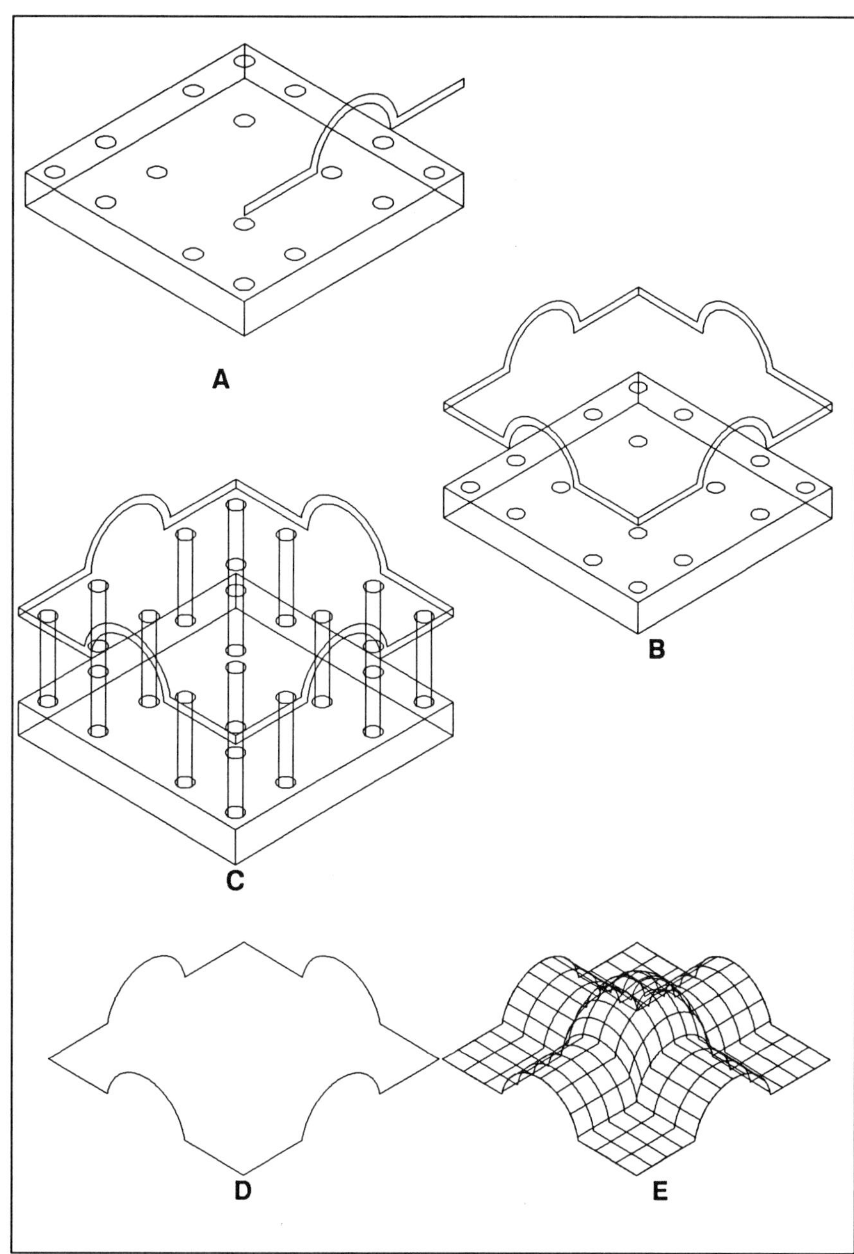

Figure 14.15 Exercise 7 Explained

Index

D

E

F

G

H

R

S

MicroStation Software Order Form

CGM2DGN - Convert CGM to DGN

CGM2DGN is a directly file conversion software from CGM (Computer Graphic Metafile) into MicroStation design files. Now you can get reports and diagrams from Harvard Graphics, Freelance or any other software that generates CGM output directly into MicroStation.

For a FREE evaluation of the conversion capabilities, send us your CGM file on a DOS floppy disk. US$1,000 for single CPU copy, available on DOS only.

REFMAN - Reference File Manager

REFMAN is the ultimate reference file manipulation tool! REFMAN has a user friendly, screen based interface and allows users to view an manipulate reference file attachments.

REFMAN lets you keep track of your reference files. View and/or edit reference file attachments including levels, display flags and symbology. Reference files can be deleted and deleted reference file attachments can even be reattached. Copy or move reference files to other directoris, even other disks and REFMAN will take care of the reference files. Use REFMAN to obtain reports of all aspects of the reference files attached to your design files. REFMAN has powerful facilities for changing the path(s) of reference files. Be free at last to move your design files. All this and more with REFMAN.

REFMAN is the design and reference file management tool you have been looking for! Site licences and demonstration versions available. US$300.00 for single CPU copy. Available on DOS only.

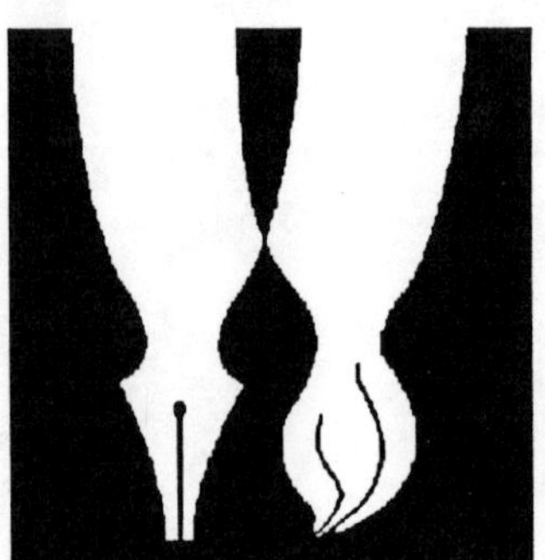

Pen And Brush Publishers
2nd Floor, 94 Flinders Street
Melbourne Vic 3000
AUSTRALIA

Tel: (03) 818 6226
Fax: (03) 818 3704
International Fax: +61 3 818 3704

Check Out These Other MicroStation Titles Available from OnWord Press

OnWord Press Products are Available Directly From

1. Your Local MicroStation Dealer or Intergraph Education Center
2. Your Local Bookseller
3. In Australia, New Zealand, and Southeast Asia from:
   Pen & Brush Publishers
   2nd Floor, 94 Flinders Street
   Melbourne Victoria 3000
   Australia
   Phone 61 (0)3 818 6226
   Fax 61(0)3 818 3704

4. .... Or Directly From OnWord Press
   (see ordering information on the last page.)

Upgrading to MicroStation 4.X from an earlier version? OnWord Press has Upgrade Tools!

The MicroStation 4.X Delta Book

Frank Conforti
A Quick Guide To Upgrading to 4.X on all Platforms from Earlier Versions of MicroStation

This short and sweet book takes you from 3.X versions of MicroStation into the new world of MicroStation 4.X.

Did you know that there are now over 950 commands in MicroStation? Have you seen the new graphical user interface (GUI)? Are you ready for dialog boxes? Associative dimensioning? Palettes? The new MicroStation Manager?

Intergraph's release notes don't tell the whole story. Sure they let you know about a few new features.

But the MicroStation 4.X Delta Book gets you up and running with these new productivity tools fast.

Highly graphic, the MicroStation 4.X Delta Book doesn't just tell you, it shows you!

Written by Frank Conforti, author of best-selling INSIDE MicroStation, the MicroStation 4.X Delta Book will have you turning out MicroStation 4.X productivity in no time.

Price: **US$19.95** (Australia A$29.95)
Pages: 110
Illustrations: 75+
ISBN: 0-934605-34-3

INSIDE MicroStation takes you beyond stand-alone commands. Using real world examples it gives you practical and effective methods for building good working habits with MicroStation.

Frank Conforti is an independent MicroStation consultant based in Delray Beach Florida. Formerly CAD Manager at Keith and Schnarrs Engineers, Frank managed a network of VAX-based IGDS and MicroStation systems running workstations, PCs and MACs. Conforti has been using CAD, IGDS and MicroStation for 16 years. He is an avid writer, trainer, and user group sponsor. He is co-author of the Macintosh CAD/CAM book.

- Optional INSIDE MicroStation Disk: **US$14.95** (Australia A$29.95) Includes the tutorial example design files, menus, listings, examples and more.

Price: **US$29.95** (Australia A$54.95)
Pages: 550
Illustrations: 220+
ISBN: 0-934605-49-1

The INSIDE MicroStation Companion Workbook

Michael Ward and Support From Frank Conforti
32 Steps to MicroStation -- A self-paced tutorial workbook for individual or classroom use.
First Edition, Release 4.X, Supports all 4.X and 3.X Platforms including DOS, Unix, Mac, and VMS

This highly readable hands-on text is your guide to learning MicroStation. As a companion to the INSIDE MicroStation book, the workbook gives you practical exercises for developing MicroStation skills.

The INSIDE MicroStation Companion Workbook is set up for self-paced work alone, or in classroom or group training. The workbook comes setup for three types of training scenarios. You can work through the exercises and test yourself. Or you can organize the course into a three or four day professional or 10 - 14 week semester course. Or you can make up your own grouping from the 32 basic training units.

The workbook course takes you through at least two paths to develop your proficiency with MicroStation. First, you are taught basic functions and given exercises to develop your skills. Second, you can select from an architectural, civil, or electrical engineering "real life" professional drawing that you work on during the length of the course.

- The INSIDE MicroStation Companion Workbook training Disk comes with the book (5.25" DOS standard, other formats available on request). On the disk you will find exercise tools, exercise drawings, plot files, even a diploma!
- Also included in the package are several professional blueprints used as part of the training course.

Price: **US$34.95** (Australia A$54.95)
Pages: 400 (Includes DOS 5.25" Disk and Blueprints)
Illustrations: 75+
ISBN: 0-934605-42-4

Optional INSIDE MicroStation Companion Workbook Teacher/Trainer's Guide.

This 75 page instructor's guide to the INSIDE MicroStation Companion Workbook is geared for the professional trainer, college professor, or corporate training department. Filled with teaching tricks, organization tools and more, the instructor's guide is a must for anyone in an instructor's position.

Price: **US$9.95** (Australia A$18.95)
NOTE: The Teacher/Trainer's Guide is **FREE** to any group ordering 10 copies or more of the workbook. Contact OnWord Press for ordering information.
Pages: 75
ISBN: 0-934605-39-4

The MicroStation Productivity Book

Kincaid, Steinbock, Malm
Tapping the Hidden Power of MicroStation
Second Edition, Release 4.X, Supports all 4.X and 3.X Platforms including DOS, Unix, Mac, and VMS

With this book beginning and advanced users alike can take big leaps in productivity, job security and personal satisfaction. Thirty-six step-by-step chapters show you how to take charge of MicroStation.

Completely updated for MicroStation 4.X, the MicroStation Productivity book now includes complete information on MDL, using MicroStation with Oracle, 100 pages of new 3D tools, and more.

This book is really two books in one. The first half is "The Power Users Guide to MicroStation". Use the tools and tutorials in these chapters to go beyond the basic MicroStation menus and commands. Power drawing, power editing, automation of repetitive tasks, working in 3D, this section teaches you how to get the most out of MicroStation.

Turn to the second half "The Unofficial MicroStation Installation Guide" to take charge of your MicroStation installation. Here you'll find tools to supercharge your software installation, customize command environment, build menus, manage your files and make DXF transfers. There is even a chapter on the undocumented EDG editor that tells you how to fix corrupted design files.

Learn how to write user commands and immediately put to work the "Ten User Commands for Everybody" and "Our Ten Favorite User Commands". Take advantage of attribute data with links to database programs including dBASE and ORACLE. This book shows you how.

Appendices include listing of all TCB variables, command names, and the syntax you need to know to be a power user.

John Kincaid and Bill Steinbock are MicroStation installation managers for the U.S. Army Corps of Engineers, Rock Island and Louisville districts respectively. They are avid writers and MicroStation bulletin board aficionados. Rich Malm managed the Corp's Intergraph/MicroStation procurement efforts until he recently retired to devote his time to consulting and MicroStation database applications. Between them, the authors have over 40 years of CAD, IGDS and MicroStation experience.

- Optional Productivity Disk: **US$49.95** (Australia A$79.95) Includes all of the user commands in the book, design file and tutorial examples, menus for user commands, custom batch files, and the exclusive MicroStation 3D menu.

Price: **US$39.95** (Australia A$69.95)
Pages: 600
Illustrations: 220+
ISBN: 0-934605-53-X

MicroStation Reference Guide, Pocket Edition

John Leavy
Everything you want to know about MicroStation -- Fast!
Second Edition, Release 4.X, Supports all 4.X and 3.X Platforms including DOS, Unix, Mac, and VMS

Finally, all of MicroStation's commands are in one easy to use reference guide. Important information on every MicroStation command is at your fingertips (including some that even Intergraph/Bentley don't tell you about!).

This book gives you everything you need to know about a command including:

How to find it
How to use it
What it does
What happens in an error situation

The book includes all commands, key-ins, ACTIVES, and environmental settings. Also you'll find background on such important concepts as coordinate entry, PLACE commands, and drawing construction techniques. Actual screens and examples help you get up and drawing now.

Now completely updated for 4.X, the MicroStation Reference Guide covers over 950 commands (The last edition of the book had only 400 commands!). Every user and every workstation should have one of these books handy.

The new edition features hundreds of command illustrations including the new 4.X GUI palettes for command operations.

John Leavy is president and chief Intergraph/MicroStation consultant with Computer Graphic Solutions, Inc. His specialty is MicroStation training and applications development. Leavy spent 12 years with Intergraph before moving into the private consulting world.

⊛ Optional MicroStation Reference Guide, Pocket Edition Disk: **US$14.95** (Australia A$19.95) The MicroStation Online Reference Guide disk puts this book online with MicroStation.

Price: **US$18.95** (Australia A$26.95)
Pages: 320
Illustrations: 200+
ISBN: 0-934605-55-6

The Complete Guide To MicroStation 3D

David Wilkinson
First Edition, Release 4.X, Supports all 4.X and 3.X Platforms including DOS, Unix, Mac, and VMS
(Published by Pen & Brush Publishers, Distributed by OnWord Press)

If you are a MicroStation 2D user wanting to advance into the "real-world" of 3D, then this is the book for you. It is both a tutorial and a reference guide. Written especially for MicroStation 2D users, it teaches you how to use the 3D capabilities of MicroStation. Diagrams (over 220 in total) are used extensively throughout this book to illustrate the various points and topics.

Chapters 1 and 2 incorporate the basic tutorial section. Here you are introduced to the 3D environment of MicroStation and shown how to place elements in 3D. This is explained with simple exercises, graphically illustrated to avoid any confusion.

From this basic introduction you can advance, step-by-step, through complex problem solving techniques. This can be accomplished at your own pace. Having learned to create a 3D model, you are shown methods for extracting drawings and how to create rendered (shaded) images.

MicroStation version 4 tools are covered comprehensively, along with those for version 3.3 (PC version for 286 machines).

Among topics covered are the following:

- Basic placement and manipulation of elements
- Views and view rotation
- Projection and Surface of Revolution
- B-Spline Surface Constructions (version 4)
- Rendering (for both versions 3.3 and 4)

⊛ Optional Complete Guide to MicroStation 3D Disk: **US$19.95** (Australia A$29.95) Includes examples from the book as well as tips and tricks for 3D work, model building, shading and rendering.

Price: **US$39.95** (Australia A$69.95)
Pages: 400
Illustrations: 200+
ISBN: 0-934605-66-1 (Australia ISBN: 0-646-01678-4)

101 MDL Commands

Bill Steinbock
First Edition, Release 4.X, Supports all 4.X and 3.X Platforms including DOS, Unix, Mac, and VMS

This is the book you need to get started with MDL!

With MicroStation 4.0 comes the MicroStation Development Language or MDL. MDL is a powerful programming language built right in to MicroStation.

MDL can be used to add productivity to MicroStation or to develop complete applications using MicroStation tools. Virtually all of the 3rd party applications vendors are already using MDL for their development.

Now you can too, with 101 MDL Commands.

The first part of this book is a 100 page introduction to MDL including a guide to how source code is created, compiled, linked, and run. This section includes full discussion of Resource Files, Source Codes, Include Files, Make Files, dependencies, conditionals, interference rules, command line options and more.

Learn to control MicroStation's new GUI with dialog boxes, state functions, element displays and file control.

The second part of the book is 101 actual working MDL commands ready-to-go. Here you will find about 45 applications with over 101 MDL tools. Some of these MDL commands replace user commands costing over $100 a piece in the 3.0 market!

Here's a sampling of the MDL applications in the book and on the book:

MATCH - existing element parameters
Creation - create all the new element types from MDL
Multi-line - convert existing lines and linestrings to multi-line elements
CALC - Dialog box calculator
DATSTMP - Places and updates filename and in-drawing date stamp
PREVIEW - previews a design file within a dialog box
Text - complete text control - underline, rotate, resize, upper/lower, locate text
string, import ASCII columns, extract text
Fence - complete fence manipulations including patterning, group control,
circular fence and more
3D surfaces - complete projection and surface of revolution control
Cell routines - place along, place view dependent cell, scale cell, extract to
cell library
Dialog boxes - make your own using these templates!
Search Criteria - delete, fence, copy, etc based on extensive search criteria.

Use these MDL commands to get you started with the power of MicroStation MDL. You can put these tools to work immediately, or use the listings to learn about MDL and develop your own applications.

⊛ Optional 101 MDL Commands Disk: **US$101.00** (Australia A$155.00) Includes all of the MDL commands from the book in executable form, ready to be loaded and used.

Price: **US$49.95** (Australia A$85.00)
Pages: 680
Illustrations: 75+
ISBN: 0-934605-61-0

Bill Steinbock's

Pocket MDL Programmers Guide

Bill Steinbock
First Edition, Release 4.X, Supports all 4.X and 3.X Platforms including DOS, Unix, Mac, and VMS

Intergraph/Bentley's MDL documentation is over 1000 pages!

Bill's Steinbock's Pocket MDL Programmers Guide gives you all the MDL tools you need to know for most applications in a brief, easy-to-read format.

All the MDL tools, all the parameters, all the definitions, all the ranges -- in a short and sweet pocket guide.

If you're serious about MDL, put the power of MicroStation MDL in your hands with this complete quick guide.

Includes all the MDL commands, tables, indexes, and a quick guide to completing MDL source for MDL compilation.

Price: **US$24.95** (Australia A$39.95)
Pages: 256
Illustrations: 75+
ISBN: 0-934605-32-7

MDL-Guides

CAD Perfect
First Edition, Release 4.X, Supports all MDL 4.X Platforms Runs under DOS only.

1000 Pages of Intergraph MDL Documentation On-Line at Your Fingertips!

The Intergraph MDL Documentation is voluminous, to say the least. MDL-GUIDES puts the MDL and MicroCSL documentation in a hypertext TSR for reference access while you are programming or debugging in MDL.

This terminate-and-stay-ready program was sanctioned by Intergraph as the practical way to find out all the MDL information you need in an easy format.

The program is environment friendly, works with high DOS memory space to leave room for your other applications, and is quick.

The package includes the hypertext software, the complete set of MDL and MicroCSL Intergraph documentation in hypertext format, and a proper set of installation instructions.

MDL-GUIDES
For DOS Formats only (Other formats available on request from CAD Perfect)
Includes Disk and User's Manual
Disk Includes all MDL Documentation formatted for use on-line
Price: **US$295.00**
ISBN 0-934605-71-8

Programming With User Commands

Mach Dinh-Vu
Second Edition
For Intergraph IGDS and MicroStation
Release 3.X, Supports all 4.X and 3.X Platforms including DOS, Unix, Mac, and VMS. Will work with all 4.X versions of MicroStation.
(Published by Pen & Brush Publishers, Distributed by OnWord Press)

Programming With User Commands is an indispensable tool for user command newcomers and programmers alike. This book serves as both a tutorial guide and handbook to the ins-and-outs of UCMs.

Step-by-step explanations and examples help you create menus and tutorials to speed the design and drafting process. Learn how to attach "intelligence" to the drawing with or without database links. Take control of your menu and command environment and customize it for your own application.

Learn how to add your own functions to MicroStation's built in commands. You can do things like: save the current cell library, attach another, and then restore the previous one; or locate, add, move and modify elements -- all from User Commands.

The User Command language is already built into your copy of MicroStation or IGDS. Put it to work for you today with Programming With User Commands. Most users think UCMs are too complicated, too much like "programming". This book shatters that myth and makes User Commands accessible to every user.

Mach Dinh-Vu is an Intergraph and MicroStation CAD specialist in Engineering, Architectural, and Public Utility applications. His background includes six years of Intergraph experience, first on the VAX, but now on MicroStation and MicroStation DOS, VAX, and UNIX networks.

- Programming With User Commands Disk: **$40.00** (Australia A$55.00) Includes all of the User Command examples in the book in a ready-to-use form. Use them as they are, or modify them with your own editor to get a jump-start on User Command programming.

Price: **US$65.00** (Australia A$80.00)
Pages: 320
Illustrations, Tables, Examples: 120+
ISBN: 0-934605-45-9 (Australia ISBN: 0-7316-5883-3)

101 User Commands

Brockway, Dinh-Vu, Steinbock
Putting user commands to work on all Intergraph MicroStation and IGDS Platforms.
For Intergraph MicroStation & IGDS Users
Supports all versions of MicroStation and IGDS under DOS, Mac OS, UNIX and VMS

With this book user and programmers alike can jump ahead with MicroStation or IGDS productivity. 101 User Commands gives you 101 programs to automate your CAD environment.

Never programmed with user commands before?
This book shows you how. With the program listings, input and output variables, prompting sequences and more, you will be using user commands in no time.
Or copy the program listings from the optional disk into your word processor or line editor and you will be programming in no time.

Already an experienced user command programmer?
Here you'll find some of the finest in the business. Each user command is built from basic building blocks to help you organize your programs.
Mix and Match programs or subroutines to put together your own set of user commands.

Order MicroStation 4.X Tools From OnWord Press Now!

Ordering Information:

OnWord Press Products are Available From

1. Your Local MicroStation Dealer
2. Your Local Bookseller
3. In Australia, New Zealand, and Southeast Asia from:
   Pen & Brush Publishers
   2nd Floor, 94 Flinders Street
   Melbourne Victoria 3000
   Australia
   Phone 61(0)3 818 6226
   Fax 61(0)3 818 3704

4. .... Or Directly From OnWord Press:

To Order From OnWord Press:

Three Ways To Order from OnWord Press

1. Order by **FAX** 505/587-1015
2. Order by **PHONE:** 1-800-CAD NEWS™ Outside the U.S. and Canada call 505/587-1010.
3. Order by **MAIL:** OnWord Press/CAD NEWS Bookstore, P.O. Box 500, Chamisal NM 87521-0500 USA.

Shipping and Handling Charges apply to all orders: 48 States: $4.50 for the first item, $2.25 each additional item. Canada, Hawaii, Alaska, Puerto Rico: $8.00 for the first item, $4.00 for each additional item. International: $46.00 for the first item, $15.00 each additional item. **Diskettes are counted as additional items.** New Mexico delivery address, please add 5.625% state sales tax.

Rush orders or special handling can be arranged, please phone or write for details. Government and Educational Institution POs accepted. Corporate accounts available.

MicroStation Books and Tools

Use This Form If Ordering Directly From OnWord Press

Quantity	Title	Price	Extension
	MicroStation 4.X Delta Book	$19.95	
	MicroStation 4.X Upgrade Video Series	$149.00	
	INSIDE MicroStation	$29.95	
	INSIDE MicroStation Disk	$14.95	
	INSIDE MicroStation Companion Workbook	$34.95	
	Instructor's Guide: INSIDE MicroStation Companion Workbook	$9.95	
	MicroStation Productivity Book	$39.95	
	MicroStation Productivity Disk	$49.95	
	MicroStation Reference Guide	$18.95	
	MicroStation Reference Disk	$14.95	
	The Complete Guide to MicroStation 3D	$39.95	
	The Complete Guide to MicroStation 3D Disk	$19.95	
	101 MDL Commands	$49.95	
	101 MDL Commands Disk	$101.00	
	Bill Steinbock's MDL Pocket Programmer's Guide	$24.95	
	MDL-GUIDES	$295.00	
	Programming With User Commands	$65.00	
	Programming With User Commands Disk	$40.00	
	101 User Commands	$49.95	
	101 User Commands Disk	$101.00	
	Teaching Assistant for Micro~Station	$449.95	
	The MicroStation Evaluator	$149.00	
	Shipping & Handling*		
	5.625% Tax - State of New Mexico Delivery Only		
	Total		

AD CODE M47

Name______________________________

Company______________________________

Street ______________________________
(No P.O. Boxes Please)

____

City, State______________________________

Country, Postal Code ____________________

Phone ______________________________

Fax______________________________

If Ordering Disks, Please Note

Disk Type ______________________________

Payment Method

___ **Cash** ___ **Check** ___ **Amex**

___ **MasterCARD** ___**VISA**

____
Card Number

Expiration Date ____________________

____
Signature

FAX TO: 505/587-1015

or MAIL TO:

OnWord Press
Box 500
Chamisal NM 87521 USA

MicroStation Books and Tools

Use This Form If Ordering Directly From OnWord Press

Quantity	Title	Price	Extension
	MicroStation 4.X Delta Book	$19.95	
	MicroStation 4.X Upgrade Video Series	$149.00	
	INSIDE MicroStation	$29.95	
	INSIDE MicroStation Disk	$14.95	
	INSIDE MicroStation Companion Workbook	$34.95	
	Instructor's Guide: INSIDE MicroStation Companion Workbook	$9.95	
	MicroStation Productivity Book	$39.95	
	MicroStation Productivity Disk	$49.95	
	MicroStation Reference Guide	$18.95	
	MicroStation Reference Disk	$14.95	
	The Complete Guide to MicroStation 3D	$39.95	
	The Complete Guide to MicroStation 3D Disk	$19.95	
	101 MDL Commands	$49.95	
	101 MDL Commands Disk	$101.00	
	Bill Steinbock's MDL Pocket Programmer's Guide	$24.95	
	MDL-GUIDES	$295.00	
	Programming With User Commands	$65.00	
	Programming With User Commands Disk	$40.00	
	101 User Commands	$49.95	
	101 User Commands Disk	$101.00	
	Teaching Assistant for Micro~Station	$449.95	
	The MicroStation Evaluator	$149.00	
	Shipping & Handling*		
	5.625% Tax - State of New Mexico Delivery Only		
	Total		

AD CODE M47

Name______________________________

Company______________________________

Street ______________________________
(No P.O. Boxes Please)

____

City, State______________________________

Country, Postal Code ______________________

Phone ______________________________

Fax______________________________

If Ordering Disks, Please Note

Disk Type ______________________________

Payment Method

___ Cash ___ Check ___ Amex

___ MasterCARD ___VISA

____
Card Number

Expiration Date ______________________

____
Signature

FAX TO: 505/587-1015

or MAIL TO:

OnWord Press
Box 500
Chamisal NM 87521 USA